International Vocational Education Bilingual Textbook Series
国际化职业教育双语系列教材

Introduction of Metallurgy
冶 金 概 论

Gong Na
宫 娜 编

Beijing
Metallurgical Industry Press
2022

内 容 提 要

本书共分绪论和4个项目。绪论主要介绍冶金基本概念以及冶金方法和分类；项目1主要介绍铁冶金，包含2个任务，内容主要涉及炼铁原料和高炉炼铁；项目2主要介绍钢冶金，包含6个任务，内容主要涉及炼钢基础理论、炼钢原材料、转炉炼钢工艺、电弧炉炼钢工艺、炉外精炼和钢的连续浇注；项目3主要介绍金属压力加工，包含3个任务，内容主要涉及金属成型方法、金属压力加工的基本原理和轧钢生产工艺；项目4主要介绍环境保护与资源综合利用，包含3个任务，内容主要涉及冶金工业废气的污染与治理、冶金工业废水的污染与治理和冶金工业固体废物的污染与治理。

本书可作为职业院校冶金相关专业的国际化教学用书，也可作为冶金企业员工的培训教材和有关专业人员的参考书。

图书在版编目(CIP)数据

冶金概论 = Introduction of Metallurgy：中文、英文／宫娜编．—北京：冶金工业出版社，2020.9（2022.8重印）
国际化职业教育双语系列教材
ISBN 978-7-5024-8536-8

Ⅰ.①冶… Ⅱ.①宫… Ⅲ.①冶金—高等职业教育—双语教学—教材—汉、英 Ⅳ.①TF

中国版本图书馆 CIP 数据核字（2020）第 170779 号

Introduction of Metallurgy 冶金概论

出版发行	冶金工业出版社	电 话	(010)64027926
地 址	北京市东城区嵩祝院北巷39号	邮 编	100009
网 址	www.mip1953.com	电子信箱	service@mip1953.com

责任编辑 俞跃春 刘林烨 美术编辑 郑小利 版式设计 孙跃红 禹蕊
责任校对 李 娜 责任印制 李玉山
三河市双峰印刷装订有限公司印刷
2020年9月第1版，2022年8月第2次印刷
787mm×1092mm 1/16；20.75 印张；496 千字；309 页
定价 59.00 元

投稿电话 (010)64027932 投稿信箱 tougao@cnmip.com.cn
营销中心电话 (010)64044283
冶金工业出版社天猫旗舰店 yjgycbs.tmall.com
(本书如有印装质量问题，本社营销中心负责退换)

Editorial Board of International Vocational Education Bilingual Textbook Series

Director　Kong Weijun（Party Secretary and Dean of Tianjin Polytechnic College）

Deputy Director　Zhang Zhigang（Chairman of Tiantang Group, Sino-Uganda Mbale Industrial Park）

Committee Members　Li Guiyun, Li Wenchao, Zhao Zhichao, Liu Jie, Zhang Xiufang, Tan Qibing, Liang Guoyong, Zhang Tao, Li Meihong, Lin Lei, Ge Huijie, Wang Zhixue, Wang Xiaoxia, Li Rui, Yu Wansong, Wang Lei, Gong Na, Li Xiujuan, Zhang Zhichao, Yue Gang, Xuan Jie, Liang Luan, Chen Hong, Jia Yanlu, Chen Baoling

国际化职业教育双语系列教材编委会

主　　任　孔维军（天津工业职业学院党委书记、院长）

副主任　张志刚（中乌姆巴莱工业园天唐集团董事长）

委　　员　李桂云　李文潮　赵志超　刘　洁　张秀芳

　　　　　　谭起兵　梁国勇　张　涛　李梅红　林　磊

　　　　　　葛慧杰　王治学　王晓霞　李　蕊　于万松

　　　　　　王　磊　宫　娜　李秀娟　张志超　岳　刚

　　　　　　玄　洁　梁　娈　陈　红　贾燕璐　陈宝玲

Foreword

With the proposal of the 'Belt and Road Initiative', the Ministry of Education of China issued *Promoting Education Action for Building the Belt and Road Initiative* in 2016, proposing cooperation in education, including 'cooperation in human resources training'. At the Forum on China-Africa Cooperation (FOCAC) in 2018, President Xi proposed to focus on the implementation of the 'Eight Actions', which put forward the plan to establish 10 Luban Workshops to provide skills training to African youth. Draw lessons from foreign advanced experience of vocational education mode, China's vocational education has continuously explored and formed the new mode of vocational education with Chinese characteristics. Tianjin, as a demonstration zone for reform and innovation of modern vocational education in China, has started the construction of 'Luban Workshop' along the 'Belt and Road Initiative', to export high-quality vocational education achievements.

The compilation of these series of textbooks is in response to the times and it's also the beginning of Tianjin Polytechnic College to explore the internationalization of higher vocational education. It's a new model of vocational education internationalization by Tianjin, response to the 'Belt and Road Initiative' and the 'Going Out' of Chinese enterprises. Tianjin Polytechnic College and Uganda Technical College-Elgon reached a cooperation intention to establish the Luban Workshop to carry out vocational education cooperation on mechatronics technology and ferrous metallurgy technology major in 2019. The establishment of Luban Workshop is conducive to strengthen the cooperation between China and Uganda in vocational education, promote the export of high-quality higher vocational education resources, and serve Chinese enterprises in Uganda and Ugandan local enterprises. Exploring and standardizing the overseas operation of Chinese colleges, the expansion of international influences of China's higher vocational education is also one of the purposes.

The construction of 'Luban Workshop' in Uganda is mainly based on the EPIP (Engineering, Practice, Innovation, Project) project, and is committed to cultivating high-quality talents with innovative spirit, creative ability and entrepreneurial spirit. To meet the learning needs of local teachers and students accurately, the compilation of these international vocational skills bilingual textbooks is based on the talent demand of Uganda and the specialty and characteristics of Tianjin Polytechnic.

These textbooks are supporting teaching material, referring to Chinese national professional standards and developing international professional teaching standards. The internationalization of the curriculums takes into account the technical skills and cognitive characteristics of local students, to promote students' communication and learning ability. At the same time, these textbooks focus on the enhancement of vocational ability, rely on professional standards, and integrate the teaching concept of equal emphasis on skills and quality. These textbooks also adopted project-based, modular, task-driven teaching model and followed the requirements of enterprise posts for employees.

In the process of writing the series of textbooks, Wang Xiaoxia, Li Rui, Wang Zhixue, Ge Huijie, Yu Wansong, Wang Lei, Li Xiujuan, Gong Na, Zhang Zhichao, Jia Yanlu, Chen Baoling and other chief teachers, professional teams, English teaching and research office have made great efforts, receiving strong support from leaders of Tianjin Polytechnic College. During the compilation, the series of textbooks referred to a large number of research findings of scholars in the field, and we would like to thank them for their contributions.

Finally, we sincerely hope that the series of textbooks can contribute to the internationalization of China's higher vocational education, especially to the development of higher vocational education in Africa.

Principal of Tianjin Polytechnic College　Kong Weijun

May, 2020

序

随着"一带一路"倡议的提出，2016年中华人民共和国教育部发布了《推进共建"一带一路"教育行动》，提出了包括"开展人才培养培训合作"在内的教育合作。2018年习近平主席在中非合作论坛上提出，要重点实施"八大行动"，明确要求在非洲设立10个鲁班工坊，向非洲青年提供技能培训。中国职业教育在吸收和借鉴发达国家先进职教发展模式的基础上，不断探索和形成了中国特色职业教育办学模式。天津市作为中国现代职业教育改革创新示范区，开启了"鲁班工坊"建设工作，在"一带一路"沿线国家搭建"鲁班工坊"平台，致力于把优秀职业教育成果输出国门与世界分享。

本系列教材的编写，契合时代大背景，是天津工业职业学院探索高职教育国际化的开端。"鲁班工坊"是由天津率先探索和构建的一种职业教育国际化发展新模式，是响应国家"一带一路"倡议和中国企业"走出去"，创建职业教育国际合作交流的新窗口。2019年天津工业职业学院与乌干达埃尔贡技术学院达成合作意向，共同建立"鲁班工坊"，就机电一体化技术专业、黑色冶金技术专业开展职业教育合作。此举旨在加强中乌职业教育交流与合作，推动中国优质高等职业教育资源"走出去"，服务在乌中资企业和乌干达当地企业，探索和规范我国职业院校"鲁班工坊"建设和境外办学，扩大中国高等职业教育的国际影响力。

中乌"鲁班工坊"的建设主要以工程实践创新项目（EPIP：Engineering，Practice，Innovation，Project）为载体，致力于培养具有创新精神、创造能力和创业精神的"三创"复合型高素质技能人才。国际化职业教育双语系列教材的编写，立足于乌干达人才需求和天津工业职业学院专业特色，是为了更好满足当地师生学习需求。

本系列教材采用中英双语相结合的方式，主要参照中国专业标准，开发国际化专业教学标准，课程内容国际化是在专业课程设置上，结合本地学生的技术能力水平与认知特点，合理设置双语教学环节，加强学生的学习与交流能力。同时，教材以提升职业能力为核心，以职业标准为依托，体现技能与质量并重的教学理念，主要采用项目化、模块化、任务驱动的教学模式，并结合企业岗位对员工的要求来撰写。

本系列教材在撰写过程中，王晓霞、李蕊、王治学、葛慧杰、于万松、王磊、李秀娟、宫娜、张志超、贾燕璐、陈宝玲等主编老师、专业团队、英语教研室付出了辛勤劳动，并得到了学院各级领导的大力支持，同时本系列教材借鉴和参考了业界有关学者的研究成果，在此一并致谢！

最后，衷心希望本系列教材能为我国高等职业教育国际化，尤其是高等职业教育走进非洲、支援非洲高等职业教育发展尽绵薄之力。

天津工业职业学院书记、院长　孔维军

2020 年 5 月

Preface

Tianjin Polytechnic College and Uganda Technical College – Elgon reached a cooperation intention to establish the Luban Workshop to carry out vocational education cooperation on mechatronics technology and ferrous metallurgy technology major in 2019. In order to strengthen the cooperation between China and Uganda in vocational education, the two colleges plan to compile a series of international vocational skills bilingual textbooks.

This book is one of bilingual textbooks for international vocational education. This book mainly aims at the modern steel production process, from raw materials, blast furnace ironmaking, converter steelmaking, off – furnace refining, metal pressure processing and metallurgical environmental protection and other aspects. It also introduces the current steel production process in China and the world commonly used technology. From the technical principle to the use of equipment, the book is based on the current production technology used by enterprises. In addition, in combination with the high quality index of advanced production enterprises, the relevant production process, as well as the smelting and processing process of characteristic steel are introduced.

The editors of this book are all teachers of Tianjin Polytechnic College, with Gong Na in charge of the planning and drafting of this book, with Zhou Fan as the co – editor. The specific compilation division is as follows: Introduction, Project 1, and Task 2.1 and Task 2.2 of Project 2 are written by Zhou Fan; Tasks 2.3 ~ 2.6 of Project 2, Project 3 and Project 4 are written by Gong Na. In the process of compiling this book, I have referred to relevant materials and literature, and hereby express my gratitude to the author concerned.

Due to the limited level of the editor, there is something wrong in the book. I hope readers to criticize and correct.

The editor

May, 2020

前 言

2019年天津工业职业学院与乌干达埃尔贡技术学院达成合作意向,共同建立"鲁班工坊",就机电一体化技术专业、黑色冶金技术专业开展职业教育合作,双方计划编撰国际化职业教育双语系列教材。

本书是国际化职业教育双语系列教材之一。本书主要针对现代钢铁生产过程,从原料、高炉炼铁、转炉炼钢、炉外精炼、金属压力加工及冶金环境保护等方面,介绍了目前中国及世界钢铁生产过程中的常用技术,从技术原理到设备使用,都是依据目前企业所用的生产技术编写。另外,在结合先进生产企业的优质指标中,介绍了相关生产流程,以及特色钢种的冶炼及加工过程。

本书由宫娜负责策划和统稿,周凡参编。具体编写分工为:绪论、项目1和项目2的任务2.1、任务2.2由周凡编写,项目2的任务2.3~任务2.6、项目3和项目4由宫娜编写。在本书编写过程中,参考了有关资料和文献,在此对有关作者表示感谢。

由于编者水平所限,书中不妥之处,希望读者批评指正。

编 者
2020年5月

Contents

0 Introduction ········· 1

    0.1 Basic Metallurgical Concepts ········· 1

    0.2 Metallurgical Method ········· 2

    0.3 Brief Introduction of Main Metallurgical Processes ········· 3

    0.4 Classification of Metallurgy ········· 6

Project 1 Iron Metallurgy ········· 7

    Task 1.1 Ironmaking Raw Materials ········· 7

        1.1.1 Iron Ore ········· 7

        1.1.2 Flux ········· 24

        1.1.3 Fuel ········· 26

        1.1.4 Sinter Production ········· 30

        1.1.5 Pellet Production ········· 48

    Task 1.2 Blast Furnace Ironmaking ········· 59

        1.2.1 Overview of Blast Furnace Ironmaking ········· 59

        1.2.2 Basic Principles of Blast Furnace Ironmaking ········· 65

        1.2.3 Equipment of Blast Furnace Ironmaking ········· 94

Project 2 Metallurgy of Steel ········· 117

    Task 2.1 Basic Theory of Steelmaking ········· 117

        2.1.1 Physical Properties of Liquid Steel ········· 118

        2.1.2 Physical and Chemical Properties of Slag ········· 121

        2.1.3 Decarburization of Molten Steel ········· 123

2.1.4	Dephosphorization of Liquid Steel	126
2.1.5	Desulfurization of Molten Steel	131
2.1.6	Deoxidation of Liquid Steel	135
2.1.7	Reactions of Hydrogen and Nitrogen	139

Task 2.2 Steel Making Raw Materials 140

- 2.2.1 Metal Materials 141
- 2.2.2 Slagging Materials 145
- 2.2.3 Oxidants, Coolants and Carburizers 148

Task 2.3 Converter Steelmaking Process 150

- 2.3.1 Change Rules of Metal Composition in Converter Converting Process 150
- 2.3.2 Variation of Slag Composition and Bath Temperature during Converter Blowing 154
- 2.3.3 Oxygen Top Blown Converter Steelmaking Operation System 156

Task 2.4 EAF Steelmaking Process 174

- 2.4.1 Size and Classification of Electric Arc Furnace 175
- 2.4.2 Traditional EAF Steelmaking Process 176

Task 2.5 Refining Outside the Furnace 187

- 2.5.1 Basic Means of Refining Outside the Furnace 187
- 2.5.2 Main Methods of Off-furnace Refining 194

Task 2.6 Continuous Pouring of Steel 205

- 2.6.1 Overview of Steel Pouring 205
- 2.6.2 Models and Characteristics of Continuous Casting Machine 205
- 2.6.3 Main Equipment of Continuous Casting Machine 209
- 2.6.4 Solidification and Crystallization Theory of Molten Steel 214
- 2.6.5 Determination of Drawing Speed of Continuous Casting 219
- 2.6.6 Quality of Continuous Casting Billet 223

Project 3 Metal Pressure Machining 225

Task 3.1 Metal Forming Methods 225

- 3.1.1 Casting 225

3.1.2	Pressure Processing	230
3.1.3	New Material Forming Technology	236

Task 3.2 Fundamentals of Metal Pressure Processing ... 243
- 3.2.1 Physical Mature of Plastic Deformation of Metals ... 243
- 3.2.2 Effects of Metal Plastic Deformation on Microstructure and Properties of Materials ... 246
- 3.2.3 Plastic Flow Law of Metal ... 258

Task 3.3 Rolling Process ... 261
- 3.3.1 Classification of Steel Products ... 261
- 3.3.2 Production Process of Strip Steel ... 263
- 3.3.3 Section Steel Production Process ... 269
- 3.3.4 Hot Rolled Seamless Steel Tube Production Process ... 272

Project 4 Environmental Protection and Utilization of Resources ... 276

Task 4.1 Pollution and Treatment of Waste Gas from Metallurgical Industry ... 276
- 4.1.1 Classification of Waste Gases from Metallurgical Industry ... 276
- 4.1.2 Characteristics of Waste Gas from Metallurgical Industry ... 277
- 4.1.3 Treatment of Waste Gas from Metallurgical Industry ... 278

Task 4.2 Pollution and Treatment of Metallurgical Industrial Waste Water ... 285
- 4.2.1 Classification of Metallurgical Industrial Waste Water ... 286
- 4.2.2 Pollution of Metallurgical Industrial Waste Water ... 286
- 4.2.3 Treatments of Metallurgical Industrial Waste Water ... 287

Task 4.3 Pollution and Treatment of Solid Waste in Metallurgical Industry ... 301
- 4.3.1 Classification of Solid Waste in Metallurgical Industry ... 301
- 4.3.2 Principles of Solid Waste Treatment in Metallurgical Industry and Significance of Comprehensive Utilization ... 303
- 4.3.3 Management of Solid Waste in Metallurgical Industry ... 303

References ... 309

目 录

0 绪论 ……………………………………………………………………………………… 1

  0.1 冶金基本概念 ……………………………………………………………………… 1

  0.2 冶金方法 …………………………………………………………………………… 2

  0.3 主要冶金过程简介 ………………………………………………………………… 3

  0.4 冶金的分类 ………………………………………………………………………… 6

项目1 铁冶金 …………………………………………………………………………… 7

  任务1.1 炼铁原料 …………………………………………………………………… 7

    1.1.1 铁矿石 ……………………………………………………………………… 7

    1.1.2 熔剂 ………………………………………………………………………… 24

    1.1.3 燃料 ………………………………………………………………………… 26

    1.1.4 烧结矿生产 ………………………………………………………………… 30

    1.1.5 球团矿生产 ………………………………………………………………… 48

  任务1.2 高炉炼铁 …………………………………………………………………… 59

    1.2.1 高炉炼铁概述 ……………………………………………………………… 59

    1.2.2 高炉炼铁基本原理 ………………………………………………………… 65

    1.2.3 高炉炼铁设备 ……………………………………………………………… 94

项目2 钢冶金 …………………………………………………………………………… 117

  任务2.1 炼钢基础理论 ……………………………………………………………… 117

    2.1.1 钢液的物理性质 …………………………………………………………… 118

    2.1.2 炉渣的物理化学性质 ……………………………………………………… 121

    2.1.3 钢液的脱碳 ………………………………………………………………… 123

2.1.4 钢液的脱磷 …… 126

2.1.5 钢液的脱硫 …… 131

2.1.6 钢液的脱氧 …… 135

2.1.7 氢、氮的反应 …… 139

任务 2.2　炼钢原材料 …… 140

2.2.1 金属料 …… 141

2.2.2 造渣材料 …… 145

2.2.3 氧化剂、冷却剂和增碳剂 …… 148

任务 2.3　转炉炼钢工艺 …… 150

2.3.1 转炉吹炼过程中金属成分的变化规律 …… 150

2.3.2 转炉吹炼过程中熔渣成分和熔池温度的变化规律 …… 154

2.3.3 氧气顶吹转炉炼钢操作制度 …… 156

任务 2.4　电弧炉炼钢工艺 …… 174

2.4.1 电弧炉的大小与分类 …… 175

2.4.2 传统电炉炼钢工艺 …… 176

任务 2.5　炉外精炼 …… 187

2.5.1 炉外精炼的基本手段 …… 187

2.5.2 炉外精炼的主要方法 …… 194

任务 2.6　钢的连续浇注 …… 205

2.6.1 钢的浇注概述 …… 205

2.6.2 连铸机的机型及特点 …… 205

2.6.3 连铸机的主要设备 …… 209

2.6.4 钢液凝固结晶理论 …… 214

2.6.5 连铸拉速的确定 …… 219

2.6.6 连铸坯的质量 …… 223

项目 3　金属压力加工 …… 225

任务 3.1　金属成型方法 …… 225

3.1.1 铸造 …… 225

3.1.2 压力加工 …… 230

####### 3.1.3 材料成型新技术 ……………………………………………………………… 236

任务 3.2 金属压力加工的基本原理 ………………………………………………… 243
####### 3.2.1 金属塑性变形的物理本质 …………………………………………………… 243
####### 3.2.2 金属塑性变形对材料组织性能的影响 ……………………………………… 246
####### 3.2.3 金属塑性流动规律 …………………………………………………………… 258

任务 3.3 轧钢生产工艺 ………………………………………………………………… 261
####### 3.3.1 钢材品种的分类 ……………………………………………………………… 261
####### 3.3.2 板带钢生产工艺 ……………………………………………………………… 263
####### 3.3.3 型钢生产工艺 ………………………………………………………………… 269
####### 3.3.4 热轧无缝钢管生产工艺 ……………………………………………………… 272

项目 4 环境保护与资源综合利用 ……………………………………………………… 276

任务 4.1 冶金工业废气的污染与治理 ………………………………………………… 276
####### 4.1.1 冶金工业废气的分类 ………………………………………………………… 276
####### 4.1.2 冶金工业废气的特点 ………………………………………………………… 277
####### 4.1.3 冶金工业废气的治理 ………………………………………………………… 278

任务 4.2 冶金工业废水的污染与治理 ………………………………………………… 285
####### 4.2.1 冶金工业废水的分类 ………………………………………………………… 286
####### 4.2.2 冶金工业废水的污染 ………………………………………………………… 286
####### 4.2.3 冶金工业废水的治理 ………………………………………………………… 287

任务 4.3 冶金工业固体废物和污染与治理 …………………………………………… 301
####### 4.3.1 冶金工业固体废物的分类 …………………………………………………… 301
####### 4.3.2 冶金工业固体废物处理的原则和综合利用的意义 ………………………… 303
####### 4.3.3 冶金工业固体废物的治理 …………………………………………………… 303

参考文献 …………………………………………………………………………………… 309

0 Introduction
0 绪 论

This part mainly introduces the basic concepts of metallurgy, the history of metallurgical development, the current situation and future planning of metallurgical industry, and analyzes the important position and role of metallurgical industry in the national economy, so that beginners have a preliminary understanding of the metallurgical industry.

本部分主要介绍了冶金基本概念、冶金发展史以及冶金工业的发展现状和未来规划等内容，同时还分析了冶金行业在国民经济中的重要地位和作用，能够使初学者对冶金行业有初步了解。

0.1 Basic Metallurgical Concepts
0.1 冶金基本概念

Metallurgy is a science that studies how to economically extract metals or metal compounds from ores or other raw materials, and make metal materials with certain properties by various processing methods.

冶金是一门研究如何经济的从矿石或其他原料中提取金属或金属化合物，并用各种加工方法制成具有一定性能的金属材料的科学。

The ore used to extract various metals has different properties. So different production processes and equipment should be used to extract metals according to different principles, thus forming a specialized discipline metallurgy.

冶金用于提取各种具有不同性质的金属矿石，故提取金属要根据不同的原理，采用不同的生产工艺过程和设备，从而形成了冶金的专门学科——冶金学。

Metallurgy is an important part of the research on the preparation, processing and improvement of metal properties. It has developed into the research on metal composition, structure, properties and related basic theories. In terms of its research field, metallurgy credits include extraction metallurgy and physical metallurgy.

冶金学以研究金属的制取、加工和改进金属性能的各种技术为重要内容，现发展为对金属成分、组织结构、性能和有关基础理论进行研究。就其研究领域而言，冶金学分为提取冶金和物理冶金两门学科。

Extraction metallurgy is a production process that studies how to extract metals or metal compounds from ores. Because this process is accompanied by chemical reactions, it is also called chemical metallurgy.

提取冶金学是研究如何从矿石中提取金属或金属化合物的生产过程，由于该过程伴有化学反应，又称为化学冶金。

Physical metallurgy is the preparation of metal or alloy materials with certain properties by forming and processing. It studies the internal relations of its composition and structure as well as the change rules under various conditions, so as to serve the effective use and development of metal materials with specific properties. It includes metallography, powder metallurgy, metal casting, metal pressure processing, etc.

物理冶金学是通过成形加工制备有一定性能的金属或合金材料，研究其组成、结构的内在联系，以及在各种条件下的变化规律，为有效地使用和发展具有特定性能的金属材料服务。它包括金属学、粉末冶金、金属铸造、金属压力加工等。

0.2　Metallurgical Method
0.2　冶金方法

There are many ways to extract metals from ores or other raw materials, which can be summed up as follows：

从矿石或其他原料中提取金属的方法很多，可归结为以下三种：

(1) Pyrometallurgy. It refers to the process of smelting, refining reaction and melting operation of ore under high temperature to separate metal and impurity and obtain pure metal. The whole process can be divided into three processes: raw material preparation, smelting and refining. The energy needed in the process is mainly supplied by fuel combustion, and also by the chemical reaction heat in the process.

(1) 火法冶金。它是指在高温下矿石经熔炼与精炼反应及熔化作业，使其中的金属和杂质分开，获得较纯金属的过程。整个过程可分为原料准备、冶炼和精炼三个工序。过程所需能源主要靠燃料燃烧供给，也有依靠过程中的化学反应热来提供的。

(2) Hydrometallurgy. It refers to the process of treating ore or concentrate with solvent at room temperature or below 100℃, so that the metal to be extracted is dissolved in the solution while other impurities are not dissolved. And then the metal is extracted and separated from the solution. Because most of the solvent is aqueous solution, it is also called aqueous metallurgy. The process includes leaching, separation, enrichment and extraction.

(2) 湿法冶金。它是指在常温或低于100℃下，用溶剂处理矿石或精矿，使所要提取的金属溶解于溶液中，而其他杂质不溶解，然后再从溶液中将金属提取和分离出来的过程。由于绝大部分溶剂为水溶液，也称为水法冶金。该方法包括浸出、分离、富集和提取等工序。

(3) Electrometallurgy. It can be divided into electrothermal metallurgy and electrochemical metallurgy according to the form of electric energy, as follows：

(3) 电冶金。它是利用电能提取和精炼金属的方法，按电能形式可分为电热冶金和电化学冶金两类，即：

1) Electrothermal metallurgy. It uses electric energy to transform into heat energy to refine metals at high temperature, which is essentially the same as pyrometallurgy.

1)电热冶金。它是利用电能转变成热能,在高温下提炼金属。其本质与火法冶金相同。

2) Electrochemical metallurgy. The former is called solution electrolysis, such as the electrolytic refining of copper, which can be classified as hydrometallurgy. The latter is called molten salt electrolysis, such as electrolytic aluminum, which can be classified as pyrometallurgy.

2)电化学冶金。利用电化学反应使金属从含金属的盐类水溶液或熔体中析出,前者称为溶液电解,如铜的电解精炼,可归入湿法冶金;后者称为熔盐电解,如电解铝,可列入火法冶金。

Which method to extract metals and in what order depends largely on the raw materials used and the products required. Fire method and wet method are widely used in metallurgy. Fire method is mainly used in iron and steel metallurgy, while fire method and wet method are both used in non-ferrous metal extraction.

采用哪种方法提取金属,按怎样的顺序进行,在很大程度上取决于所用的原料以及要求的产品。冶金方法中以火法和湿法的应用较为普遍,钢铁冶金主要采用火法,有色金属的提取则火法和湿法兼有。

0.3 Brief Introduction of Main Metallurgical Processes
0.3 主要冶金过程简介

In production practice, various metallurgical methods often include many metallurgical processes, such as beneficiation, drying, roasting, calcination, sintering, pellet, smelting, refining and other processes in pyrometallurgy. This section focuses on the following processes:

在生产实践中,各种冶金方法往往包括许多冶金工序,如火法冶金中有选矿、干燥、焙烧、煅烧、烧结、球团、熔炼、精炼等工序。本节重点介绍以下工序:

(1) Drying. Drying refers to the removal of moisture from raw materials. The drying temperature is generally 400~600℃.

(1)干燥。干燥是指除去原料中的水分,干燥温度一般为400~600℃。

(2) Roasting. Roasting refers to the metallurgical process in which the ore or concentrate is heated to a temperature lower than their melting point in an appropriate atmosphere, oxidation, reduction or other chemical changes occur. Its purpose is to change the chemical composition of the extracted object in the raw material and meet the requirements of smelting. According to the different control atmosphere of roasting process, it can be divided into oxidation roasting, reduction roasting, sulfuric acid roasting, chlorination roasting, etc.

(2)焙烧。焙烧是指将矿石或精矿置于适当的气氛下,加热至低于它们的熔点温度,发生氧化、还原或其他化学变化的冶金过程。其目的是为了改变原料中提取对象的化学组成,满足熔炼的要求。按焙烧过程控制气氛的不同,可分为氧化焙烧、还原焙烧、硫酸化焙烧、氯化焙烧等。

(3) Calcination. Calcination refers to the process of heating and decomposing the mineral raw materials of carbonate or hydroxide in the air to remove carbon dioxide or water and make them

become oxides, also known as calcination. For example, the limestone is calcined into lime as a steel-making flux.

（3）煅烧。煅烧是指将碳酸盐或氢氧化物的矿物原料在空气中加热分解，除去二氧化碳或水分，使其变成氧化物的过程，也称焙解。如将石灰石煅烧成石灰，作为炼钢熔剂。

（4）Sintering and pelletizing. Sintering and pelletizing are the main methods of making powder ore block, which are to mix different ore powder or heat and bake it after pelletizing, and then consolidate it into porous block or spherical material.

（4）烧结和球团。烧结和球团是将不同矿粉混匀或造球后加热焙烧，固结成多孔块状或球状的物料，是粉矿造块的主要方法。

（5）Smelting. Smelting refers to the process of separating metals and impurities in ores into two liquid layers (i.e. molten metal and slag) by oxidation-reduction reaction of processed ores or other raw materials at high temperature, also known as smelting. According to smelting conditions, it can be divided into reduction smelting, matte making smelting, oxidation converting, etc.

（5）熔炼。熔炼是指将处理好的矿石或其他原料在高温下通过氧化还原反应，使矿石中金属和杂质分离为两个液相层（即金属液和熔渣）的过程，也称冶炼。按冶炼条件，其可分为还原熔炼、造锍熔炼、氧化吹炼等。

（6）Refining. Refining refers to the further processing of the crude metal containing a small amount of impurities to improve its purity. For example, pig iron is obtained by smelting iron ore and then refined into steel by oxidation. There are many refining methods, such as steelmaking, vacuum metallurgy, jet metallurgy, molten salt electrolysis, etc.

（6）精炼。精炼是指进一步处理熔炼所得的含有少量杂质的粗金属，以提高其纯度。如熔炼铁矿石得到生铁，再经氧化精炼成钢。精炼方法很多，如炼钢、真空冶金、喷射冶金、熔盐电解等。

（7）Blowing. Blowing is an important process of pyrometallurgy, which refers to the oxidation smelting in converter. During converting, air, industrial pure oxygen or other oxidizing gases are blown into the molten matte or coarse metal to oxidize the impurities into gases to escape or become oxide slag, so as to obtain pure metal or high matte (such as white matte). Blowing is widely used in the smelting of steel, copper and nickel.

（7）吹炼。吹炼是火法冶金的一个重要过程，是指在转炉中所进行的氧化熔炼。吹炼时，向熔融的锍或粗金属鼓入空气、工业纯氧或其他氧化性气体，使杂质氧化成气体逸出或成为氧化物造渣，以获得较纯的金属或高锍（如白冰铜）。吹炼广泛用于钢、铜、镍的冶炼。

（8）Distillation. Distillation refers to the method of separating some components of smelting materials under the condition of indirect heating by using the characteristics of different volatilities of various substances at a certain temperature.

（8）蒸馏。蒸馏是指将冶炼的物料在间接加热的条件下，利用某一温度下各种物质挥发度不同的特点，使冶炼物料中某些组分分离的方法。

（9）Leaching. The so-called leaching (some are also called dissolution) is to add solid materials (such as ore, concentrate, etc.) to the liquid solvent, so that one or several valuable metals in the solid materials are dissolved in the solution, while gangue and some non main metals are put into the slag, so as to separate the extracted metals from gangue and some impurities.

（9）浸出。浸出（有的也称为溶出）是指将固体物料（例如矿石、精矿等）加到液体溶剂中，使固体物料中的一种或几种有价金属溶解于溶液中，而脉石和某些非主体金属入渣，使提取金属与脉石和某些杂质分离的过程。

（10） Purification. Purification is a process used to deal with leaching solution or other solution containing excessive impurities and remove the impurities to reach the standard. It is also an effective method for comprehensive utilization of resources, improvement of economic benefits and prevention of environmental pollution. Due to the different properties of various elements in the solution, the purification methods used are also different. Therefore, it is not possible to remove all impurities by one method at a time, but to use different methods for multiple purification. Common purification methods include ion precipitation, displacement precipitation and coprecipitation.

（10）净化。净化是指利用处理浸出溶液或其他含有杂质超标的溶液，除去其中杂质以达标的过程。它也是综合利用资源、提高经济效益、防止污染环境的有效方法。由于溶液中各种元素的性质不同，采用的净化方法也不同，这样就不能试图采用一种方法将所有的杂质一次除去，而是采用不同方法，多次净化才能完成。一般常用的净化方法有离子沉淀法、置换沉淀法和共沉淀法等。

（11） Electrolysis of aqueous solution. Aqueous solution electrolysis is a process in which two electrodes (cathode and anode) are inserted into the aqueous solution electrolyte and direct current is applied to make the aqueous solution electrolyte undergo oxidation-reduction reaction. Due to the different anodes used in aqueous solution electrolysis, there are two kinds of anodes: soluble anode and insoluble anode. The former is called electrolytic refining, and the latter is called electrolytic deposition.

（11）水溶液电解。水溶液电解是指在水溶液电解质中插入两个电极（阴极与阳极），通入直流电，使水溶液电解质发生氧化—还原反应的过程。水溶液电解时因使用的阳极不同，有可溶阳极与不可溶阳极之分，前者称为电解精炼，后者称为电解沉积。

（12） Molten salt electrolysis. Molten salt electrolysis refers to the electrolysis process using molten salt as electrolyte. It is mainly used to extract light metals such as aluminum and magnesium. This is due to the high chemical activity of these metals, which are not available in the aqueous solution for electrolysis of these metals. In order to make solid electrolyte melt, the process should be carried out at high temperature.

（12）熔盐电解。熔盐电解是指用熔融盐作为电解质的电解过程。其主要用于提取轻金属，如铝、镁等。这是由于这些金属的化学活性很大，电解这些金属的水溶液得不到金属。为了使固态电解质成为熔融体，需要在高温条件下进行。

It can be seen that the metallurgical process is a process of using various chemical and physical methods to separate the main metals in the raw materials from other metals or non-metallic elements, so as to obtain metals with high purity.

由此可见，冶金过程是应用各种化学和物理的方法，使原料中的主要金属与其他金属或非金属元素分离，以获得纯度较高的金属的过程。

Metallurgy is a multi-disciplinary comprehensive applied science. On the one hand, metallurgy constantly absorbs the new achievements of other disciplines, especially physics, chemistry, mechanics, physical chemistry, hydromechanics, etc., to guide the development of metallurgical production technology to a new breadth and depth. On the other hand, metallurgical production enriches the contents of metallurgy with rich practical experience, and also provides suggestions for other disciplines for new metal materials and new research topics. The development and application of electronic technology and computer have had a profound impact on metallurgical production, promoted the continuous production of new metals and new alloy materials, and further adapted to the needs of the development of high, precision and advanced science and technology.

冶金学是一门多学科的综合应用科学：一方面，冶金学不断吸收其他学科，特别是物理学、化学、力学、物理化学、流体力学等方面的新成果，指导冶金生产技术向新的广度和深度发展；另一方面，冶金生产又以丰富的实践经验充实冶金学的内容，也为其他学科提供新的金属材料和新的研究课题。电子技术和电子计算机的发展及应用，对冶金生产产生深刻的影响，促进新金属和新合金材料的不断产出，进一步适应高、精、尖科学技术发展的需要。

0.4　Classification of Metallurgy
0.4　冶金的分类

In modern industry, metals are usually divided into ferrous metals and non-ferrous metals. Iron, chromium and manganese belong to ferrous metals, while the rest belong to non-ferrous metals. Therefore, the metallurgical industry is generally divided into ferrous metallurgy industry and non-ferrous metallurgy industry according to the two major categories of metals. The former includes the production of iron, steel and ferroalloys (such as ferromanganese and ferrochrome), so it is also called iron and steel metallurgy. The latter includes the production of various non-ferrous metals, collectively known as non-ferrous metallurgy.

现代工业中习惯把金属分为黑色金属和有色金属两大类。铁、铬、锰三种金属属于黑色金属，其余金属属于有色金属。因此，冶金工业按照金属的两大类别通常分为黑色金属冶金工业和有色金属冶金工业。前者包括铁、钢及铁合金（如锰铁、铬铁）的生产，故又称钢铁冶金；后者包括各种有色金属的生产，统称为有色金属冶金。

Project 1　Iron Metallurgy
项目1　铁冶金

This chapter firstly introduces the types, functions, components and processing methods of ironmaking raw materials, and summarizes the classification and characteristics of iron ore, the key points of iron ore quality evaluation, the functions, classification characteristics, quality requirements of fluxes in blast furnace ironmaking, the basic principles of preparation of coke, sinter and pellet required for ironmaking, production process and main equipment needed for production, etc. Secondly, the basic principle of blast furnace ironmaking, the function and structure of blast furnace ironmaking equipment are introduced.

本章首先介绍了炼铁原料的种类、作用、成分和加工方法，概述了铁矿石的分类和特性，铁矿石的质量评价要点，熔剂在高炉炼铁中的作用、分类特性和质量要求，以及炼铁所需的焦炭、烧结矿、球团矿的制备基本原理、生产工艺流程和生产需使用的主要设备等。其次，介绍了高炉炼铁的基本原理，以及高炉炼铁设备的作用和结构。

Task 1.1　Ironmaking Raw Materials
任务1.1　炼铁原料

Blast furnace ironmaking raw materials are mainly composed of iron ore, flux and fuel.
高炉炼铁原料主要由铁矿石、熔剂和燃料组成。

1.1.1　Iron Ore
1.1.1　铁矿石

1.1.1.1　Minerals, Ores and Rocks
1.1.1.1　矿物、矿石和岩石

Minerals refer to natural compounds or elements with uniform internal crystal structure, chemical composition and certain physical and chemical properties in the crust. Minerals that can be used by human beings are called useful minerals. In addition to useful minerals in the ore, there are almost some minerals or rocks that are not valuable in industry, which are called gangue minerals. Gangue minerals, which are unfavorable to smelting, should be removed as much as possible during beneficiation and other processing.

矿物是指地壳中具有均一内部结晶结构、化学组成以及一定物理、化学性质的天然化合物或自然元素。这些能够被人类利用的矿物，称为有用矿物。在矿石中除了有用矿物外，还存在一些在工业上没有提炼价值的矿物或岩石，称为脉石矿物。对冶炼不利的脉石

矿物，应在选矿和其他处理过程中尽量去除。

Ore and rock are aggregates of minerals. But under the current conditions of science and technology, minerals that can extract metals economically and reasonably are called ores. The concept of ore is relative. For example, with the development of mineral processing and smelting technology, Panzhihua vanadic titanomagnetite, which cannot be smelted, has become an important raw material for iron making.

矿石和岩石是矿物的集合体。但是在当前科学技术条件下，能从中经济合理地提炼出的金属矿物才称为矿石。矿石的概念是相对的。例如，随着选矿和冶炼技术的发展，不能冶炼的攀枝花钒钛磁铁矿已成为重要的炼铁原料。

1.1.1.2　Classification and Characteristics of Iron Ore
1.1.1.2　铁矿石的分类及特性

There are a lot of iron minerals in nature. It is rare to find pure iron only in the state of metal, generally in the form of compounds composed of iron and other elements. At present, there are more than 300 kinds of iron bearing minerals, but only 20 minerals can be used as raw materials for iron making. They are mainly composed of one or several iron bearing minerals and gangue. According to the properties of iron bearing minerals, they are mainly divided into magnetite, hematite, limonite and siderite. The classification and characteristics of iron ore is shown in Table 1−1.

自然界中含有很多铁矿物，但仅以金属状态存在的单质铁是很少见的，一般都是以铁元素与其他元素组成的化合物形式存在。目前已经知道的含铁矿物有300多种，但是能作为炼铁原料的只有20多种，它们主要由一种或几种含铁矿物和脉石组成。根据含铁矿物的性质，其主要分为磁铁矿、赤铁矿、褐铁矿和菱铁矿四类铁矿石。铁矿石的分类及特性见表1−1。

Table 1−1　Classification and characteristics of iron ore
表1−1　铁矿石的分类及特性

Mineral 矿石名称	Name of iron bearing mineral 含铁矿物名称及化学式	Theoretical iron content in minerals (by mass) /% 矿物中理论铁含量（质量分数）/%	Ore density /t·m^{-3} 矿物密度 /t·m^{-3}	Color 颜色	Streak 条痕	Actual iron content in minerals (by mass)/% 矿物中实际铁含量（质量分数）/%	Harmful impurities 有害杂质	Strength and reducibility 强度及还原性
Hematite 赤铁矿	Fe_2O_3	70.0	4.9~5.3	Red to light gray or even black 红色至淡灰色甚至黑色	Dull red 暗红色	55~60	Less 少	Easy to be broken, soft and reducible 较易破碎，软，易还原

Continued Table 1-1

Mineral 矿石名称	Name of iron bearing mineral 含铁矿物名称及化学式	Theoretical iron content in minerals (by mass) /% 矿物中理论铁含量(质量分数)/%	Ore density /t·m^{-3} 矿物密度 /t·m^{-3}	Color 颜色	Streak 条痕	Actual iron content in minerals (by mass)/% 矿物中实际铁含量(质量分数)/%	Harmful impurities 有害杂质	Strength and reducibility 强度及还原性
Magnetite 磁铁矿	Fe_3O_4	72.4	5.2	Black or gray 黑色或灰色	Black 黑色	45~70	High S and P content S、P含量高	Hard, dense, and hard to restore 坚硬，致密，难还原
Limonite 褐铁矿	Hematite 水赤铁矿 $2Fe_2O_3 \cdot H_2O$	66.1	4.0~5.0	Tan, or tan to black 黄褐色、黄褐色至黑色	Yellow or brown 黄色或褐色	37~55	High content of P P含量高	Loose, mostly soft ore, and easy to reduce 疏松，大部分属于软矿石，易还原
	Needle hematite 针赤铁矿 $Fe_2O_3 \cdot H_2O$	62.9	4.0~4.5					
	Water needle iron ore 水针赤铁矿 $3Fe_2O_3 \cdot 4H_2O$	60.9	3.0~4.4					
	Limonite 褐铁矿 $2Fe_2O_3 \cdot 4H_2O$	60.0	3.0~4.2					
	Yellow needle iron ore 黄针铁矿 $Fe_2O_3 \cdot 2H_2O$	57.2	3.0~4.0					
	Yellow ochre 黄赭石 $Fe_2O_3 \cdot 3H_2O$	55.2	2.5~4.0					
Siderite 菱铁矿	$FeCO_3$	48.2	3.8	Gray or tan 灰色或黄褐色	Gray or yellow 灰色或带黄色	30~40	Less 少	Easy to be broken and reduced after roasting 焙烧后，易破碎，最易还原

Due to different chemical composition, crystal structure and geological conditions, various iron ores have different external morphology and physical characteristics.

由于化学成分、结晶构造及生成的地质条件不同，各种铁矿石具有不同的外部形态和物理特征。

Hematite

赤铁矿

Hematite, also known as 'Red Ore', has a variety of organizational structures, ranging from very dense crystals to loose and dispersed powders. The mineral structural components also have a variety of forms, with lamellar and tabular crystal forms. It is called specular iron ore if its surface is flaky, metallic luster and bright as mirror; it is called mica like hematite if its surface is mica like and its luster is not as good as the former; it is called red earth like hematite (also called iron ochre) if its texture is soft, lusterless and contains impurities of mellow soil; and it is generally hard to form oolitic, pisiform and kidney like hematite by colloidal deposition reality.

赤铁矿又称"红矿",其组织结构多种多样,从非常致密的结晶体到疏松分散的粉体。矿物结构成分也具有多种形态,晶形为片状和板状。外表呈片状、具有金属光泽、明亮如镜的称为镜铁矿砂;外表呈云母片状、光泽度不如前者的称为云母状赤铁矿;质地松软、无光泽、含有熟土杂质的称为红色土状赤铁矿(又称为铁赭石);以胶体沉积形成鲕状、豆状和肾形集合体的赤铁矿,其结构一般较坚实。

The color of hematite crystal is steel gray or iron black, and others are dark red. But all hematite is dark red. The hardness of hematite depends on its type, the vickers hardness of crystalline hematite is $5.5 \sim 6.0HV$, and the hardness of other forms is lower. The content of sulfur and phosphorus in hematite is less than that in magnetite. For low-grade hematite, flotation is generally used to improve its iron content, and the obtained concentrate is used for sintering and pelletizing. The reserves of hematite account for 48.3% of the world's total, which is the most abundant ore and the main source of iron ore for ironmaking.

结晶的赤铁矿外表颜色为钢灰色或铁黑色,其他为暗红色。但所有赤铁矿的条痕检测皆为暗红色。赤铁矿的硬度视其类型而不同,结晶赤铁矿的维氏硬度为$5.5 \sim 6.0HV$,其他形态的硬度较低。赤铁矿中杂质硫和磷的含量比磁铁矿中少。对低品位赤铁矿一般用浮选法提高其含铁品位,所获得的精矿供烧结、球团造块。赤铁矿储量占世界总量的48.3%,为储量最多的矿石,是炼铁所需的主要铁矿石来源。

Magnetite

磁铁矿

Magnetite is also known as 'Black Ore'. The crystal is octahedral, with dense and hard structure, generally massive. Its appearance is steel gray or black gray, with black streaks, which is difficult to be reduced and broken. It is magnetic and easy to be separated and enriched by electromagnetic separation.

磁铁矿又称"黑矿",晶体呈八面体,组织结构比较致密、坚硬,一般呈块状。其外表呈钢灰色或黑灰色,具有黑色条痕,难以还原和破碎。其显著特性是具有磁性,易用电磁选矿方法分选富集。

In nature, due to oxidation, parts of magnetite can be oxidized into hematite, which can become the ore containing both Fe_2O_3 and Fe_3O_4. But it still keeps the crystal form of the original magnetite. This phenomenon is called pseudomorphism. This kind of ore is mostly called pseudomorphism or semimorphism hematite. In general, they can be distinguished by the ratio of

$w(\mathrm{TFe})/w(\mathrm{FeO})$：

在自然界中，由于氧化作用，可使部分磁铁矿氧化成赤铁矿，成为既含 $\mathrm{Fe_2O_3}$、又含 $\mathrm{Fe_3O_4}$ 的矿石，但仍保持原磁铁矿的结晶形态，这种现象称为假象化，这种矿石多称为假象赤铁矿或半假象赤铁矿。它们一般可用 $w(\mathrm{TFe})/w(\mathrm{FeO})$ 的比值来区分：

(1) $w(\mathrm{TFe})/w(\mathrm{FeO}) = 2.33$，which is pure magnetite（为纯磁铁矿）；

(2) $w(\mathrm{TFe})/w(\mathrm{FeO}) < 3.5$，which is magnetite（为磁铁矿）；

(3) $w(\mathrm{TFe})/w(\mathrm{FeO}) = 3.5 \sim 7.0$，which is semi semimorphism hematite（为半假象赤铁矿）；

(4) $w(\mathrm{TFe})/w(\mathrm{FeO}) > 7.0$，which is semimorphism hematite（为假象赤铁矿）.

Among them, $w(\mathrm{TFe})$ is the mass fraction of total iron (also known as total iron) in the ore in %; and $w(\mathrm{FeO})$ is the mass fraction of FeO in the ore in %.

其中，$w(\mathrm{TFe})$ 为矿石中总铁（又称全铁）的含量（质量分数），%；$w(\mathrm{FeO})$ 为矿石中 FeO 的含量（质量分数），%。

The main gangues in magnetite are quartz, silicate and carbonate, and sometimes a small amount of clay. In addition, the ore may also contain pyrite and apatite, and even chalcopyrite sphalerite.

磁铁矿中的主要脉石有石英、硅酸盐和碳酸盐，有时还含有少量黏土。此外，矿石中还可能含有黄铁矿和磷灰石，甚至含有黄铜矿利闪锌矿等。

Generally, the content of iron (by mass) in the magnetite mined is 30%~60%. When the content of iron (by mass) is more than 45%, if the grain size is more than 5mm, it can be directly used for smelting in blast furnace, which is called rich ore; and if the grain size is less than 5mm, it is called rich ore powder, which can be sent to sintering and block making. When the content of iron (by mass) is less than 45% or the content of harmful impurities exceeds the specification value, it is necessary to obtain concentrate through beneficiation and block after impurity removal.

一般开采出来的磁铁矿中铁的含量（质量分数）为 30%~60%。当铁的含量（质量分数）大于 45% 时，若粒度大于 5mm，可直接供高炉冶炼，称为富矿；若粒度小于 5mm，则称为富矿粉，可送烧结造块。当铁的含量（质量分数）小于 45%，或有害杂质含量超过规格值时，则须经过选矿获得精矿和去杂后造块。

Limonite

褐铁矿

Limonite is hematite with crystal water ($m\mathrm{Fe_2O_3} \cdot n\mathrm{H_2O}$). Due to the different content of crystal water, the limonite can be divided into 6 kinds: turgite ($2\mathrm{Fe_2O_3} \cdot \mathrm{H_2O}$), raphisiderite ($\mathrm{Fe_2O_3} \cdot \mathrm{H_2O}$), hydrogoethite ($3\mathrm{Fe_2O_3} \cdot 4\mathrm{H_2O}$), limonite ($2\mathrm{Fe_2O_3} \cdot 4\mathrm{H_2O}$), yellow goethite ($\mathrm{Fe_2O_3} \cdot 2\mathrm{H_2O}$) and yellow ochre ($\mathrm{Fe_2O_3} \cdot 3\mathrm{H_2O}$). Most of limonite in nature exists in the form of limonite ($2\mathrm{Fe_2O_3} \cdot 4\mathrm{H_2O}$), and its theoretical iron content (by mass) is 59.8%.

褐铁矿为含结晶水的赤铁矿（$m\mathrm{Fe_2O_3} \cdot n\mathrm{H_2O}$）。因结晶水含量不同，褐铁矿石可分为 6 种，即水赤铁矿（$2\mathrm{Fe_2O_3} \cdot \mathrm{H_2O}$）、针赤铁矿（$\mathrm{Fe_2O_3} \cdot \mathrm{H_2O}$）、水针铁矿（$3\mathrm{Fe_2O_3} \cdot 4\mathrm{H_2O}$）、褐铁矿（$2\mathrm{Fe_2O_3} \cdot 4\mathrm{H_2O}$）、黄针铁矿（$\mathrm{Fe_2O_3} \cdot 2\mathrm{H_2O}$）、黄赭石（$\mathrm{Fe_2O_3} \cdot 3\mathrm{H_2O}$）。自然界中的褐铁矿绝大部分以褐铁矿（$2\mathrm{Fe_2O_3} \cdot 4\mathrm{H_2O}$）形态存在，其理论铁含量（质量

分数）为 59.8%。

The appearance of limonite is yellowish brown and dark brown to black. The stripes are yellow or brown, and nonmagnetic. It is formed by weathering of other iron ores. Its structure is soft, density is small, water content is large, and there are many pores. When the temperature rises, the crystal water will be removed and new pores will be left. So the reducibility is better than the former two kinds of iron ores.

褐铁矿的外观呈黄褐色、暗褐色至黑色，条痕呈黄色或褐色，无磁性。它是由其他铁矿风化而成，其结构松软、密度小、含水量大、气孔多，且在温度升高结晶水脱除后留下新的气孔，所以还原性比前两种铁矿好。

In nature, limonite is rich in very few minerals. Generally, the content of iron (by mass) is 37%~55%. The gangue is mainly clay and quartz, but the content of sulfur and phosphorus is relatively high. When the iron content grade is lower than 35%, beneficiation is needed. At present, limonite is mainly treated by gravity beneficiation and magnetization roasting magnetic separation.

自然界中褐铁矿富矿很少，一般铁的含量（质量分数）为 37%~55%，其脉石主要为黏土、石英等，但杂质硫、磷的质量分数较高。当含铁品位低于 35% 时，需进行选矿处理。目前，褐铁矿主要用重力选矿和磁化焙烧—磁选联合法处理。

Siderite
菱铁矿

The content of FeO (by mass) in siderite is 62.1%. A part of iron in carbonate can be mixed with other metals to form complex salts, such as $(Ca, Fe)CO_3$, $(Mg, Fe)CO_3$ etc. Under the action of water and oxygen, it is easy to transform into limonite and cover the surface of siderite deposit. Clay siderite is widely distributed in nature, and its inclusions are clay and sediment.

菱铁矿中 FeO 的含量（质量分数）为 62.1%。碳酸盐内的一部分铁可被其他金属混入而生成复盐，如 $(Ca, Fe)CO_3$、$(Mg, Fe)CO_3$ 等。在水和氧的作用下，其易转变成褐铁矿而覆盖在菱铁矿矿床的表面。在自然界中分布最广的是黏土质菱铁矿，其夹杂物为黏土和泥沙。

The common dense and hard siderite is gray or yellowish brown in appearance. After weathering, it turns to dark brown, with gray or yellow stripes, glass luster and non magnetism.

常见的致密、坚硬的菱铁矿，其外表颜色呈灰色或黄褐色，风化后则转交为深褐色，具有灰色或带黄色条痕，有玻璃光泽，无磁性。

The siderite with low iron content can be treated by gravity separation, magnetization roasting magnetic separation and magnetic separation flotation. Due to the decomposition of carbonate at high temperature, the content of iron can be greatly increased.

对含铁品位低的菱铁矿，可用重选法和磁化焙烧—磁选联合法，也可用磁选—浮选联合法处理。这类矿石因在高温下碳酸盐分解，可使产品铁的含量大大提高。

1.1.1.3 Pretreatment of Iron Ore
1.1.1.3 铁矿石入炉前处理

According to the quality requirements of iron ore, general iron ore is difficult to fully meet. So necessary preparation must be carried out before entering the furnace.

根据铁矿石质量要求,一般的铁矿石很难完全满足,必须在入炉前进行必要的准备处理。

For natural rich ores [such as iron content (by mass) more than 50%], they must be crushed and screened to obtain appropriate and uniform particle size. For limonite, siderite and dense magnetite, roasting treatment should be carried out to remove crystal water and CO_2, improve grade, loose structure, and improve reducibility and smelting effect.

对于天然富矿[如铁含量(质量分数)为50%以上],必须经破碎、筛分,以获得合适而均匀的粒度。对于褐铁矿、菱铁矿和致密磁铁矿还应进行焙烧处理,去除其结晶水和CO_2,提高品位,疏松组织,改善还原性,提高冶炼效果。

The treatment of poor iron ore is much more complicated. Generally, the concentrate powder containing (by mass) more than 60% iron must be obtained through crushing, screening, fine grinding and cleaning. After being mixed, it will be agglomerated to become artificial rich ore, and then it will be properly crushed and screened according to the blast furnace particle size requirements before entering the furnace.

对于贫铁矿的处理要复杂得多。一般都必须经过破碎、筛分、细磨、精选,得到含铁(质量分数)60%以上的精矿粉,经混匀后进行造块,变成人造富矿,再按高炉粒度要求进行适当破碎、筛分后入炉。

Due to the limited natural rich ore resources, and its metallurgical performance is not superior to man-made rich ore, most modern blast furnaces use man-made rich ore or most of them use man-made rich ore, mixed with a few natural rich ore smelting. In this case, the iron and steel plant have two kinds of treatment processes: man-made rich block ore and natural rich ore. The flow of preparation and treatment of iron ore before entering the furnace is shown in Figure 1-1.

由于天然富矿资源有限,其冶金性能不如人造富矿优越,所以绝大多数现代高炉都采用人造富矿或用人造富矿兑加少数天然富矿冶炼。在这种情况下,钢铁厂便兼有人造富块矿和天然富矿两种处理流程。铁矿石入炉前准备处理的流程如图1-1所示。

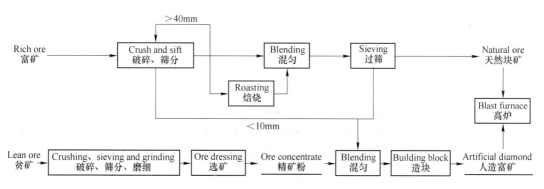

Figure 1-1 Flow chart of preparation and treatment of iron ore before entering the furnace

图1-1 铁矿石入炉前准备处理流程图

Crushing and Screening
破碎和筛分

Crushing and screening are the basic links in the preparation and treatment of iron ore. Through crushing and screening, the particle size of iron ore can reach the standard of 'Small, Hook and Net'. For lean ore, crushing separates the iron mineral from gangue monomer for beneficiation. The finer the iron ore is embedded, the finer the crushing accuracy is required.

破碎和筛分是铁矿石准备处理工作中的基本环节,通过破碎和筛分使铁矿石的粒度达到"小、匀、净"的标准。对贫矿而言,破碎使铁矿物与脉石单体分离,以便选矿。铁矿物嵌布越细密,破碎精度要求越细。

According to the different levels of product granularity requirements, crushing operations are divided into coarse crushing, medium crushing, fine crushing and crushing. Commonly used crushing equipment include jaw crusher, cone crusher, hammer crusher, roller crusher, ball mill and rod mill. The particle size range and crushing equipment of crushing operation are shown in Table 1-2.

根据对产品粒度要求的级别不同,破碎作业分为粗碎、中碎、细碎和粉碎。破碎的常用设备有颚式、圆锥、锤式、辊式破碎机、球磨机和棒磨机。破碎作业粒度范围及破碎设备见表1-2。

Table 1-2 Particle size range and crushing equipment of crushing operation
表1-2 破碎作业粒度范围及破碎设备

Crushing operation 破碎作业	Granularity of ore feed/mm 给矿粒度/mm	Ore discharge granularity/mm 排矿粒度/mm	Crushing equipment 破碎设备
Coarse crushing 粗碎	1000	100	Jaw and cone crusher 颚式、圆锥破碎机
Medium crushing 中碎	100	30	Jaw and cone crusher 颚式、圆锥破碎机
Finely broken 细碎	30	5	Hammer and roller crusher 锤式、辊式破碎机
Smash 粉碎	5	<1	Ball mill, rod mill 球磨机、棒磨机

In order to screen out large blocks and powders, and to classify the ores within the qualified particle size range, screening is required. Screening is a process that the mixed materials with different particle sizes are divided into several different particle sizes through a single or multi-layer screen surface. The common screening equipment includes fixed bar screen, cylinder screen, vibrating screen, etc.

为了筛出大块和粉末,并对合格粒度范围内的矿石进行分级,需要进行筛分。筛分是将颗粒大小不同的混合物料通过单层或多层筛面，分成若干不同粒度级别的过程。筛分的常用设备有固定条筛、圆筒筛、振动筛等。

Roasting

焙烧

Roasting is a process in which iron ore is heated to a temperature lower than its melting point in a proper atmosphere, and physical and chemical changes occur in the solid state. The common roasting methods are oxidation roasting, reduction magnetization roasting and chlorination roasting.

焙烧是在适当的气氛中，使铁矿石加热到低于其熔点的温度，并在固态下发生物理化学变化的过程。常见的焙烧方法有氧化焙烧、还原磁化焙烧和氯化焙烧。

Oxidation roasting is carried out in an oxidizing atmosphere with sufficient air to ensure complete combustion of fuel and oxidation of ore. It is mainly used to remove CO_2, H_2O and S (decomposition of carbonate and crystal water, and oxidation of sulfide), making the structure of dense ore loose and easy to reduce. For example, the roasting of siderite is decomposed between 500~900℃ according to reaction (1-1):

氧化焙烧在空气充足的氧化性气氛中进行，以保证燃料的完全燃烧和矿石的氧化。其多用于去除 CO_2、H_2O 和 S（碳酸盐和结晶水分解、硫化物氧化），使致密矿石的组织变得疏松而易于还原。例如菱铁矿的焙烧，在500~900℃之间按反应（1-1）分解：

$$4FeCO_3 + O_2 = 2Fe_2O_3 + 4CO_2(g) \tag{1-1}$$

During the dehydration of limonite, the reaction (1-2) ~ reaction (1-5) take place between 250~500℃:

在褐铁矿的脱水过程中，250~500℃之间可发生反应（1-2）~反应（1-5）：

$$2Fe_2O_3 \cdot 3H_2O = 2Fe_2O_3 + 3H_2O(g) \tag{1-2}$$

Oxidation roasting can also oxidize sulfur in ore. The reaction is given by:

氧化焙烧还可使矿石中的硫氧化，其反应式为：

$$3FeS_2 + 8O_2 = Fe_3O_4 + 6SO_2(g) \tag{1-3}$$

The reduction magnetization roasting is carried out in the reduction atmosphere, the main purpose of which is to transform Fe_2O_3 in the poor hematite into Fe_3O_4 with magnetism, so as to facilitate magnetic separation. The reactions are given by:

还原磁化焙烧则在还原气氛中进行，其主要目的是使贫赤铁矿中的 Fe_2O_3 转变为具有磁性的 Fe_3O_4，以便磁选，其反应式为：

$$3Fe_2O_3 + CO = 2Fe_3O_4 + CO_2(g) \tag{1-4}$$

$$3Fe_2O_3 + H_2 = 2Fe_3O_4 + H_2O(g) \tag{1-5}$$

The purpose of chlorination roasting is to recover non-ferrous metals (such as zinc, copper, tin, etc.) from hematite or remove other harmful impurities.

氯化焙烧的目的是为了回收赤铁矿中的有色金属（如锌、铜、锡等），或去除其他有害杂质。

Blending

混匀

It is necessary to mix the iron ore in order to stabilize the chemical composition of the iron ore entering the blast furnace and the operation of the blast furnace, keeping the furnace running smoothly and improving the smelting index.

进入高炉的铁矿石,为了稳定其化学成分,从而稳定高炉操作、保持炉况顺行、改善冶炼指标,需要对铁矿石进行混匀作业。

The mixing operation follows the principle of 'Paving Straight', that firstly, the incoming materials shall be overlapped one thin layer by one in order, and laid into a stack with a certain height and size, and then cut and transported out vertically along the cross section of the stack, one section by one.

混匀作业遵行"平铺直取"的原则,即先将来料按顺序一薄层一薄层地往复重叠,铺成具有一定高度和大小的条堆,然后再沿料堆横断面一个截面一个截面地垂直切取运出。

At present, because the main raw material of blast furnace is sinter or pellet, the mixing work of natural iron ore is very little. But the main raw materials of sintering and pelletizing which include iron concentrate and iron powder, also need to be mixed to ensure the stability of sinter and pelletizing composition.

目前由于高炉普遍使用的主体原料是烧结矿或球团矿,天然铁矿石的混匀工作已很少。但是烧结、球团原料的主体原料铁精矿和铁粉矿同样需要混匀,以保证烧结矿和球团矿成分的稳定。

Beneficiation

选矿

In order to improve the ore grade, to remove some harmful impurities, to recover some useful elements in the composite ore, and to make the poor ore resources be effectively utilized, it is necessary to separate the ore (beneficiation). Mineral processing is a process of enriching useful minerals by mechanically separating useful minerals and gangues according to the properties of ores and adopting appropriate methods.

为了提高矿石品位,去除部分有害杂质,回收复合矿中的一些有用元素,使贫矿资源得到有效利用,需要对矿石进行选别(即选矿)。选矿是指依据矿石的性质,采用适当的方法,把有用矿物和脉石机械地分开,从而使有用矿物富集的过程。

Concentrate refers to the useful mineral concentrate obtained through beneficiation, such as iron concentrate, iron vanadium concentrate, etc.. While the rest mainly composed of gangue is called tailings, which is generally abandoned. However, in some complex iron ores, some useful elements are often enriched in tailings (such as titanium in vanadic titanomagnetite, rare earth elements in Baotou ore, etc.), which must be further selected. Intermediate products with useful mineral content between concentrate and tailings are called middling, and further separation is needed to improve metal recovery. There are three methods commonly used to clean iron ore in modern times:

精矿是指通过选矿获得的有用矿物富集品,如铁精矿、铁钒精矿等。而主要由脉石组成的其余部分则称为尾矿,一般废弃。但在一些复合铁矿石中,常有一些有用元素富集于尾矿中(如钒钛磁铁矿中的钛、包头矿中的稀土元素等),必须将它们进一步精选出来。有用矿物含量介于精矿和尾矿之间的中间产品称为中矿,也需进一步选分以提高金属回收

率。现代常用的精选铁矿石的方法主要有以下三种:

(1) Re-election. Gravity separation uses the difference of density between iron bearing useful minerals and gangue minerals. When the size of the two minerals is similar and the minerals sink in the medium, the dense ore will settle rapidly and separate from gangue. The commonly used medium is water. Sometimes a liquid with a density greater than water is used as the medium, which is called heavy liquid separation. At present, the commonly used equipment of iron ore gravity separation in China are jig, heavy medium chute, plane or centrifugal shaker and centrifugal concentrator.

(1) 重选。重选是利用含铁有用矿物与脉石矿物密度的差异来选别。当两者粒度相近且在介质中沉落时,密度大的含铁矿物将迅速沉降并与脉石分开。常用的介质为水,有时还用密度大于水的液体作介质,称为重液选。目前中国铁矿石常用的重选设备有跳汰机、重介质溜槽、平面或离心式摇床和离心选矿机。

(2) Magnetic separation. Magnetic separation uses the different characteristics of magnetic conductivity of useful minerals and gangue minerals for separation. If the permeability of pure iron is 100%, the magnetite with strong magnetism, the ilmenite with medium magnetism and the hematite with weak magnetism are 40.2%, 24.7% and 1.32% respectively, and the gangue with non-magnetic pyrite is less than 0.5%. Under the action of magnetic field, strong magnetic particles (such as Fe_3O_4) are separated from weak magnetic particles (such as Fe_2O_3) or non-magnetic particles (such as quartz). If hematite is processed by magnetic separation, it needs to be magnetized and roasted in advance. Generally, dry magnetic separator is used to process coarse ore and wet magnetic separator is used to process fine ore. According to the magnetic field strength, the magnetic separator with the magnetic field strength higher than 320000A/m is called strong magnetic separator, and the magnetic separator with the magnetic field strength between 72000A/m and 320000A/m is called weak magnetic separator. The commonly used magnetic separators are electromagnetic separators and permanent magnetic separators.

(2) 磁选。磁选是利用有用矿物与脉石矿物导磁性特点的不同进行选分。如纯铁的磁导率为100%,则强磁性的磁铁矿为40.2%,中磁性的钛铁矿为24.7%,弱磁性的赤铁矿为1.32%,无磁性的黄铁矿石英脉石等在0.5%以下。在磁场作用下,强磁性的颗粒(如Fe_3O_4)便与弱磁性(如Fe_2O_3)或无磁性(如石英)的颗粒分开。赤铁矿若采用磁选则需事先进行磁化焙烧。一般用干式磁选机处理粗粒矿石,用湿式磁选机处理细粒矿石。按磁场强度,高于320000A/m的磁选机称为强磁选机,在72000~320000A/m之间的称为弱磁选机。常用的磁选机有电磁选机和永磁矿机。

(3) Flotation. Flotation uses the surface of minerals with different hydrophilicity for separation. Before flotation, minerals should be ground to a certain size, so that useful minerals and gangue minerals can basically achieve monomer separation. When air stirring is carried out in fine grinding pulp, the particles with strong hydrophilicity are easy to be wetted by water and sink, while the particles with weak hydrophilicity are difficult to be wetted by water and float up, so as to separate useful minerals from gangue. In order to improve the flotation effect, various flotation reagents are often used to regulate and control the flotation process, such as forming thin

films on the surface of ore particles, controlling wetting, promoting flotation agents, forming bubbles and stabilizing foam, and ensuring the floating agent does not sink. Due to the various functions of flotation agents, minerals can be separated according to the needs, so flotation is particularly suitable for the treatment of composite ore and non-ferrous metal ore.

(3) 浮选。浮选利用矿物表面具有不同的亲水性进行选分。浮选前矿物要磨碎到一定粒度，使有用矿物与脉石矿物基本达到单体分离。在细磨矿浆中进行充气搅拌时，亲水性强者的颗粒表面易被水润湿而下沉，亲水性弱者的颗粒表面难以被水润湿而浮起，从而使有用矿物与脉石分离。为了提高浮选效果，常使用各种浮选药剂来调节和控制浮选过程，如在矿粒表面形成薄膜、控制润湿、促进浮起的捕集剂，形成气泡和稳定泡沫、保证浮起者不下沉的气泡剂等。由于浮选剂的多种作用，可以根据需要来选别矿物，因此浮选特别适用于处理复合矿和有色金属矿石。

Because of the complexity of some ore properties, it is often necessary to combine several methods for beneficiation in order to maximize the comprehensive recovery and utilization of useful metal elements.

有些矿石性质复杂，往往需要将几种方法联合起来选矿，以最大限度地综合回收利用其中的有用金属元素。

1.1.1.4 Quality Evaluation of Iron Ore
1.1.1.4 对铁矿石的质量评价

Iron ore is the main raw material for blast furnace ironmaking, which affects the smelting process and technical economic indicators. The factors that determine the quality of iron ore mainly include chemical composition, physical properties and metallurgical properties, which are generally evaluated from the following aspects.

铁矿石是高炉炼铁的主要原料，它直接影响高炉的冶炼过程和技术经济指标。决定铁矿石质量的因素主要包括化学成分、物理性质和冶金性能，通常从以下几方面评价。

Grade of Iron Ore

铁矿石的品位

The grade of iron ore refers to its iron content, which is the main indicator to measure iron ore. The higher the grade is, the lower the coke ratio and the higher the output will be, so as to improve the economic benefits. Experience shows that if the content of iron (by mass) in ore is increased by 1%, the coke ratio is reduced by 2% and the output is increased by 3%. Because the increase of grade means that the acid gangue is greatly reduced and less limestone can be added to make slag during smelting, the slag amount is greatly reduced, which not only saves heat, but also promotes the smooth operation of the furnace. For example, the content of iron (by mass) in the acid lean iron ore in Anshan area is 30%, and the content of SiO_2 (by mass) is 50%; the grade of concentrate after enrichment reaches 60%, the content of SiO_2 (by mass) is reduced to 14%, the content of iron is doubled, the content of SiO_2 is reduced by nearly 3/4, and the amount of slag and flux used to produce 1t pig iron is reduced to 1/8 of the original. It can be seen that improving the taste has a great influence on smelting.

铁矿石的品位是指其铁含量,是衡量铁矿石的主要指标。品位越高,越有利于降低焦比和提高产量,从而提高经济效益。经验表明,若矿石中铁含量(质量分数)提高1%,则焦比降低2%,产量增加3%。因为品位提高意味着酸性脉石大幅度减少,冶炼时可少加石灰石造渣,因而渣量大大减少,这样既节省热量,又促进炉况顺行。例如,鞍山地区的酸性贫铁矿中铁含量(质量分数)为30%,SiO_2含量(质量分数)为50%;富选后精矿的品位达到60%,SiO_2含量(质量分数)降低到14%,铁含量提高一倍,SiO_2含量降低近3/4,而每生产1t生铁的渣量和熔剂用量减少到原来的1/8。由此可见,提高品位对冶炼的影响是很大的。

The grade of iron ore is also related to gangue composition, impurity content and ore type. For example, the requirements for iron content of limonite, siderite and alkaline gangue can be appropriately relaxed, because the grade of limonite and siderite will be improved after decomposition of H_2O and CO_2 by heating. The content of CaO in alkaline gangue is high, so limestone can be added less or not during smelting, and the grade should be evaluated by deducting the iron content of CaO.

铁矿石的品位与矿石的脉石成分、杂质含量和矿石类型等因素有关。如对褐铁矿、菱铁矿和碱性脉石矿铁含量的要求可适当放宽,因为褐铁矿、菱铁矿受热分解出H_2O和CO_2后品位会提高。碱性脉石矿CaO含量高,冶炼时可少加或不加石灰石,其品位应按扣去CaO的铁含量来评价。

Gangue Composition
脉石成分

Gangue is divided into alkaline gangue and acid gangue. The main components of alkaline gangue are Cao and MgO, and the main components of acid gangue are SiO_2 and Al_2O_3. In general, most of the iron ores contain acid gangue that the content of SiO_2 is high. A considerable amount of limestone needs to be added to make slag with basicity of $w(CaO)/w(SiO_2) \approx 1.0$, so as to meet the needs of smelting process. Therefore, it is hoped that the less of the content of acid gangue is better. The alkaline gangue with high CaO content has high smelting value. If the composition of an iron ore is: $w(Fe) = 45.30\%$, $w(CaO) = 10.05\%$, $w(MgO) = 3.34\%$, $w(SiO_2) = 11.20\%$, then the natural basicity of the iron ore is $w(CaO)/w(SiO_2) = 0.9$, $[w(CaO)+w(MgO)]/w(SiO_2) = 1.2$, which is close to the normal range of basicity of slag and belongs to self melting rich ore. MgO in gangue can also improve slag performance, but this kind of ore is rare. The content of Al_2O_3 in gangue should also be controlled. If the content of Al_2O_3 is too high, and the content of Al_2O_3 (by mass) in slag exceeds 20%, the slag is difficult to melt and flow, which makes smelting difficult. Therefore, the problem of slag fluidity should be solved by increasing MgO content.

脉石可分为碱性脉石和酸性脉石。碱性脉石的主要成分为CaO和MgO,酸性脉石的主要成分为SiO_2和Al_2O_3。一般铁矿石中含酸性脉石者居多,即其中SiO_2含量高,需加入相当数量的石灰石造成碱度$w(CaO)/w(SiO_2) \approx 1.0$的炉渣,以满足冶炼工艺的需求。因此希望酸性脉石含量越少越好,而CaO含量高的碱性脉石则具有较高的冶炼价值。如某铁矿成分为:$w(Fe) = 45.30\%$,$w(CaO) = 10.05\%$,$w(MgO) = 3.34\%$,$w(SiO_2) = 11.20\%$,

则该铁矿的自然碱度 $w(CaO)/w(SiO_2)= 0.9$，$[w(CaO)+w(MgO)]/w(SiO_2)= 1.2$，接近炉渣碱度的正常范围，属于自熔性富矿。脉石中的 MgO 还有改善炉渣性能的作用，但这类矿石不多见。脉石中 Al_2O_3 含量也应控制，若 Al_2O_3 含量过高，如使炉渣中 Al_2O_3 含量（质量分数）超过20%时，炉渣难熔而不易流动，从而给冶炼造成困难。因此，应采取提高 MgO 含量的方法来解决炉渣流动性的问题。

Content of Harmful Impurities and Beneficial Elements
有害杂质和有益元素的含量

Harmful impurities usually refer to S, P, Pb, Zn, As and so on. The lower their content, the better. Cu is sometimes harmful and sometimes beneficial, depending on the circumstances.

有害杂质通常指 S、P、Pb、Zn、As 等，它们的含量越低越好。Cu 有时有害，有时有益，视具体情况而定。

S is the most harmful element to steel, which makes steel hot brittle. The so-called hot embrittlement refers to that sulfur is almost insoluble in solid iron and forms FeS with it. The melting point of the eutectic formed by FeS and Fe is 988℃, which is lower than the starting temperature of steel hot working (1150~1200℃). During hot working, the eutectic distributed in the grain boundary melts first and causes cracking. Therefore, the lower the sulfur content, the better. According to the national standard, $w(S) \leqslant 0.07\%$ in pig iron and $w(S) \leqslant 0.03\%$ in high-quality pig iron, that the sulfur content in steel should be strictly controlled. More than 90% sulfur can be removed in the process of blast furnace ironmaking. However, the basicity of slag needs to be improved to increase the slag quantity, which leads to the increase of coke ratio and the decrease of output. For every 0.1% increase of sulfur content (by mass) in ore, the coke ratio increases by 5%. Generally, it is specified that the ore with $w(S) \leqslant 0.06\%$ is primary ore, the ore with $w(S) \leqslant 0.2\%$ is secondary ore, and the ore with $w(S) > 0.3\%$ is high sulfur ore. For high sulfur ores, the sulfur content can be reduced by mineral processing and sintering. Sulfur can improve the cutting performance of steel. In free cutting steel, sulfur content (by mass) can reach 0.15%~0.3%.

S 是对钢铁危害最大的元素，它使钢材具有热脆性。热脆是指硫几乎不溶于固态铁而与其形成 FeS。FeS 与 Fe 形成的共晶体的熔点为 988℃，低于钢材热加工的开始温度（1150~1200℃）。热加工时，分布于晶界的共晶体先行熔化而导致开裂，因此矿石硫含量越低越好。国家标准规定生铁中 $w(S) \leqslant 0.07\%$，优质生铁中 $w(S) \leqslant 0.03\%$，即应严格控制钢中硫含量。高炉炼铁过程可除去 90% 以上的硫。但脱硫需要提高炉渣碱度，使渣量增加，导致焦比增加而产量降低。矿石中硫含量（质量分数）每增加 0.1%，焦比升高 5%。一般规定 $w(S) \leqslant 0.06\%$ 的矿石为一级矿，$w(S) \leqslant 0.2\%$ 的为二级矿，$w(S) > 0.3\%$ 的为高硫矿。对于高硫矿石，可以通过选矿和烧结的方法降低硫含量。硫可改善钢材的切削性能，在易切削钢中，硫含量（质量分数）可达 0.15%~0.3%。

P is a harmful component in steel, which makes steel cold brittle. S can be dissolved in α-Fe (up to 1.2%). The strength of ferrites between grains is greatly increased by the phosphorus atoms which are dissolved and enriched in the grain boundary, so that the room temperature strength and brittleness of steel increase, which is called cold brittleness. S is easy to segregate in the process of steel crystallization, and it is difficult to eliminate it by heat treatment, which will increase the risk of steel cold embrittlement. However, the liquid iron containing phosphorus has good fluidity and filling property, which is beneficial to the manufacture of deformed and complex castings. S can also improve the cutting performance of steel, so the content of S (by mass) in free cutting steel can reach 0.08%~0.15%. S in ore is not easy to be removed in the process of beneficiation and sintering, and almost all phosphorus enters pig iron in the process of blast furnace smelting. Therefore, the content of S in pig iron depends on the content of S in ore, and the lower the content of S in iron ore, the better.

P 是钢材中的有害成分,使钢具有冷脆性。P 能溶于 α-Fe 中(可达 1.2%),固溶并富集在晶粒边界的磷原子使铁素体在晶粒间的强度大大增高,从而使钢材的室温强度提高而脆性增加(称为冷脆)。磷在钢的结晶过程中容易偏析,而且很难用热处理的方法来消除,也会使钢材冷脆的危险性增加。但含磷铁水的流动性和充填性好,对制造畸形复杂铸件有利。P 也可改善钢材的切削性能,所以在易切削钢中 P 含量(质量分数)可达 0.08%~0.15%。矿石中的 P 在选矿和烧结过程中不易除去,在高炉冶炼过程中 P 几乎全部进入生铁。因此,生铁中 P 含量取决于矿石中的 P 含量,要求铁矿石中的 P 含量越低越好。

Pb, Zn and As are easy to be reduced in blast furnace. Pb is insoluble in iron and its density is higher than that of iron. After reduction, it is deposited at the bottom of furnace, which is very destructive. When it boils at 1750℃, the volatilized lead vapor circulates in the furnace, forming furnace lump. After zinc reduction, a large amount of zinc vapor volatilises and rises in the high temperature area, and part of it is deposited in the furnace wall with ZnO, which makes the furnace wall expand and crack and form furnace tumor. As can be completely reduced into pig iron, which can reduce the weldability of steel and make it cold brittle. The content of As in pig iron shall be less than 1%, and high quality pig iron shall be free of arsenic. Pb, Zn and As in iron ores are usually in the form of sulfide, such as galena (PbS), sphalerite (ZnS) and arsenopyrite (FeAsS). It is difficult to exclude Pb and Zn in the sintering process, so the lower the content is, the better the content is. Generally, the content of each of them is less than 0.1%. The iron ore with high Pb can be separated by chlorination roasting and flotation. The ore with high content of Zn can not be smelted directly alone. It should be mixed with the ore with low content of Zn or processed by roasting and beneficiation to reduce the content of Zn in the iron ore. In the sintering process, As in ore can be partly removed, which can be eliminated by chlorination roasting. Generally, the content of As (by mass) in iron ore should not exceed 0.07%.

Pb、Zn 和 As 在高炉内均易还原。Pb 不溶于铁且密度比铁大,还原后沉积于炉底,破坏性很大。其在 1750℃ 时沸腾,挥发的 Pb 蒸气在炉内循环,形成炉瘤。Zn 还原后在高温区以锌蒸气大量挥发上升,部分以 ZnO 沉积于炉墙,使炉墙胀裂并形成炉瘤。As 可全部还原进入生铁,它可降低钢材的焊接性并使之冷脆。生铁中 As 含量(质量分数)应小于 1%,优质生铁应不含 As。铁矿石中的 Pb、Zn、As 常以硫化物形态存在,如方铅矿(PbS)、闪锌矿(ZnS)和毒砂(FeAsS)。烧结过程中很难排除 Pb 和 Zn,因此要求其含量越低越好,一般要求各自的含量(质量分数)不超过 0.1%。Pb 含量高的铁矿石可以通过氯化焙烧和浮选的方法使 Pb、Fe 分离。Zn 含量高的矿石不能单独直接冶炼,应该与含

Zn 少的矿石混合使用或进行焙烧、选矿等处理，降低铁矿石中的 Zn 含量。烧结过程中能部分去除矿石中的砷，可以采用氯化焙烧方法排除。通常要求铁矿石中砷含量（质量分数）不超过 0.07%。

If the content of Cu (by mass) in the steel is less than 0.3%, the corrosion resistance of the steel can be increased; if it is more than 0.3%, the weldability will be reduced, and there will be hot embrittlement. Generally, Cu can not be removed in sintering, and all Cu is reduced into pig iron in blast furnace, so the content of Cu in steel depends on that in raw material. Generally, the content of Cu (by mass) in iron ore is not more than 0.2%. For some refractory high oxide ores of Cu, the chlorination roasting method can be used to recover Cu, and the high copper $[w(Cu)>1.0\%]$ cast pig iron can be smelted at the same time. It has good mechanical properties and corrosion resistance.

钢中 Cu 含量（质量分数）若不超过 0.3%，可增加钢材抗蚀性；若超过 0.3%，则降低其焊接性，并有热脆现象。Cu 在烧结中一般不能去除，但在高炉中能全部还原进入生铁，所以钢铁中 Cu 含量取决于原料中的 Cu 含量。一般铁矿石允许的 Cu 含量（质量分数）不超过 0.2%。对于一些难选的高铜氧化矿，可采用氯化焙烧法回收 Cu，同时可炼高 Cu $[w(Cu)>1.0\%]$ 铸造生铁，具有很好的力学性能和耐腐蚀性能。

In addition, some iron ores also contain alkali metals: K and Na. Most of them are volatilized after reduction in the high temperature area of the lower part of the blast furnace, and oxidized in the upper part. Then enter the charge, causing circulation accumulation and furnace wall nodulation. Therefore, the content of alkali metal in ore must be strictly controlled. The quantity of alkali metal (K_2O+Na_2O) in China is limited to 5~7kg/t, while that in foreign countries is less than 3.5kg/t.

此外，一些铁矿石还含有碱金属 K、Na，它们在高炉下部高温区大部分被还原后挥发，到上部又氧化进入炉料中，从而造成循环累积，使炉墙结瘤。因此，矿石中碱金属含量必须严格控制。中国普通高炉碱金属（K_2O+Na_2O）入炉量限制为 5~7kg/t，国外高炉碱金属（K_2O+Na_2O）入炉限制量为低于 3.5kg/t。

F enters the slag in the form of CaF_2 during smelting. CaF_2 can reduce the melting point of slag and increase the fluidity of slag. When the content of F in iron ore is high, the slag is formed early in the blast furnace, which is not conductive to ore reduction. When the content of F in ore is less than 1%, it has no effect on smelting; when it reaches 4%~5%, it is necessary to control the fluidity of slag. In addition, the volatilization of F at high temperature has certain corrosion effect on refractories and metal components.

F 在冶炼过程中以 CaF_2 形态进入渣中。CaF_2 能降低炉渣的熔点，增加炉渣的流动性。当铁矿石中 F 含量高时，炉渣在高炉内过早形成，不利于矿石还原。矿石中 F 含量（质量分数）不超过 1%时，对冶炼无影响；达到 4%~5%时，需要注意控制炉渣的流动性。此外，高温下 F 的挥发对耐火材料和金属构件有一定的腐蚀作用。

Mn, Cr, Ni, Co, V, Ti and Mo are often found in iron ores, and Nb, Ta, Ce and La are also found in Baotou Bayan Obo Iron Ore. These elements can improve the properties of steel, so they are called beneficial elements. When their contents in the ore reaches a certain value, such as $w(Mn) \geqslant 5\%$, $w(Cr) \geqslant 0.06\%$, $w(Ni) \geqslant 0.2\%$, $w(Co) > 0.03\%$, $w(V) > 0.1\% \sim 0.15\%$, $w(Mo) > 0.3\%$, $w(Cu) \geqslant 0.3\%$, the ore is called composite ore, which has great economic value and should be considered for comprehensive utilization.

铁矿石中常共生有 Mn、Cr、Ni、Co、V、Ti 和 Mo，包头白云鄂博铁矿还含有 Nb、Ta 及稀土元素 Ce、La 等。这些元素有改善钢铁性能的作用，故称为有益元素。当它们在矿石中的含量达一定数值时，如 $w(Mn) \geqslant 5\%$、$w(Cr) \geqslant 0.06\%$、$w(Ni) \geqslant 0.2\%$、$w(Co) > 0.03\%$、$w(V) > 0.1\% \sim 0.15\%$、$w(Mo) > 0.3\%$、$w(Cu) \geqslant 0.3\%$，则称该矿石为复合矿石，其经济价值很大，应考虑综合利用。

For some harmful impurities in iron ore, if the content is high, such as $w(Pb) \geqslant 0.5\%$, $w(Zn) \geqslant 0.7\%$, $w(Sn) \geqslant 0.2\%$, it should be regarded as comprehensive utilization of composite ore. Because these impurities themselves are also important metals.

对于铁矿石中一些有害杂质，如果含量较高，如 $w(Pb) \geqslant 0.5\%$、$w(Zn) \geqslant 0.7\%$、$w(Sn) \geqslant 0.2\%$ 时，应视其为复合矿石综合利用，因为这些杂质本身也是重要的金属。

Particle Size and Strength of Iron Ore
铁矿石的粒度和强度

The iron ore entering the furnace shall have appropriate particle size and sufficient strength. If the particle size is too large, it will reduce the contact area between gas and iron ore and make iron ore hard to reduce; if the particle size is too small, it will increase the air flow resistance and be blown out of the furnace easily to form furnace dust loss; and if the particle size is uneven, it will seriously affect the permeability of the material column. Therefore, the bulk should be broken, the powder should be sieved, and the particle size should be suitable and uniform. Generally, the ore particle size is required to be within 5~40mm, and efforts are made to reduce the upper and lower limit particle size difference.

入炉铁矿石应具有适宜的粒度和足够的强度。粒度过大，会减小煤气与铁矿石的接触面积，使铁矿石不易还原；粒度过小，会增加气流阻力，同时易被吹出炉外形成炉尘损失；粒度大小不均，则严重影响料柱透气性。因此，大块应破碎，粉末应筛除，粒度应适宜而均匀。一般要求矿石粒度在 5~40mm 范围内，并力求缩小上下限粒度差。

The strength of iron ore refers to its impact resistance and friction resistance. With the continuous expansion of the blast furnace volume, the strength of the iron ore into the furnace should also be increased accordingly. Otherwise, it is easy to generate powder and fragments, that it will increase the loss of furnace dust, and make the permeability of the blast furnace charging column bad, and causing the furnace condition not smooth.

铁矿石的强度是指铁矿石耐冲击、耐摩擦的强弱程度。随着高炉容积不断扩大，入炉铁矿石的强度也要相应提高，否则易生成粉末和碎块，增加炉尘损失，使高炉料柱透气性变坏，引起炉况不顺。

Reduction of Iron Ore
铁矿石的还原性

The reducibility of iron ore refers to the ease of reduction of iron ore by reducing gas CO or H_2, which is an important index to evaluate the quality of iron ore. The better the reducibility is, the better the coke ratio is and the higher the output is. Improving the reducibility of iron ore (or adopting easily reducible iron ore) is one of the important measures to strengthen blast furnace smelting. The main factors affecting the reducibility of iron ore are mineral composition, density of ore structure, particle size and porosity.

铁矿石的还原性是指铁矿石被还原性气体 CO 或 H_2 还原的难易程度,是评价铁矿石质量的重要指标。还原性越好,越有利于降低焦比,提高产量。改善铁矿石的还原性(或采用易还原铁矿石)是强化高炉冶炼的重要措施之一。影响铁矿石还原性的因素主要有矿物组成、矿石结构的致密程度、粒度和孔隙率等。

Stability of Chemical Composition of Iron Ore
铁矿石化学成分的稳定性

The fluctuation of chemical composition of iron ore will cause the fluctuation of furnace temperature, basicity and properties of slag and pig iron quality, which will cause the furnace condition to be not smooth, increase the coke ratio and decrease the output. At the same time, the frequent fluctuation of furnace condition will make the automatic control of blast furnace difficult to realize. Therefore, the fluctuation range of charge composition is strictly controlled at home and abroad. The effective way to stabilize the ore composition is to mix the ore evenly.

铁矿石化学成分的波动会引起炉温、炉渣碱度和性质以及生铁质量的波动,造成炉况不顺,从而使焦比升高,产量下降。同时,炉况的频繁波动还会使高炉自动控制难以实现。因此,国内外都严格控制炉料成分的波动范围。稳定矿石成分的有效方法是对矿石进行混匀处理。

1.1.2 Flux
1.1.2 熔剂

In addition to iron ore and coke, a certain amount of flux should be added in the blast furnace smelting.

高炉冶炼中除主要加入铁矿石和焦炭外,还要加入一定量的助熔物质,即熔剂。

1.1.2.1 Function of Flux
1.1.2.1 熔剂的作用

Since the gangue and coke ash in the ore are mostly composed of acid oxide SiO_2 and Al_2O_3, they are all high melting point compounds. The melting point of SiO_2 is 1713℃, and the melting point of Al_2O_3 is 2050℃. These high melting point substances are difficult to melt at the smelting temperature of blast furnace. Therefore, the role of adding flux lies in two aspects. On the one hand, the flux can generate low melting point compounds and co-melts with gangue in ore and high melting point substances in coke ash, forming slag that is easy to flow out of the hearth and separated from molten iron. On the other hand, a certain number of slag with certain physical and chemical properties can be formed, which can remove some harmful impurities (such as S) so as to improve the quality of pig iron.

由于矿石中脉石和焦炭灰分的组成大多为酸性氧化物 SiO_2 和 Al_2O_3,它们都属于高熔点的化合物，SiO_2 的熔点为 1713℃，Al_2O_3 的熔点为 2050℃，这些高熔点物质在高炉冶炼的温度下很难熔化。因此，加入熔剂的作用在于两个方面，一方面，熔剂能与矿石中的脉石、焦炭灰分中的高熔点物质生成低熔点的化合物和共熔体，形成易从炉缸流出的炉渣，与铁水分离；另一方面，形成一定数量和具有一定物理和化学性能的炉渣，能够去除部分有害杂质（如 S），从而起到改善生铁质量的作用。

1.1.2.2 Classification and Characteristics of Flux
1.1.2.2 熔剂的分类及特性

Since the gangue of iron ore is mainly composed of SiO_2, basic fluxes containing CaO and MgO are often used to make gangue slag.

由于铁矿石的脉石成分绝大多数以 SiO_2 为主，所以常用含有 CaO 和 MgO 的碱性熔剂使矿物中的脉石造渣。

Limestone and dolomite are the most commonly used basic fluxes for blast furnace, and their chemical composition is shown in Table 1-3.

高炉最常用的碱性熔剂是石灰石和白云石，其化学成分见表 1-3。

Table 1-3 Chemical composition of common fluxes for blast furnace
表 1-3 高炉常用熔剂的化学成分

Types of flux 溶剂类型	Components (mass fraction)/% 化学成分（质量分数）/%					
	CaO	MgO	SiO_2	Al_2O_3	S	Burning loss 烧损率
Dolomite 白云石	32.22	18.37	3.71	1.79	0.094	43.67
Limestone 石灰石	52.26	1.58	2.58	1.71	0.139	41.51

When the content of basic oxide in gangue is high, acid flux is used (SiO_2 is commonly used). In recent years, in order to increase the output of cast iron, SiO_2 is often added to the high basicity sinter. In order to make full use of iron and steel industry waste, some blast furnaces use converter slag with high basicity instead of basic flux. At present, most blast furnaces use the charge structure with appropriate proportion of basic charge (high basicity sinter) and acid charge (pellet and lump ore), so no or less flux can be added. When the slag is thick and the furnace condition is abnormal, part of fluorite (CaF_2) is used for short time to dilute the slag and eliminate the accumulation (or bond).

当脉石中碱性氧化物含量较高时，则用酸性熔剂（常用的有硅石等）。近年来，为提高生铁的产量，在高碱度烧结矿冶炼铸造生铁时，常加入 SiO_2。为充分利用钢铁工业废弃物，有些高炉用高碱度的转炉钢渣代替碱性熔剂。目前高炉大多使用碱性炉料（高碱度烧结矿）和酸性炉料（球团矿和块矿）比例合适的炉料结构，故可不加或少加熔剂。高炉直接加入熔剂只作为临时调剂措施,在炉渣黏稠、炉况失常时,短期使用部分萤石(CaF_2)来稀释炉渣,消除堆积(或黏结物)。

1.1.2.3 Quality Requirements for Flux
1.1.2.3 对熔剂的质量要求

The quality requirements for flux are as follows:
对熔剂的质量要求如下:

(1) Basic oxide (CaO+MgO) has a high content of active ingredients, and the less acidic oxide ($SiO_2+Al_2O_3$). Therefore, high effective alkalinity is required for limestone and dolomite. The effective alkalinity is the percentage of the remaining alkaline oxide in the flux after deducting the alkaline oxide required by its own acidic oxide for slagging.

(1) 碱性氧化物(CaO+MgO)的有效成分含量要高,酸性氧化物($SiO_2+Al_2O_3$)越少越好。所以,对石灰石与白云石来说,要求有效碱度高。有效碱度是指熔剂含有的碱性氧化物扣除其本身酸性氧化物造渣所需的碱性氧化物后,剩余部分的含量(质量分数)。

(2) The content of harmful impurities such as S and P should be less. There are few S and P impurities in limestone. The content of S (by mass) is only 0.01%~0.08%, and the content of P (by mass) is only 0.001%~0.03% in limestone used in iron and steel plants in China.

(2) 有害杂质S、P等含量要少。石灰石中一般S、P杂质都较少。在中国各钢铁厂使用的石灰石中,S含量(质量分数)只有0.01%~0.08%,P含量(质量分数)只有0.001%~0.03%。

1.1.3 Fuel
1.1.3 燃料

Fuel is one of the essential raw materials for blast furnace smelting, and almost all blast furnaces use coke as fuel. Due to the shortage of coking coal resources, the technology of injecting fuel from tuyere has developed rapidly to replace expensive coke. At present, the amount of injected fuel accounts for 10%~40% of the total fuel. The main fuels used for injection are coal powder, heavy oil and natural gas.

燃料是高炉冶炼不可缺少的基本原料之一,几乎所有高炉都使用焦炭作燃料。由于焦煤资源的紧缺,使风口喷吹燃料的技术迅速发展,以代替昂贵的焦炭。目前喷吹燃料用量已占全部燃料的10%~40%。用作喷吹的燃料主要有煤粉、重油和天然气等。

1.1.3.1 Function of Coke in Blast Furnace
1.1.3.1 焦炭在高炉内的作用

Coke plays the role of heating agent, reducing agent and column framework in blast furnace. Its combustion in front of tuyere produces high temperature and reducing gas containing CO and H_2, which provides reductant and heat needed in blast furnace smelting process. 70%~80% of heat in blast furnace smelting process comes from coke combustion. Coke accounts for 1/3~1/2 of the volume in the charge column. In the high temperature zone, coke is the only charge existing in solid state after ore soft melting. So it plays a role of skeleton supporting the charge column up to tens of meters. The permeability of the charge column at the lower part of the blast furnace is completely maintained by coke. At present, there is no other fuel to replace the role of coke.

焦炭在高炉内起到发热剂、还原剂及料柱骨架的作用。其在风口前燃烧产生高温和含有 CO 和 H_2 的还原性气体，并提供高炉冶炼过程所需的还原剂和热量。高炉冶炼过程中的热量有 70%~80%来自焦炭的燃烧。焦炭在料柱中占 1/3~1/2 的体积，在高温区，矿石软熔后焦炭是唯一以固态存在的炉料，故起着支撑高达数十米料柱的骨架作用。高炉下部料柱的透气性完全由焦炭来维持，焦炭的这一作用目前还没有其他燃料能够代替。

In addition, coke is the carburizing agent of pig iron, and the combustion of coke also provides a free space for the furnace charge to drop.

另外，焦炭是生铁的渗碳剂，焦炭的燃烧还为炉料下降提供了自由空间。

1.1.3.2 Requirements for Coke Quality in Blast Furnace Smelting
1.1.3.2 高炉冶炼对焦炭质量的要求

The quality of coke is generally measured from its chemical and physical properties. Chemical properties are often expressed by industrial analysis of coke, such as fixed carbon, ash, volatile, moisture and sulfur content. The physical properties of coke mainly include mechanical strength, particle size and porosity. The quality of coke has a direct impact on the blast furnace smelting process and various technical economic indicators. Therefore, there are certain quality requirements for coke entering the furnace：

衡量焦炭质量一般从其化学性质和物理性质两方面来分析。化学性质常以焦炭的工业分析来表示，即固定碳、灰分、挥发分、水分和硫的含量。焦炭的物理性质主要包括机械强度、粒度和孔隙率等。焦炭质量的好坏直接影响高炉冶炼的进行和各项技术经济指标。因此，对入炉焦炭有一定的质量要求，其中包括：

（1）Fixed carbon and ash. The content of fixed carbon and ash in coke is increasing and decreasing. And high ash content means low fixed carbon content.

（1）固定碳和灰分。焦炭中固定碳与灰分的含量互为消长，灰分含量高则意味着固定碳含量低。

The ash content of coke is mainly composed of acid oxides (SiO_2 and Al_2O_3), so it is necessary to add basic oxides with the same amount of ash to make slag. If the ash content is high, the amount of slag will increase, which will lead to the increase of coke ratio and the decrease of output. The blast furnace smelting practice shows that the ash content of coke increases by 1%, the coke ratio increases by 2%, and the output decreases by 3%. Therefore, the ash content of coke is required to be as low as possible. The ash content of coke (by mass) in China is generally between 11% and 15%. The ash in coke comes from raw coal, so coal should be washed and blended reasonably before coking to reduce the ash content.

焦炭灰分主要由酸性氧化物（SiO_2 和 Al_2O_3）构成，故在冶炼中必须配加与灰分数量大体相等的碱性氧化物来造渣。灰分含量高，渣量就会增加，会导致焦比升高、产量下降。高炉冶炼实践证明，焦炭灰分增加 1%，焦比升高 2%，产量下降 3%。因此，要求焦炭灰分含量尽量低。中国焦炭灰分含量（质量分数）一般在11%~15%之间。焦炭中的灰分来自原煤，因此在炼焦前应洗煤，并进行合理配煤，从而降低灰分含量。

(2) S. Generally, the amount of S brought in by coke accounts for 80% of the total amount of S in the charge. Therefore, reducing the content of S in coke is very important to improve the quality of pig iron. The practice shows that the sparse content of coke is increased by 0.1%, and the coke ratio is increased by 1.2%~2%. A part of S can be removed in the coking process, but most of the S remains in the coke. Therefore, the basic way to reduce the content of S in coke is through coal washing and reasonable coal blending.

(2) S。一般焦炭带入的S含量占入炉料总S含量的80%，因此，降低焦炭S含量对提高生铁质量极为重要。实践证明，焦炭中S含量（质量分数）提高0.1%，焦比升高1.2%~2%。在炼焦过程中能够去除一部分S，但是仍有大部分S留在焦炭中。因此，降低焦炭S含量的基本途径是通过洗煤和合理配煤。

(3) Moisture. The fluctuation of coke composition and performance will lead to the instability of blast furnace smelting process. Especially the fluctuation of moisture content will cause the fluctuation of coke weight, which will affect the furnace temperature and lead to the fluctuation of heat regime. The moisture content of wet quenching coke (by mass) is generally 2%~6%, which requires the moisture content in coke to be stable.

(3) 水分。焦炭成分与性能波动会导致高炉冶炼过程不稳定，特别是水分波动会引起入炉焦炭的质量波动，从而影响炉温，导致热制度波动。湿法熄焦含水量（质量分数）一般为2%~6%，要求焦炭中的水分含量要稳定。

(4) Volatile matter. Volatile matter refers to the organic matter (H_2, CH_4, N_2, etc.) that has not been completely decomposed and volatilized in the coking process, which is the main indicator to identify the maturity of coke. Under normal conditions, the volatile content is generally 0.7%~1.2%. if the content is too high, it means that the maturity of coke is poor, there are many coke, and the strength is not enough. And the powder will be produced due to brittle fracture in smelting process, which will affect the permeability of material column; and if the content is too low, it means that the coke is coking too high and fragile. So the volatile content is required to be appropriate.

(4) 挥发分。挥发分是指炼焦过程中未分解挥发完的有机物（H_2、CH_4、N_2等），是鉴别焦炭成熟程度的主要标志。正常情况下，挥发分含量一般为0.7%~1.2%。若含量过高，则表明焦炭成熟程度差，生焦多，强度不够，在冶炼过程中易碎裂产生粉末而影响料柱透气性；若含量过低，则表明焦炭结焦过高且易碎，故要求挥发分含量适当。

(5) Mechanical strength and particle size. The mechanical strength refers to the ability of coke to resist fragmentation and wear under the action of mechanical force and thermal stress. Coke must be transported for many times before entering the furnace. In the process of falling down in the furnace, it is affected by high temperature, gravity and friction between the charges. If the strength is not good, a large amount of powder will be produced. Entering the primary slag will cause the slag to become viscous, resulting in the furnace condition not smooth. The strength of coke is usually measured by small drum test. In China, it is stipulated to use a small drum (Micombe turns drums), which is a closed drum with a diameter and a length of 1000mm. The

four 100mm×50mm×10mm angle steel baffles are welded on the inner wall, which are arranged 90° to each other. During the test, take 50kg of coke with particle size greater than 60mm and put it into the drum, rotate it for 4min at the speed of 25r/min, and then sieve the sample with ϕ40mm and ϕ10mm round hole sieve. The percentage of coke with particle size greater than 40mm in the mass of the sample is called the impact strength (crushing strength) index of coke, which is expressed by M_{40}; and the percentage of coke with particle size less than 10mm is called the friction strength (grinding). The index of damage strength is represented by M_{10}. In China, the specified drum index of coke strength is: Grade I, $M_{40} \geq 75\%$, $M_{10} \leq 9.0\%$; Grade II, $M_{40} = 64\% \sim 68\%$, $M_{10} \leq 11.5\%$. The size of coke is required to be uniform and appropriate. The size of large blast furnace is 40~60mm, medium blast furnace is 25~40mm and small blast furnace is 15~25mm.

（5）机械强度与粒度。机械强度是指焦炭在机械力和热应力作用下抵抗碎裂和磨损的能力。焦炭在入炉前要经过多次转运，在炉内下降过程中受高温和炉料间重力及摩擦力的作用，如果强度不好会产生大量的粉末，进入初渣会导致炉渣变得黏稠，造成炉况不顺。常通过小转鼓试验测定焦炭强度。中国规定采用小转鼓（米库姆转鼓），它是一个直径和长度均为 1000mm 的封闭转鼓，内壁焊有 4 条 100mm×50mm×10mm 的角钢挡板，互成 90°布置。进行试验时，取粒度大于 60mm 的焦炭 50kg 装入转鼓内，以 25r/min 的速度旋转 4min，然后将试样用 ϕ40mm 和 ϕ10mm 的圆孔筛筛分，大于 40mm 的焦炭占试样质量的百分数称为焦炭的抗冲击强度（破碎强度）指标，用 M_{40} 表示；小于 10mm 的碎焦所占的质量百分数称为焦炭的抗摩擦强度（磨损强度）的指标，用 M_{10} 表示。中国规定焦炭强度转鼓指标为：一级品，$M_{40} \geq 75\%$，$M_{10} \leq 9.0\%$；二级品，$M_{40} = 64\% \sim 68\%$，$M_{10} \leq 11.5\%$。焦炭的粒度要求均匀、大小合适，大型高炉为 40~60mm，中型高炉为 25~40mm，小型高炉为 15~25mm。

（6）Reactivity and combustibility of coke. Reactivity refers to the rate at which coke reacts with CO_2 to form CO at a certain temperature. The reaction formula is：

（6）焦炭的反应性和燃烧性。反应性是指焦炭在一定温度下与 CO_2 作用生成 CO 的速度，其反应式为：

$$C+CO_2 = 2CO \qquad (1-6)$$

Flammability refers to the rate at which coke reacts with oxygen to generate CO_2 at a certain temperature. The reaction formula is：

燃烧性是指焦炭在一定温度下与氧反应生成 CO_2 的速度，其反应式为：

$$C+O_2 = CO_2 \qquad (1-7)$$

If the above reaction rate is fast, it shows that the combustion and reactivity of coke is good. It is generally believed that in order to expand the combustion zone, make the hearth temperature and gas flow distribution more reasonable and the charge drop smoothly, it is hoped that the combustibility of coke will be worse. In order to increase the CO_2 content in the top gas and improve the utilization degree of gas, it is hoped that the reactivity of coke will be worse when the temperature is lower.

若反应（1-7）速度快，则表明焦炭的燃烧性和反应性好。一般认为，为了扩大燃烧带，使炉缸温度和煤气流分布更为合理，并使炉料顺利下降，则希望焦炭的燃烧性差一些。为了提高炉顶煤气中的 CO_2 含量，改善煤气利用程度，在温度较低时则希望焦炭的反应性差一些。

1.1.4 Sinter Production
1.1.4 烧结矿生产

Sintering is a method of sintering powder materials (such as powder ore and concentrate) into blocks under the condition of incomplete melting by high temperature heating. The product is called sinter, and its shape is irregular and porous. The heat energy needed for sintering is provided by burning the carbon mixed into the sinter and the excess air, so it is also called oxidation sintering. Sinter mainly depends on liquid phase bonding (also known as melting sintering), and solid phase bonding only plays a secondary role.

烧结是指将粉状物料（如粉矿和精矿）进行高温加热，在不完全熔化的条件下烧结成块的方法。所得产品称为烧结矿，其外形是不规则多孔状。烧结所需热能由配入烧结料内的碳与通入过剩的空气经燃烧提供，故又称氧化烧结。烧结矿主要依靠液相黏结（又称熔化烧结），固相黏结仅起次要作用。

1.1.4.1 Sintering Principle
1.1.4.1 烧结原理

The sintering process starts from the surface of the material layer to the bottom, and there is obvious stratification along the height of the material layer. According to the change of temperature in the sinter layer and the difference of physical and chemical changes in the sintering process, the sinter layer can be divided into five layers from top to bottom, which are sinter layer, combustion layer, preheating layer, drying layer and over wet layer. After ignition, five layers appear one after another, and move down continuously. And finally all become sinter layer. The distribution and main reaction of each material layer in the sintering process is shown in Figure 1-2.

由于烧结过程是由料层表面开始逐渐向下进行的，沿料层高度方向有明显的分层性。按照烧结料层中温度的变化和烧结过程中所发生物理化学变化的不同，可以将正在烧结的料层自上而下分为五层，依次出现烧结矿层、燃烧层、预热层、干燥层和过湿层。点火后五层相继出现，不断向下移动，最后全部变为烧结矿层。烧结过程各料层分布和主要反应如图1-2所示。

Sinter Bed
烧结矿层

The combustion of fuel in the sinter bed has ended, forming porous sinter cake. The main change of this layer is that the high temperature melt solidifies into sinter, accompanied by crystallization and precipitation of new minerals. At the same time, the cold air pumped in is preheated, the sinter is cooled, and the low-cost oxide in contact with the air may be re-oxidized. The temperature of this layer is below 1100℃. With the moving down of combustion layer and the passing of cold air, the temperature of material decreases gradually. The molten phase is cooled and solidified into porous sinter. The sinter layer is gradually thickened, which makes the permeability of the whole layer better and the vacuum degree lower. The thickness of this layer is 40~50mm.

烧结矿层中燃料燃烧已结束，形成多孔的烧结矿饼。此层的主要变化是高温熔融物凝

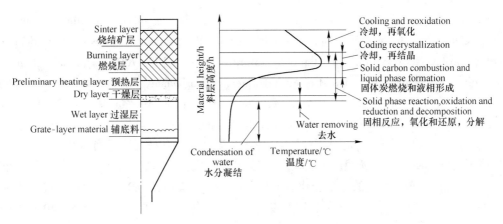

Figure 1-2 Distribution and main reaction of each material layer in sintering process
图 1-2 烧结过程各料层分布及主要反应

固成烧结矿,伴随着结晶和析出新矿物。同时抽入的冷空气被预热,烧结矿被冷却,与空气接触的低价氧化物可能被再氧化。这一层的温度在 1100℃ 以下,随着燃烧层的下移和冷空气的通过,物料温度逐渐下降,熔融液相被冷却,凝固成多孔结构的烧结矿。烧结矿层逐渐增厚,使整个料层透气性变好,真空度变低。该层厚度为 40~50mm。

Combustion Layer
燃烧层

When the air preheated by the sinter layer enters into the layer and contacts with the solid carbon, a large amount of heat will be released and a high temperature of 1300~1500℃ will be generated to form a certain composition of gas phase. Under this condition, a series of complex changes take place in the material layer, mainly including: the low melting point material continuing to form and melt, and forming a certain amount of liquid phase; the reduction, oxidation and decomposition of some oxide, the decomposition of sulfide, sulfate and carbonate etc. The combustion layer has a certain thickness, generally 15~50mm. Because of the high temperature and liquid phase melt in the combustion layer, there are many influences on the sintering process. If the combustion layer is too wide and the permeability of the material layer is poor, the output will decrease; and if the combustion layer is too thin, the liquid phase will not stick well and the strength will be low.

被烧结矿层预热的空气进入此层,与固体碳接触时发生燃烧反应,放出大量的热,产生 1300~1500℃ 的高温,形成一定成分的气相组成。在此条件下,料层中发生一系列复杂的变化,主要有:低熔点物质继续生成并熔化,形成一定数量的液相;部分氧化物分解、还原和氧化,硫化物、硫酸盐和碳酸盐分解等。燃烧层有一定厚度,一般为 15~50mm。由于燃烧层出现液相熔融物,且存在很高的温度,所以对烧结过程有多方面的影响。燃烧层过宽,则料层透气性差,导致产量下降;燃烧层太薄,则液相黏结不好、强度低。

Preheating Layer
预热层

Due to the heating effect of the high temperature exhaust gas from the combustion layer, the temperature quickly rises to close to the ignition point of the solid fuel, thus forming the preheating layer. As the heat exchange is very intense and the temperature of exhaust gas decreases rapidly, the layer is very thin, and the temperature is between 150~700℃. The main changes of this layer are: decomposition of part of crystal water and carbonate, decomposition and oxidation of sulfide and high valent iron oxide, reduction of part of iron oxide, and solid - state reaction. The thickness of this layer is generally 20~40mm.

受到来自燃烧层产生的高温废气的加热作用,温度很快升高到接近固体燃料着火点,从而形成预热层。由于热交换很剧烈,废气温度很快降低,所以此层很薄,其所处的温度在150~700℃之间。该层发生的主要变化有:部分结晶水、碳酸盐分解、硫化物、高价铁氧化物分解、氧化,部分铁氧化物还原,以及发生固相反应等。此层厚度一般为20~40mm。

Drying Layer
干燥层

The waste gas from the preheating layer heats the sinter and the free water in the sinter layer evaporates rapidly. Due to the good thermal conductivity of the wet material, the material temperature quickly rises above 100℃, and when it rises to 120~150℃, the moisture completely evaporates. Due to the rapid heating rate, it is difficult to separate the dry coal seam and the preheating layer completely, so it is sometimes referred to as the dry preheating layer, whose thickness is only 20~40mm. When the thermal stability of the ball in the mixture is not good, it will produce damage phenomenon in the process of intense temperature rise and water evaporation, which will affect the permeability of the material layer.

从预热层下来的废气将烧结料加热,料层中的游离水迅速蒸发。由于湿料的导热性好,料温很快升高到100℃以上,升至120~150℃时水分完全蒸发。由于升温速度太快,干煤层和预热层很难截然分开,所以有时又统称为干燥预热层,其厚度只有20~40mm。当混合料中料球的热稳定性不好时,会在剧烈升温和水分蒸发过程中产生破坏现象,影响料层透气性。

Over Wet Layer
过湿层

From the sintering of the surface sinter, the moisture in the sinter layer begins to evaporate into water vapor. A large amount of water vapor flows with the waste gas. If the temperature of the raw material is low and the temperature of the waste gas when it contacts the cold material falls below the corresponding dew point (generally 60~65℃), the water vapor will condense and make the water content of the sintering material exceed the appropriate value to form a super wet layer. During sintering, it is found that there is a serious over wetting phenomenon in the lower layer of the sinter, which is due to the strong air flow and gravity, and the high sintering water content. The original structure of the sinter is destroyed and the water in the sinter layer is transferred down mechanically, especially for those materials with small wet capacity. The air permeability of the material layer is greatly deteriorated by water vapor condensation, which has a great influence on the sintering process. Therefore, measures must be taken to reduce or eliminate the over wet layer.

从表层烧结料烧结开始，料层中的水分就开始蒸发成水汽。大量水汽随着废气流动，若原始料温较低，废气与冷料接触时其温度降到与之相应的露点（一般为 60~65℃）以下，则水蒸气凝结下来，使烧结料的含水量超过适宜值而形成过湿层。烧结时发现在烧结料下层有严重的过湿现象，这是由于在强大的气流和重力作用下，且烧结水分比较高，烧结料的原始结构被破坏，料层中的水分向下机械转移，特别是那些湿容量较小的物料容易发生这种现象。水汽冷凝使得料层的透气性大大恶化，对烧结过程产生很大的影响。所以，必须采取措施减少或消除过湿层。

1.1.4.2　Sintering Process Flow
1.1.4.2　烧结生产工艺流程

According to the different sintering equipment and air supply mode, the sintering methods can be divided into:

按照烧结设备和供风方式的不同，烧结方法可分为：

(1) Blast sintering. The blast sintering adopts the sintering pot and the flat blowing method, which is the local sintering method of small-scale factory and has been gradually eliminated.

(1) 鼓风烧结。鼓风烧结采用烧结锅和平地吹方式，是小型厂的土法烧结，已逐渐被淘汰。

(2) Suction sintering. Suction sintering is divided into continuous type and intermittent type. Continuous sintering equipment includes belt sintering machine and ring sintering machine. The intermittent sintering equipment includes fixed sintering machine and mobile sintering machine, fixed sintering machine such as disk sintering machine and box sintering machine, and mobile sintering machine such as step sintering machine.

(2) 抽风烧结。抽风烧结分为连续式和间歇式。连续式烧结设备有带式烧结机和环式烧结机等。间歇式烧结设备有固定式烧结机和移动式烧结机，固定式烧结机如盘式烧结机和箱式烧结机，移动式烧结机如步进式烧结机。

(3) Sintering in flue gas. Sintering in flue gas includes rotary kiln sintering and suspension sintering.

(3) 在烟气中烧结。在烟气中烧结包括回转窑烧结和悬浮烧结。

The belt type suction sintering machine is objective to widely used, because it has the advantages of high productivity, strong adaptability of raw materials, high degree of mechanization, good working conditions, and convenient for large-scale and automation. So more than 90% of the world's sinters are produced by this method.

广泛采用带式抽风烧结机是因为它具有生产率高、原料适应性强、机械化程度高、劳动条件好和便于大型化、自动化等优点，所以世界上有 90% 以上的烧结矿采用这种方法生产。

The process of sintering with belt suction mainly includes the preparation of raw materials, batching, mixing, sintering and product processing. Firstly, according to the production quality requirements, all kinds of iron ore powder are mixed into neutralization powder, the flux and fuel are crushed and screened to make the particle size meet the requirements of production, and the ingredients are calculated according to the chemical composition of raw materials. Then prepare the batching in the batching room. The batching method is to continuously distribute the material layer by layer on the running belt. The layered raw materials are mixed with water twice and then distributed on the sintering machine trolley. After being blown and ignited, the fuel is burned and sintering begins. The sinter is discharged at the tail of the machine. And after cooling, crushing and screening, the finished sinter, return ore and bottom material are obtained. The cold return ore and bottom material are returned to participate in the sintering process. The sintering process flow is shown in Figure 1-3.

带式抽风烧结过程主要包括烧结原料的准备、配料、混合、烧结和产品处理等工序。烧结生产首先按生产质量要求，将各种含铁矿粉混匀成中和粉，将熔剂和燃料进行破碎、筛分，从而使粒度达到生产所需要求，并根据原料化学成分进行配料计算。然后在配料室准备配料，配料方式是往运转的皮带上分层连续布料。分层的各种原料经两次加水混合后布在烧结机台车上，经抽风、点火后燃料燃烧，开始进行烧结。烧结矿在机尾排出，经冷却、破碎、筛分后获得成品烧结矿、返矿和捕底料，冷返矿和铺底料再返回参加烧结过程。烧结生产工艺流程如图 1-3 所示。

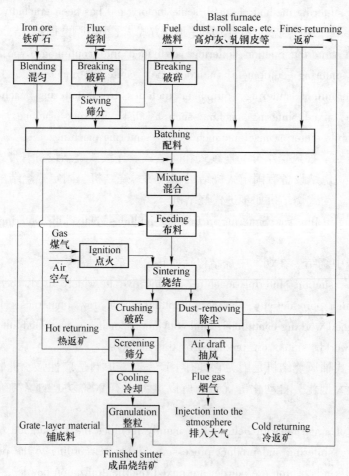

Figure 1-3 Sintering process flow
图 1-3 烧结生产工艺流程

Preparation of Raw Materials for Sintering
烧结原料的准备

There are many kinds of raw materials used in sintering production. In order to ensure the smooth production process and the output and quality of sinter, there are certain requirements for the raw materials and fuels used.

烧结生产所用原料品种较多,为了保证生产过程顺利进行,并保证烧结矿的产量和质量,应对所用原燃料有一定要求。

- Ferrous Material
- 含铁原料

Iron ore and iron concentrate are the main raw materials for sintering. The ore with high iron content is crushed and screened, and the qualified ore is directly sent to blast furnace for ironmaking. The part of ore powder with less than 10mm under the sieve is used as the raw material for sintering.

铁矿石和铁精矿是烧结的主要含铁原料。铁含量较高的矿石经破碎、筛分后,将合格矿直接送到高炉炼铁。将其筛下物小于10mm的矿粉作为烧结原料。

Generally, there are four sources of iron raw materials:

通常,含铁原料的来源有:

(1) Powdered ore. The 0~10mm iron ore formed in the process of mining and crushing is often called powder ore.

(1) 粉矿。开采、破碎过程中形成的0~10mm的铁矿石,常称为粉矿。

(2) Concentrate. The fine-grained iron ore obtained from the deep grinding and fine separation of lean ore is often called concentrate.

(2) 精矿。贫矿经过深磨细选后所得到的细粒铁矿石,常称为精矿。

(3) Metallurgical impurities. Metallurgical impurities include fine particles formed in smelting or other processes, as well as valuable and recyclable powders.

(3) 冶金杂料。冶金杂料包括冶炼或其他工艺过程形成的细粒以及含有价成分、可回收的粉末。

(4) Sinter return. In the process of transportation, crushing and pelletizing, the powder of less than 5mm size is returned to sintering. The chemical composition of the returned ore is basically the same as that of the sinter.

(4) 烧结返矿。烧结矿在运输、破碎、整粒过程中形成的小于5mm粒级的粉末返回烧结,返矿的化学成分基本与烧结矿相同。

The main iron ores used in sintering production are magnetite, hematite, limonite and siderite. Their sintering properties are quite different. Magnetite is hard, dense and hard to reduce, but it has a good burnability. Because it is oxidized and exothermic in high temperature treatment, and FeO is easy to form low melting point compound with gangue composition. It also has good energy saving and agglomeration strength. However, hematite particles have many pores, which are easier to reduce and break than magnetite, but it is difficult to form low melting point compounds due to its high degree of iron oxidation, so its burnability is poor, and its fuel consumption is higher than that of magnetite. Limonite has a lot of crystallized water and pores, so it has a great shrinkage when used in sintering, which makes the product quality reduce. Only by extending the high temperature treatment time can the product strength improved correspondingly, but it will lead to the increase of fuel consumption and processing cost. Due to the large shrinkage of siderite during sintering, the strength of product is reduced, the production capacity of equipment is low, and the fuel consumption is also increased due to the decomposition of carbonate.

用于烧结生产的主要铁矿石有磁铁矿、赤铁矿、褐铁矿和菱铁矿，它们的烧结性能有较大差异。磁铁矿坚硬、致密、难还原，但可烧性良好，因其在高温处理时氧化放热，且FeO 易与脉石成分形成低熔点化合物，所以造块节能和结块强度好。而赤铁矿颗粒内孔隙多，比磁铁矿易还原、破碎，但因其铁氧化程度高而难形成低熔点化合物，故其可烧性较差，造块时燃料消耗比磁铁矿高。褐铁矿因含结晶水、气孔多，用于烧结时收缩性很大，使产品质量降低，只有通过延长高温处理时间的方法才能使产品强度相应提高，但会导致燃料消耗增大、加工成本提高。菱铁矿在烧结时因收缩量大，导致产品强度降低、设备生产能力低，燃料消耗也因碳酸盐分解而增加。

It is generally required that raw materials containing iron have high grade, stable composition and few impurities. In addition to iron ore, there are also some industrial by-products, such as blast furnace ash, steel mill skin, pyrite cinder, steel slag, etc., which can also be used as sintering raw materials.

一般要求含铁原料品位高，成分稳定，杂质少。除铁矿石外还有一些工业副产品，如高炉灰、轧钢皮、黄铁矿烧渣、钢渣等，也可作为烧结原料。

- Flux
- 熔剂

It is generally required that the content of effective CaO in flux is high, the impurities are few, the composition is stable, the water content is about 3%, and the particle size less than 3mm accounts for more than 90%. With the refinement of concentrate size, the flux size should be relatively smaller. In some factories, when using fine concentrate sintering, the flux particle size is controlled below 2mm, which has achieved good results. When using quicklime, the particle size can be controlled within 5mm to facilitate water absorption and digestion.

一般要求熔剂中有效 CaO 含量高，杂质少，成分稳定，含水量在 3% 左右，粒度小于 3mm 的粒级占 90% 以上。随着精矿粒度的细化，熔剂粒度也要相对缩小。有的工厂在使用细精矿烧结时，将熔剂粒度控制在 2mm 以下，具有良好的效果。使用生石灰时，粒度可以控制在 5mm 以内，以便吸水消化。

Adding a certain amount of dolomite into the sinter to make the sinter containing appropriate MgO has a good effect on the sintering process and it can also improve the quality of the sinter.

在烧结料中加入一定量的白云石，使烧结矿含有适当的 MgO，对烧结过程有良好的作用，可以提高烧结矿的质量。

- Fuel
- 燃料

The main fuels used in sintering are coke powder and anthracite. The requirements for fuel include: high fixed carbon content, low ash content, low volatile content, low sulfur content, stable composition, water content less than 10%, and particle size less than 3mm accounts for more than 95%.

烧结所用燃料主要为焦粉和无烟煤。对燃料的要求是：固定碳含量高，灰分低，挥发分低，硫含量低，成分稳定，含水量小于10%，粒度小于3mm的粒级占95%以上。

It is generally considered that coke powder is a better sintering fuel, which can not only meet the above requirements, but also utilize the powder screened from blast furnace coke. But many factories use anthracite as fuel. The production practice shows that the hardness of anthracite is small, it is easy to be broken, the ignition point is low, and it is flammable. So anthracite is also a good fuel.

一般认为焦粉作烧结燃料较好，它既能满足上述要求，同时又利用了高炉焦炭筛分后的粉末。但不少厂家采用无烟煤作燃料的生产实践表明，无烟煤硬度小、易破碎、着火点低、易燃，所以无烟煤也是可取的燃料。

The main fuel for sintering is coke powder. It is the under screen material of coke in iron making plant and coking plant (i.e. coke crushing and coke powder). The quality requirements of coke powder are generally high fixed carbon content, low ash and sulfur content, particle size of 0~3mm. There is no clear requirements for its mechanical strength and ash soft melting temperature.

用于烧结燃料的主要是焦粉，它是炼铁厂和焦化厂焦炭的筛下物（即碎焦和焦粉）。对焦粉质量的要求一般是固定碳含量高、灰分和硫含量低、粒度为0~3mm，对其机械强度和灰分软熔温度没有明确要求。

When anthracite is used for sintering as fuel, its particle size is generally broken into 0~3mm. Anthracite with high fixed carbon content with 70%~80% (by mass), low volatile content with 2%~8% (by mass) and low ash content with 6%~10% (by mass) should be selected. Its structure is dense, black, bright luster and low water content. It is often used as a substitute for coke powder to reduce production cost. It should be noted that bituminous coal must not be used in the process of air extraction sintering.

当无烟煤作烧结燃料时，粒度一般破碎成0~3mm，应选用固定碳含量（质量分数）高（70%~80%）、挥发分含量低（2%~8%）、灰分少（6%~10%）的无烟煤。其结构致密，呈黑色，具有明亮光泽，含水量很低。它常作为焦粉代用品以降低生产成本。应注意，烟煤绝不能在抽风烧结中使用。

Batching and Mixture
配料与混合

- Batching
- 配料

There are many kinds of raw materials used in sintering production. In order to obtain the sinter with stable chemical composition and physical properties and meet the requirements of blast furnace smelting, it is necessary to carry out accurate batching.

烧结生产使用的原料种类较多，为了获得化学成分和物理性质稳定的烧结矿，满足高炉冶炼的要求，必须进行精确配料。

At present, the commonly used batching methods are volume batching and quality batching. The volume batching method is based on the condition that the bulk density of the material is constant and the mass of the raw material is proportional to the volume. In fact, the bulk density of raw materials is not stable, so the accuracy of volume batching method is poor. The quality proportioning method is based on the quality of raw materials. It is more accurate than the volume proportioning method, and it is easy to realize automation.

目前，常用的配料方法有容积配料法和质量配料法。容积配料法是基于物料堆积密度不变，原料的质量与体积成比例这一条件进行的。实际上原料的堆积密度并不稳定，因此容积配料法的准确性较差。质量配料法是根据原料的质量配料，它比容积配料法准确，并且便于实现自动化。

- Mixture
- 混合

The purpose of mixing the sinter with a certain proportion is to make the composition of the sinter uniform, the moisture suitable and even, so as to obtain the sinter with a good particle size composition, and to ensure the quality of the sinter improving the output.

将按一定配比组成的烧结料进行混合，目的是使烧结料的成分均匀、水分合适且均匀，从而获得粒度组成良好的烧结混合料，以保证烧结矿的质量和产量。

Mixing operations include wetting with water, blending and granulation. According to the different properties of raw materials, two processes can be adopted: one is mixed and the other is mixed twice. The purpose of one-time mixing is to wet and mix well, and preheat the material when it is heated to return to the ore. In order to improve the permeability of sinter layer, the second mixing is mainly granulation. The mixing time is 1~3min for primary mixing and 2.5min for secondary mixing.

混合作业包括加水润湿、混匀和制粒。根据原料性质的不同，可采用一次混合或二次混合两种流程。一次混合的目的是润湿与混匀，当加热返矿时还可使物料预热。二次混合除继续混匀外，主要是制粒，以改善烧结料层透气性。采用一次混合，混合时间为1~3min；采用二次混合，混合时间一般不少于2.5min。

The sinter should be fully mixed to ensure its uniform composition. At the same time, the powder should be made into balls to improve the pelletization of the mixture. Water must be added during mixing, and the moisture content of the mixture must be appropriate.

烧结料要充分混合，确保其成分均匀，同时使粉料成球，提高混合料成球性。混合时必须加水，混合料的水分含量必须适宜。

Sintering and Product Treatment
烧结与产品处理

Sintering operation is the central link of sintering production, which includes the main processes of distribution, ignition, sintering and sinter treatment.

烧结作业是烧结生产的中心环节，它包括布料、点火、烧结、烧结矿处理等主要工序。

- Feeding
- 布料

Feeding refers to the operation of laying the bottom material and mixture on the sintering machine trolley. Whether the distribution uniform or not affects the output and quality of sintering. Uniform distribution is the basic requirement of sintering production.

布料是指将铺底料、混合料铺在烧结机台车上的作业。布料均匀与否影响烧结的产量和质量，均匀布料是烧结生产的基本要求。

When the bottom material laying process is adopted, a layer of small sinter with particle size of 10~25mm and thickness of 20~25mm shall be laid as the bottom material before the mixture distributed, the purpose of which is to protect the grate, reduce the dust-proof load, extend the service life of the fan rotor, reduce or eliminate the grate slime.

当采用铺底料工艺时，在布混合料之前，先铺一层粒度为 10~25mm、厚度为 20~25mm 的小块烧结矿作为铺底料，其目的是保护炉箅、降低防尘负荷、延长风机转子寿命、减少或消除炉箅黏料。

After the bottom material is laid, the cloth shall be distributed accordingly. In the process of distribution, it is required that the particle size and chemical composition of the mixture be evenly distributed along the vertical and horizontal direction of the trolley, with certain looseness and smooth surface. At present, the round roller spreader is widely used for distribution, but there is segregation along the height of the material layer and the length and width of the sintering machine. In order to reduce or eliminate the segregation of distribution, it is necessary to ensure that the height of the material surface in the trough is 1/2~2/3 of the height of the trough, or to adopt the combined distribution of shuttle distributor and round roller feeder, which can greatly improve the distribution effect.

铺完底料后，随后进行布料。布料时要求混合料的粒度和化学成分等沿台车纵横方向均匀分布，并且有一定的松散性，表面平整。目前采用较多的是圆辊布料机进行布料，但此布料方式存在沿料层高度及烧结机长度和宽度方向上的偏析现象。为减轻或消除布料偏析，生产中应保证料槽内料面高度为料槽高度的 1/2~2/3 或采取梭式布料器—圆辊给料机联合布料，使用联合布料方法可大大改善布料效果。

- Ignition
- 点火

Ignition operation is to ignite and burn the surface of the material layer on the trolley. Sufficient ignition temperature and appropriate high temperature holding time are required. The ignition shall be uniform along the width of the trolley to facilitate the smooth combustion of fuel in the material layer.

点火操作是对台车上的料层表面进行点燃并使之燃烧，要求有足够的点火温度和适宜的高温保持时间，沿台车宽度点火应均匀，从而使料层中燃料顺利燃烧。

The ignition temperature depends on the melting temperature of the sintered product. This temperature range is usually 1000~1200℃. In practice, the ignition temperature is usually controlled at (1150 ± 50)℃. When the ignition temperature is too high, it will make the sinter surface over melt to form a hard shell, reduce the permeability of the material layer, slow down the vertical sintering speed and reduce the productivity. When the ignition temperature is too low, the surface appears floating ash, and the amount of ore returned increases.

点火温度取决于烧结生成物的熔化温度。这一温度范围通常为1000~1200℃。实际操作中点火温度常控制在（1150±50）℃。点火温度过高时，会使烧结料表面过熔形成硬壳，降低料层透气性，减慢垂直烧结速度，降低生产率；过低时，表层出现浮灰，返矿量增加。

At a certain temperature, in order to ensure the required heat of the surface material layer, it is necessary to have enough ignition time, usually about 1min.

在一定温度下，为了保证表面料层所需热量，需要有足够的点火时间，通常为1min左右。

Ignition heat q can be given by:

点火热量q的计算公式为：

$$q = t \cdot I \tag{1-8}$$

Where t——the ignition time in min;
I——the heat per unit surface area per unit time in $kJ/(m^2 \cdot min)$.

式中 t——点火时间，min；
I——单位时间内单位表面积所获得的热量，$kJ/(m^2 \cdot min)$。

The vacuum degree of ignition affects the depth of ignition. Generally, the ignition depth is required to be 10~20mm, so that the ignition heat is concentrated in a certain thickness of the surface layer.

点火真空度影响点火深度。一般要求点火深度为10~20mm，使点火热量集中于表层一定厚度内。

Sintering

烧结

At present, almost all sintering plants adopt the continuous belt sintering machine with suction type. After the iron containing raw materials, fluxes and fuels are prepared, which are proportioned in a certain proportion in the sintering batching room, mixed and granulated to form a mixture. And then they are distributed to the sintering machine trolley (The bottom material is distributed before the mixture is distributed), and the trolley moves along the sintering machine track to the discharge end. The igniter on the trolley ignites on the surface of the sinter, so the sintering reaction begins. During ignition and after ignition, the air in the material layer and the coke in the sinter are burned, due to the forced ventilation of the lower air box. And the heat generated makes the sinter mixture undergo physical and chemical changes, forming sinter. When the trolley reaches the discharge end of sinter, the sintering process is completed.

目前，各烧结厂几乎都采用抽风式的带式连续烧结机。将含铁原料、熔剂、燃料准备好后，在烧结配料室按一定比例配料，经过混合和制粒形成混合料，然后布到烧结机台车上（在布混合料前先布铺底料），台车沿着烧结机的轨道向排料端移动。台车上的点火器在烧结料表面进行点火，开始烧结反应。点火时和点火后，由于下部风箱强制抽风，通过料层的空气与烧结料中的焦炭燃烧,所产生的热量使烧结混合料发生物理化学变化,形成烧结矿。台车到达烧结矿排料端时,则烧结过程完成。

Sinter Treatment

烧结矿处理

Most sintering plants adopt cold ore process, including crushing, screening, cooling and pelletizing. Figure 1-4 shows the main processing flow of sinter.

烧结厂大都采用冷矿流程,其包括破碎、筛分、冷却和整粒。烧结矿主要处理流程如图 1-4 所示。

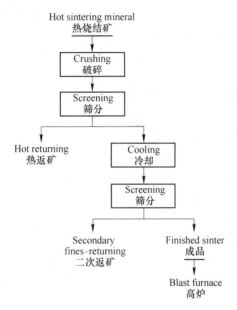

Figure 1-4 Main processing flow of sintering production

图 1-4 烧结矿主要处理流程

- Crushing and Screening of Sinter
- 烧结矿的破碎与筛分

The production practice has proved that when crushing and screening operations are not set up, the bulk sinter not only blocks the ore trough, but also enters the hearth when it is not fully reduced in the upper and middle part of the blast furnace in the smelting process, which destroys the thermal system of the hearth and increases the coke ratio. If the powder is not removed, not only the cooling of sinter will be affected, but also the permeability of the material column will be deteriorated when the powder enters into the blast furnace, resulting in uneven gas distribution, unsmooth furnace condition, increased air pressure, suspension and collapse of the material. And the output of blast furnace will be reduced. According to statistics, for every 1% increase of powder in sinter, the output of blast furnace decreases by 6%~8%, the coke ratio increases, and a large amount of blast dust will accelerate the wear of furnace top equipment and worsen the working conditions. According to production experience, for every 10% reduction of powder less than 5mm in sinter, the coke ratio can be reduced by 1.6% and the output can be increased by

7.6%. Therefore, it is necessary to set up crushing and screening operation at the end of sintering machine for sintering plant and smelter. The utility model has the following advantages:

生产实践表明,当不设置破碎和筛分作业时,大块烧结矿不仅堵塞矿槽,而且在冶炼过程中高炉的上、中部未能充分还原便进入炉缸,从而破坏了炉缸的热工制度,造成焦比升高。若不筛除粉末,不仅影响烧结矿的冷却,而且粉末进入高炉内会恶化料柱透气性,引起煤气分布不均,炉况不顺,风压升高,造成悬料、崩料,使高炉产量下降。据统计,烧结矿中的粉末每增加1%,高炉产量下降6%~8%,焦比升高,大量炉尘吹出会加速炉顶设备的磨损,恶化劳动条件。据生产经验,烧结矿中小于5mm的粉末每减少10%,可降低焦比1.6%,使产量增加7.6%。因此,在烧结机尾设置破碎和筛分作业,对烧结厂和冶炼厂都是十分必要的。其具有如下优点:

(1) Small pulverization degree and high yield in the crushing process;

(1) 破碎过程中的粉化程度小,成品率高;

(2) Simple and reliable structure, convenient use and maintenance;

(2) 结构简单、可靠,使用及维修方便;

(3) Low crushing energy consumption.

(3) 破碎能耗低。

Hot ore vibrating screen with high screening efficiency is widely used in the screening of hot sinter in China, which can effectively reduce the dust in the finished sinter, reduce the resistance of sinter bed in the cooling process and reduce the dust. Meanwhile, the obtained hot return ore can improve the particle size composition and preheating mixture of sinter mixture, which is beneficial to the increase of the output and quality of sinter. However, the hot ore screen also has disadvantages. Because of the high temperature, the vibrating screen has many accidents, which reduces the operation rate of the sintering machine. Therefore, in recent years, the large-scale sintering machine designed and put into operation has cancelled the hot ore screen, and the sinter directly enters the cooler for cooling after being broken by a single roller from the tail of the machine.

热烧结矿的筛分,国内多采用筛分效率高的热矿振动筛,它能有效地减少成品烧结矿中的粉尘,可降低冷却过程中的烧结矿层阻力和减少扬尘。同时,所获得的热返矿可改善烧结混合料的粒度组成和预热混合料,对提高烧结矿的产量和质量有好处。但热矿筛也有缺点,因在高温下工作,振动筛事故多,降低了烧结机作业率。因此,近年来设计投产的大型烧结机取消了热矿筛,烧结矿自机尾经单辊破碎后直接进入冷却机进行冷却。

Cooling of sinter. The cooling methods of sinter mainly include blast cooling, suction cooling and on-board cooling. At present, blast cooling is mainly adopted.

烧结矿的冷却。烧结矿的冷却方式主要有鼓风冷却、抽风冷却和机上冷却三种,目前主要采取鼓风冷却。

- Blast Cooling
- 鼓风冷却

The blast cooling adopts thick material layer (thickness 1500mm) and low speed. The cooling time is about 60min, cooling area is relatively small, and the ratio of cooling area to sintering area is 0.9~1.2. After cooling, the temperature of the hot exhaust gas is 300~400℃, which is higher than that of the exhaust gas cooled by exhaust air. So it is convenient for the recovery and utilization of the exhaust gas. The disadvantage of blast cooling is that the required air pressure is high, generally 2000~5000Pa. So it is necessary to select a sealing device with good sealing performance.

鼓风冷却采用厚料层（厚度为1500mm）、低转速，冷却时间长约60min，冷却面积相对较小，冷却面积与烧结面积之比为0.9~1.2。冷却后热废气温度为300~400℃，比抽风冷却废气温度高，便于废气回收利用。鼓风冷却的缺点是所需风压较高，一般为2000~5000Pa，因此必须选用密封性能好的密封装置。

- Exhaust Cooling
- 抽风冷却

Belt cooler and ring cooler are relatively mature exhaust cooling equipment, which are widely used at home and abroad. Both of them have better cooling effect. Compared with the two, the ring cooler has the advantages of small floor area and compact plant layout. The belt cooler can play a transport role in the cooling process at the same time. For plants with more than two sintering machines, the process is easy to arrange, the distribution is relatively uniform, the sealing structure is simple, and the cooling effect is good.

带式冷却机和环式冷却机是比较成熟的抽风冷却设备，在国内外获得广泛的应用。它们都有较好的冷却效果，两者相比较，环式冷却机具有占地面积小、厂房布置紧凑的优点。带式冷却机则在冷却过程中能同时起到运输作用。对于有多于两台烧结机的厂房，该工艺便于布置，而且布料较均匀，密封结构简单，冷却效果好。

- Machine Cooling
- 机上冷却

Machine cooling is the process of extending the sintering machine and cooling the sinter directly in the second half of the sintering machine. The advantages are: the single roll crusher has low working temperature, does not need hot ore screen and separate cooling machine. It can also improve the equipment operation rate, reduce the equipment maintenance cost, and facilitate the dedusting of cooling system and environment.

机上冷却是将烧结机延长后，使烧结矿直接在烧结机的后半部进行冷却的工艺。其优点是：单辊破碎机工作温度低，不需要热矿筛和单独的冷却机，可以提高设备作业率，降低设备维修费，便于冷却系统和环境的除尘。

- Final Sinter Screening
- 烧结矿的整粒

In order to meet the needs of modernization, upsizing and energy saving of blast furnace, the quality requirements of sinter are getting higher and higher. In recent years, most of the newly-built sintering plants in China have set up the final sinter screening system. In a sintering plant with a final sinter screening system, generally after sinter is discharged from the cooling machine, the sinter shall go through cold crushing, and then after 2~4 times of screening, the grain size smaller than 5mm shall be separated as the return ore, the grain size of 10~20mm (or 15~

25mm) shall be used as the bottom material, and the rest shall be finished sinter. The upper limit of the grain size of finished sinter shall generally not exceed 50mm. The sinter size is uniform and the amount of powder is small, which is beneficial to improve the smelting index of blast furnace. For example, when a blast furnace uses sinter after final sintering and sieving into the furnace, the utilization coefficient of blast furnace increases by 18%, the coke ratio of cast iron per ton decreases by 20kg, the dust blown out from the top of the furnace reduces, and the service life of the equipment on the top of the furnace is extended. The final sinter screening diverging process of sintering plant is different, large sintering plant mostly uses fixed screen and single vibrating screen as the final sinter screening diverging process of four stages, and small sintering plant mostly uses single or double vibrating screen for three stages.

为了满足高炉现代化、大型化和节能的需要，对烧结矿的质量要求越来越高。近年来国内新建的烧结厂大都设有整粒系统，一些老厂的改造也增设了较完善的整粒系统。设有整粒系统的烧结厂，一般烧结矿从冷却机卸出后要经过冷破碎，然后经2~4次筛分，分出小于5mm的粒级作为返矿，10~20mm（或15~25mm）的粒级作为铺底料，其余的为成品烧结矿，成品烧结矿的粒度上限一般不超过50mm。经过整粒的烧结矿粒度均匀、粉末量少，有利于高炉冶炼指标的改善。如某公司高炉使用整粒后的烧结矿入炉，高炉利用系数提高了18%，每吨生铁焦比降低20kg，炉顶吹出粉尘减少，并延长了炉顶设备的使用寿命。烧结厂的整粒流程各异，大型烧结厂多采用固定筛和单层振动筛作四段筛分的整粒流程，小型烧结厂则多采用单层或双层振动筛作三段筛分。

1.1.4.3 Main Sintering Equipment
1.1.4.3 烧结生产主要设备

The main equipment of sintering production includes sintering machine, suction system and feeding system.

烧结生产主要设备包括烧结机、抽风系统和供料系统等。

Sintering Machine

烧结机

At present, the belt sintering machine is mainly used in sintering production, and its structure is shown in Figure 1-5. The utility model is composed of a trolley, a transmission device, an ignition device, a sealing device, a frame, etc.

目前烧结生产中主要采用带式烧结机，其结构如图1-5所示。带式烧结机是由台车、传动装置、点火装置、密封装置和机架等组成。

- Trolley
- 台车

The trolley is a very important part of the sintering machine. It is the main equipment for carrying materials and sintering. The belt type sintering machine is a closed-circuit circulating sintering chain belt composed of many trolleys, which is composed of the body, grate bar and block plate, running wheel and clamping roller, etc.

台车是烧结机上非常重要的部件，它是载料并进行烧结的主要设备。带式烧结机是由许多台车组成的闭路循环运转的烧结链带，由本体、箅条和挡板、运行轮和卡辊等组成。

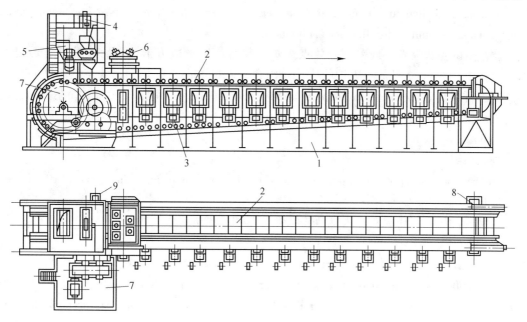

Figure 1-5 Belt sintering machine
图 1-5 带式烧结机

1—Framework of sintering machine；2—Trolley；3—Suction chamber；4—Charging；5—Bottom charging；6—Igniter；
7—Drive part of sintering machine；8—Chip outlet of discharge part；9—Chip outlet of sintering machine head
1—烧结机的骨架；2—台车；3—抽风室；4—装料；5—装铺底料；6—点火器；7—烧结机传动部分；
8—卸料部分碎屑出口处；9—烧结机头部碎屑出口处

- Drive
- 传动装置

The driving device of sintering machine is composed of adjustable speed motor, reducer, driving gear and driving star wheel. At present, flexible driving device is widely used. The big star wheel drive at the head of sintering machine is a driving mechanism of closed-circuit sintering chain belt circulation composed of trolley. There is a driven star wheel at the tail to return the trolley.

烧结机的传动装置由调速电动机、减速器、传动齿轮和传动星轮组成，目前广泛采用的是柔性传动装置。烧结机头部的大星轮驱动是由台车组成的闭路烧结链带循环运转的传动机构。机尾部位有使台车返回的从动星轮。

- Ignition Device
- 点火装置

The ignition device of sintering machine generally adopts open ignition furnace, which has semicircle or square furnace cover. And there are several rows of burners on the furnace top, about 10 in each row. The advantage of the utility model is that the ignition is uniform along the width direction of the trolley; and the disadvantage is that the material surface can not be prevented from sudden cooling, so a heat preservation cover is set up behind to prevent the material surface from sudden cooling. A burner is arranged in the igniter, and the gas and air are mixed and burned in the channel, and then the sintering process begins. Because the furnace is a high temperature area, it must be built with refractory materials.

烧结机的点火装置一般采用开放点火炉，其有半圆形或方形炉罩，在炉顶有数排烧嘴，每排10个左右。开放点火炉的优点是沿台车宽度方向点火均匀；缺点是不能防止料面急冷，所以其后常设保温罩以防料面急冷。点火器内设有烧嘴，煤气和空气在其通道内混合燃烧，随之烧结过程开始。因炉内为高温区，其内必须用耐火材料砌筑。

Exhaust System

抽风系统

The suction system of sintering is composed of air box, large flue, dust removal device and suction fan.

烧结的抽风系统由风箱、大烟道、除尘装置和抽风机组成。

- Bellows
- 风箱

The air box is in the shape of a square funnel. Its upper port is connected with the sealing slide fixed by the sintering machine, and its lower port is connected with the air box duct. The function of the air box is to gather the flue gas passing through the sinter layer and discharge it by the exhaust fan. The small sintering machine has a row of air boxes, the large sintering machine is designed with two rows of air boxes, and the flue gas is discharged from both sides.

风箱呈方漏斗形，其上口与烧结机固定的密封滑道连接，下口与风箱文管相连。风箱的作用是集聚透过烧结料层的烟气，由抽风机排出。小型烧结机有一排风箱，大型烧结机设计有两排风箱，烟气从两侧排出。

- Large Flue
- 大烟道

The main function of the large flue (main exhaust pipe) is to send the sintering waste gas to the dedusting equipment and the exhaust fan, and the other function is to separate the dust. The larger dust particles can be separated under the action of gravity. Because the large flue is long and subject to large thermal expansion, the pipeline is installed on the carriage with roller support, in order to prevent the pipeline and frame from thermal stress damage, so that it can expand freely.

大烟道（主排气管）的主要功能是将烧结废气送往除尘设备和抽风机，另一功能是分离粉尘，较大的粉尘颗粒在重力作用下可被分离出来。由于大烟道很长，受到的热膨胀量较大，为了使管道和构架不受热应力破坏，管道安装在有滚柱支撑的拖架上，使之能自由伸缩。

- Dedusting Device
- 除尘装置

According to different working principles, dust-proof devices are mainly divided into multi tube dust removal and electric dust removal devices. The multi tube dedusting device is composed of a plurality of cyclones, and the dust containing gas enters from the opening, enters into each single cyclone separately, and then generates rotation movement through the guide to reduce the dust. The electrostatic precipitator is based on the principle that the charged dust moves under the action of electric field force. The dust tends to the dust collecting electrode and falls down to discharge after discharge. The investment of multi tube dedusting device is less, but the maintenance is complex. The investment of ESP is high, but the maintenance is simple.

防尘装置按工作原理不同，主要分为多管除尘和电除尘装置。多管除尘装置由多个旋风子组成，含尘气体从开口进入，然后分别进入每个单体的旋风子中，经导向器产生旋转运动而使灰尘降下来。电除尘装置是利用带电灰尘在电场力作用下产生移动的原理工作。灰尘趋向收尘电极，放电后落下排出。多管除尘装置投资少，但检修复杂。电除尘投资高，但维修简单。

- Exhaust Fan
- 抽风机

The main exhaust fan is one of the most important process equipment in sintering production. With the development of large-scale sintering machine and thick layer sintering operation, the working pressure, working temperature and working medium of the suction fan, the requirements for the performance of the suction fan are higher and higher.

主抽风机（排风机）是烧结生产中最重要的工艺设备之一。随着烧结机的大型化和厚料层烧结操作的发展，以及抽风机的工作压力、工作温度和工作介质的状况等因素的影响，对抽风机性能要求越来越高。

Feeding System
供料系统

The feeding system mainly includes the silo and mixer.

供料系统主要包括料仓和混合机。

- Bunker
- 料仓

The bin is one of the key equipment in the batching operation. Its volume matches the production capacity of the sintering machine. Generally, the main raw material bin should meet the needs of 6~7h production of the sintering machine. In order to ensure the smooth discharge of raw materials, the discharge port shall be as large as possible. The inner wall of the bin is provided with a smooth lining plate to reduce stickiness.

料仓是配料操作的关键设备之一。其容积与烧结机的生产能力相匹配，一般主原料仓要满足烧结机生产6~7h的需要。为了使原料顺利排出，要求排料口尽量大。料仓内壁设有光滑的衬板，从而减少黏料。

- Cylinder Mixer
- 圆筒混合机

The cylinder mixer is a device for adding water, mixing and granulating the mixture. The utility model mainly comprises a cylindrical body, a ring-shaped roll ring installed at two ends of the barrel, four idlers and two stop wheels corresponding to the roll ring.

圆筒混合机是混合料加水、混匀和制粒的设备。其主要包括圆筒形本体、安装在筒体两头的圆环形辊圈以及与辊圈相对应的四个托辊和两个挡轮。

1.1.5 Pellet Production
1.1.5 球团矿生产

Pelletizing is another main method of making iron ore powder. The prepared raw materials (such as, fine grinding materials, additives or binders, etc.) are proportioned and mixed according to a certain proportion, and a certain size of green ball is formed by rolling on the pelletizer. Then a series of physical and chemical changes and consolidation are made by drying and roasting or other methods. The products produced by this method are called pellets, which are spherical in shape, uniform in size, and high in strength and reduction.

球团法是铁矿粉造块的另外一种主要方法。它是将准备好的原料（如细磨物料、添加剂或黏结剂等）按一定比例进行配料、混匀，在造球机上经滚动形成一定大小的生球，然后采用干燥和焙烧或其他方法使其发生一系列的物理化学变化而固结。这种方法生产的产品称为球团矿，其呈球形，粒度均匀，具有高强度和高还原性。

1.1.5.1 Pelletizing Principle
1.1.5.1 球团原理

Pelletizing process and roasting consolidation are two important processes in pelletizing production.

球团的成球过程和焙烧固结是球团矿生产过程中两大重要的工序。

Pelletizing of fine grinding materials is divided into continuous pelletizing and batch pelletizing. Continuous pelletizing is mainly used in production, which is divided into three stages as follows:

细磨物料造球分为连续造球和批料造球。生产中主要以连续造球为主，其分为如下三个阶段：

(1) The formation of the mother ball. As the core of pelletizing, the mother ball is a compact aggregate of particles with high capillary water content. The mixture particles used for pelletizing are in a loose state, the ore particles are covered by adsorbed water and membrane water, and the capillary water only exists in the contact point between the ore particles. Moreover, the rest space is filled by air, the contact between the ore particles is not close, and the membrane water can not play a role. In addition, because the capillary water content is less, the pore size is too large, the capillary pressure is small, and the binding force between ore particles is weak, so it is impossible to form balls. At this time, the mixture is wetted unevenly, and the mechanical force is used to make the mineral powder locally compact, resulting in smaller pores and larger capillary pressure, pulling the surrounding mineral particles to the center of the water drop, and forming a closer aggregate of particles, thus forming a mother ball.

（1）母球的形成。母球是造球的核心，是毛细水含量较高的紧密颗粒集合体。用于造球的混合料颗粒之间处于松散状态，矿粒被吸附水和薄膜水覆盖，毛细水仅存在于各矿粒间的接触点上，其余空间被空气充填，矿粒之间接触不紧密，薄膜水不能够发挥作用。此外，由于毛细水含量较少，毛细孔过大，毛细压小，矿粒间结合力较弱，不能成球。此时对混合料进行不均匀的点滴润湿，并利用机械力的作用，使矿粉得到局部紧密，造成更小的毛细孔和较大的毛细压力，将周围矿粒拉向水滴中心，形成较紧密的颗粒集合体，从而形成母球。

(2) The growth of the mother ball. The growth of the mother ball is also due to the capillary effect. When the mother ball rolls in the pelletizer, the mother ball with the original structure not too tight is compressed, and the excess capillary water inside is squeezed to the surface of the mother ball. If the mother ball continues to be wetted with water, the surrounding mineral powder will continue to stick. This rolling compaction is repeated many times, and the mother ball gradually grows up to the specification size.

(2) 母球的长大。母球的长大也是由于毛细效应。母球在造球机内滚动，原来结构不太紧密的母球压紧，内部过剩的毛细水被挤到母球表面，继续加水润湿母球表面，就会不断黏结周围矿粉。这种滚动压紧重复多次，母球便逐步长大至规格尺寸。

(3) The tightness of the ball. At this stage, the mechanical actions of rolling and rubbing force of pelletizer are the decisive factor. They will make the particles in the ball selectively arrange according to the maximum contact area, extrude all the capillary water inside the green ball and be absorbed by the surrounding mineral powder. At the same time, the mineral particles in the green ball are arranged more closely, which makes it possible for the film water layers to contact each other, moving form the hydration film shared by many mineral particles and to strengthen the water binding force. They also greatly improve the strength of the green ball. When the green ball reaches a certain size and strength, it will roll out automatically from the pelletizer by centrifugal force.

(3) 生球的紧密。此阶段造球机的滚动和搓力的机械作用为决定因素。它们将使球内颗粒选择性地按接触面积最大化排列，将生球内部的毛细水全部挤出，被周围矿粉所吸收。同时，生球内的矿粒排列更紧密，使薄膜水层有可能相互接触迁移，形成众多矿粒共有的水化膜而加强水分结合力，使生球强度大大提高。当达到一定粒度和强度后，生球依靠离心力作用从造球机自动滚出。

1.1.5.2 Pellet Production Process
1.1.5.2 球团生产工艺流程

The pelletizing process generally includes raw material preparation, batching, mixing, pelletizing, drying and roasting, cooling, finished product and ore return processing, as shown in Figure 1-6.

球团生产工艺流程一般包括原料准备、配料、混合、造球、干燥和焙烧、冷却、成品和返矿处理等工序，如图 1-6 所示。

Raw Material Preparation
原料准备

- Iron Concentrate
- 铁精矿

The main raw material used in pelletizing production is iron concentrate powder, which

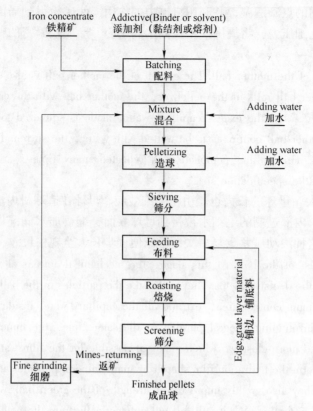

Figure 1-6 Pellet production process
图 1-6 球团生产工艺流程

accounts for more than 90% of the pelletizing mixture. The quality of iron concentrate powder is strictly required in the production of closed pellets, because it plays a decisive role in the quality of raw pellets and finished pellets. The specific quality requirements are as follows:

球团矿生产采用的原料主要是铁精矿粉,占造球混合料的90%以上。球闭矿生产对铁精矿粉的质量要求比较严格,因为其对生球与成品球团矿的质量起着决定性的作用。具体质量要求有以下几方面:

(1) Particle size: Generally, it is required that 90% of the fine sand with particle size less than 0.074mm or less than 0.044mm accounts for 60%~85%, and the proportion especially less than 20μm should not be less than 20%. But it is not that the finer the particle size is, the better the particle size is, which will increase the energy consumption of grinding.

(1) 粒度:一般要求精砂粒度小于0.074mm的部分达90%以上或者小于0.044mm的部分占60%~85%,尤其是小于20μm部分的比例不能小于20%。但是并不是矿粉粒度越细越好,粒度过细会增加磨矿的能耗。

(2) Moisture content: It has a great influence on pelletizing process, pelletizing quality, drying roasting system and pelletizing equipment to control and adjust the moisture content of concentrate. In order to stabilize pelletizing, the smaller the fluctuation of moisture content is, the better, and the fluctuation range should not exceed ±0.2%. Generally, it is required that the

moisture content of concentrate powder is between 7.5% and 10.5% (by mass). When the particle size of less than 0.044mm accounts for 65%, the appropriate moisture content is 8.5% (by mass). When the particle size of 0.044mm accounts for 90%, the appropriate moisture content can reach 11% (by mass).

(2) 水分含量：控制和调节精矿水分含量对造球过程、生球质量、干燥焙烧制度和造球设备工作影响很大。为了稳定造球，其水分含量波动越小越好，波动范围不应超过±0.2%。一般要求精矿粉水分含量在7.5%~10.5%（质量分数）之间，当小于0.044mm的粒级占65%时适宜水分含量为8.5%（质量分数），当0.044mm的粒级占90%时适宜水分含量可达到11%（质量分数）。

(3) Chemical composition: The stability and uniformity of the chemical composition directly affect the production process and product quality. The fluctuation range of $w(TFe)$ and $w(SiO_2)$ is required to be ±0.5% and ±0.5%, respectively.

(3) 化学成分。化学成分的稳定、均匀程度直接影响生产工艺过程和产品质量，要求$w(TFe)$波动范围为±0.5%，$w(SiO_2)$波动范围为±0.5%。

When the concentrate powder used to produce pellets can not meet the above requirements, it needs to be processed. Figure 1-6 shows the pellet production prcess.

当用于生产球团矿的精矿粉不能满足上述要求时，需进行加工处理。球团生产工艺流程如图1-6所示。

- Additive
- 添加剂

Adding additives to pelletizing materials is to strengthen pelletizing process and improve pellet quality. The additives are mainly binders and fluxes. The binders used in pellet production include bentonite, slaked lime, cement, etc. the commonly used fluxes in pellet production include limestone, dolomite, slaked lime, etc.

在造球物料中加入添加剂是为了强化造球过程和改善球团矿质量。添加剂主要为黏结剂和熔剂。球团矿使用的黏结剂有膨润土（皂土）、消石灰、水泥等，球团生产常用的熔剂有石灰石、白云石、消石灰等。

Bentonite is widely used as a adhesive and has the best effect. When the concentrate is pelletized, adding appropriate amount of bentonite (generally accounting for 1%~2% (by mass) of the content of the mixture) can improve the strength of the pelletizing, adjust the moisture content of the raw materials, improve the pelletizing rate of the materials, and make the size of the pelletizing small and uniform. Moreover, it can increase the bursting temperature of raw pellets, accelerate the drying speed, shorten the drying time, improve the quality of pellets, and promote the consolidation strength of finished pellets. Bentonite is a kind of clay mineral with montmorillonite as the main component. It is a kind of hydrous silicate with swelling property and layered structure. The theoretical chemical formula of bentonite is $Si_8Al_4O_{20}(OH)_4 \cdot nH_2O$. And the chemical composition is SiO_2 66.7% (by mass), and Al_2O_3 28.3% (by mass), Vickers hardness is 1~2HV, whose density changes greatly due to different water absorption, generally 1~2g/cm^3. The actual composition of bentonite is $w(SiO_2) = 60\%~70\%$, and $w(Al_2O_3) \approx$

15%. It also contains some other impurities, such as Fe_2O_3, Na_2O, K_2O, etc.

膨润土作为黏结剂使用较广泛，效果最佳。在精矿造球时加入适量（一般占混合料含量（质量分数）的 1%~2%）的膨润土可提高生球的强度，调整原料中的水分含量，提高物料的成球率，并使生球粒度小而均匀。同时，它能提高生球的爆裂温度，使干燥速度加快，缩短干燥时间，提高球团矿质量，对成品球的固结强度也有促进作用。膨润土是以蒙脱石为主要成分的黏土矿物，它是一种具有膨胀性能、呈层状结构的含水硅酸盐。膨润土的理论化学分子式为 $Si_8Al_4O_{20}(OH)_4 \cdot nH_2O$，化学成分为 SiO_2 66.7%（质量分数）、Al_2O_3 28.3%（质量分数），维氏硬度为 1~2HV，因吸水量不同其密度变化较大，一般为 1~2g/cm³。膨润土实际成分为 $w(SiO_2) = 60\%\sim70\%$，$w(Al_2O_3) \approx 15\%$，还含有一些其他杂质，如 Fe_2O_3、Na_2O、K_2O 等。

The purpose of adding flux is to adjust the composition of pellets, improve the reduction degree and softening temperature, and reduce the reduction pulverization rate and reduction expansion rate. Slaked lime has the characteristics of fine particle, large specific surface area, good hydrophilicity and strong adhesion. It is used as both binder and flux in pellet production. However, when the amount of slaked lime is too much, it will reduce the ball forming speed, lead to irregular surface of the ball, and cause the ball burst temperature to reduce. In addition, the difficulty of lime digestion and preparation limits its use. Limestone is also a kind of material with strong hydrophilicity. Because of its rough surface, it can increase the friction between the particles inside the green ball and increase the strength of the green ball. The lime bin powder added during pelletizing shall be finely ground, and the part with particle size less than 0.074mm shall account for more than 80%. However, its cohesion is not as good as slaked lime, so the main purpose of adding limestone is to improve the basicity of pellet.

添加熔剂的目的是调剂球团矿的成分，提高还原度和软化温度，降低还原粉化率和还原膨胀率等。消石灰具有颗粒细、比表面积大、亲水性好、黏结力强的特点。它在球团生产中既是黏结剂又是熔剂。但当消石灰的用量过多时会降低成球速度，导致生球表面不规则，从而引起生球爆裂温度降低。再加上石灰消化制备困难，因此限制了它的使用。石灰石也是一种亲水性较强的物料，因其颗粒表面粗糙，能增加生球内部颗粒间的摩擦力，使生球强度提高。在造球时加入的石灰仓粉须经细磨，粒度小于 0.074mm 的部分应占 80%以上。但其黏结力不如消石灰，所以加入石灰石的主要目的是提高球团矿的碱度。

Ingredients

配料

In order to stabilize the chemical composition, physical properties and metallurgical properties of the pellets, it is necessary to make accurate batching. The batching calculation should be carried out according to the type and composition of the raw materials, the chemical composition and properties of the pellets required by the blast furnace smelting, so as to ensure that the main indexes such as iron content, basicity, sulfur content and ferrous oxide content of the pellets are within the specified range. The calculation of on-site ingredients needs to be simple, convenient, accurate and rapid. After the calculation to determine the amount of various raw materials, mainly through the quality batching method, i.e. through weighing (belt scale or electronic scale) and disc feeder to determine and control the amount of material.

为了稳定球团矿的化学成分、物理性能和冶金性能，必须精确配料，根据原料的种类、成分以及高炉冶炼要求的球团矿化学成分和性质进行配料计算，从而保证将球团矿的铁含量、碱度、硫含量和氧化亚铁含量等主要指标控制在规定范围内。现场配料计算要求简单、方便、准确和迅速。经计算确定各种原料用量后，主要通过质量配料法，即通过称量（皮带秤或电子秤）利圆盘给料机来确定和控制料量。

Mixing
混合

Although there are few kinds of raw materials used in pelletizing production, the mixing operation is needed in order to make the mixture particles uniformly dispersed and well mixed with water. The commonly used mixer is cylinder mixer.

球团矿生产使用的原料种类虽然较少，但为了使混合料颗粒间能获得均匀分散的效果，并能与水良好混合，需要进行混合作业。常用的混合机为圆筒混合机。

Ball making
造球

Pelletizing is an important part of pelletizing process. The quality of raw pellets has a great influence on the quality of finished pellets, so the quality of raw pellets must be strictly required. The general requirement of green ball is that the particle size is uniform, the strength is high, and the powder content is small. The particle size shall be generally controlled within the range of 10~16mm. The compressive strength of each ball is: the wet ball is not less than 90N/piece, the dry ball is not less than 450N/piece, and the fracture temperature (the initial temperature of the structure damaged during the heating process of the raw ball) should be higher than 400℃. The dry ball shall have good wear resistance. In order to improve the productivity and pelletizing quality of the roasting equipment, the balls smaller than 10mm and larger than 16mm should be removed by sieving, and then participate in pelletizing after crushing. At present, disc pelletizer and cylinder pelletizer are widely used at home and abroad. In foreign pelletizing production, 66.7% of them use cylinder pelletizer, 29% use disc pelletizer and 4.3% use cone pelletizer.

造球是球团生产工艺的重要环节。生球的质量对成品球团矿的质量影响很大，因此必须对生球质量严格要求。对生球的一般要求包括粒度均匀，强度高，粉末含量少。粒度一般应控制在 10~16mm 范围内。每个球的抗压强度为：湿球不小于90N/个，干球不小于450N/个，破裂温度（生球在被加热过程中结构遭到破坏的初始温度）应高于400℃。干球应具有良好的耐磨性能。为了提高焙烧设备的生产率和成球质量，应将小于10mm和大于16mm的球筛除，经破碎后再参与造球。目前国内外广泛采用圆盘造球机和圆筒造球机。国外造球生产中，使用圆筒造球机的约占66.7%，使用圆盘造球机的约占29%，使用圆锥造球机的为4.3%。

Roasting
焙烧

The common methods of pelletizing production are shaft furnace, belt roaster and grate—rotary kiln pelletizing. The shaft furnace pelletizing method developed firstly and rapidly at one time. Due to the demand of raw materials and output, and the large-scale equipment, the belt roaster and grate rotary kiln pellet method have been developed successively. The comparison of pellet production by three kinds of pellet roasting equipment is shown in Table 1-4.

  球团焙烧生产应用较为普遍的方法有竖炉、带式焙烧机和链箅机—回转窑球团法。竖炉球团法发展最早，一度发展很快。由于原料和产量的要求，以及设备大型化，相继发展了带式焙烧机和链箅机—回转窑球团法。三种球团焙烧设备生产球团的比较见表1-4。

Table 1-4 Comparison of pellet production by three kinds of pellet roasting equipment
表1-4 三种球团焙烧设备生产球团的比较

Equipment 设备	Shaft furnace 竖炉	Belt roaster 带式焙烧机	Grate—rotary kiln 链箅机—回转窑
Advantages 优点	(1) Simple structure; (1) 结构简单; (2) No special requirements for material; (2) 材质无特殊要求; (3) Good heat utilization in the furnace (3) 炉内热利用好	(1) Easy to operate, manage and maintain; (1) 便于操作、管理和维护; (2) It can handle all kinds of ores; (2) 可以处理各种矿石; (3) The baking cycle is short and the length of each section is easy to control (3) 焙烧周期短，各段长度易控制	(1) Simple equipment structure; (1) 设备结构简单; (2) The roasting is even and the product quality is good; (2) 焙烧均匀，产品质量好; (3) It can handle all kinds of ores; (3) 可处理各种矿石; (4) Heat resistant alloy material is not required (4) 不需耐热合金材料
Disadvantages 缺点	(1) The roasting is not uniform enough; (1) 焙烧不够均匀; (2) Single machine production capacity is limited; (2) 单机生产能力受限制; (3) Single ore treatment (3) 处理矿石单一	(1) The quality of the balls in the upper and lower layers is uneven; (1) 上下层球团质量不均; (2) High temperature resistant alloy shall be used for trolley and calculation bar; (2) 台车需用耐高温合金; (3) Complicated process of edge and bottom material laying (3) 铺边、铺底料流程复杂	(1) Easy to form rings in the kiln; (1) 窑内易结圈; (2) Heavy maintenance work (2) 维修工作量大
Throughput 生产能力	The single machine production capacity is small, with a maximum capacity of 2000t/d, suitable for the production of small and medium-sized enterprises 单机生产能力小，最大为2000t/d，适于中小型企业生产	Large single machine production capacity with 6000~6500t/d, suitable for large enterprises 单机生产能力大，最大为6000~6500t/d，适于大型企业生产	Large single machine production capacity with 6000~12000t/d, suitable for large enterprises 单机生产能力大，最大为6000~12000t/d，适于大型企业生产

Continued Table 1-4

Equipment 设备	Shaft furnace 竖炉	Belt roaster 带式焙烧机	Grate—rotary kiln 链算机—回转窑
Product quality 产品质量	Slightly worse 稍差	Good 良好	Good 良好
Infrastructure investment 基建投资	Low 低	Higher 较高	Higher 较高
Operating expenses 经营费用	Commonly 一般	Slightly higher 稍高	Low 低
Electricity consumption 电耗	High 高	Medium 中	Slightly lower 稍低

1.1.5.3 Main Equipment for Pellet Production
1.1.5.3 球团生产主要设备

The main equipment used in pellet production includes pelletizing equipment and roasting equipment.

球团生产使用的主要设备有造球设备和焙烧设备。

Pelletizing Equipment

造球设备

Disc pelletizer widely used at home and abroad, can be divided into bevel gear driven disc pelletizer and internal ring gear driven disc pelletizer according to the structure. The structure of the disc pelletizer driven by bevel gear is shown in Figure 1-7, which is mainly composed of disc, scraper, scraper frame, large bevel gear, small bevel gear, main shaft, angle adjusting mechanism, reducer, motor, base, etc. The rotating speed and the inclination of the disc of the pelletizer can be adjusted.

圆盘造球机是目前国内外广泛采用的造球设备，按结构可分为伞齿轮传动的圆盘造球机和内齿圈传动的圆盘造球机。伞齿轮传动的圆盘造球机的构造如图1-7所示，它主要是由圆盘、刮刀、刮刀架、大伞齿轮、小圆锥齿轮、主轴、倾角调节机构、减速机、电动机、底座等组成。该造球机的转速和圆盘倾角可调。

The advantages of the disc pelletizer are: the pelletizing particle size is uniform, and there is no cycle load; when the solid fuel is used for baking, a ring groove is added at the edge of the disc to add solid fuel to the surface of the pelletizing, and no special equipment is required; and in addition, the equipment is light in weight, low in power consumption, and easy to operate. The disadvantage is low output of single machine.

圆盘造球机的优点是：造出的生球粒度均匀，没有循环负荷；采用固体燃料焙烧时，在圆盘边缘加一环形槽就能向生球表面附加固体燃料，不必另置专门设备；另外，设备质

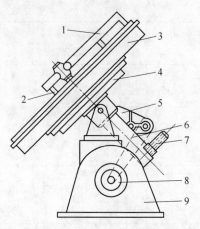

Figure 1-7 Disc pelletizer driven by bevel gear
图 1-7 伞齿轮传动的圆盘造球机

1—Scraper frame; 2—Scraper; 3—Disc; 4—Bevel gear; 5—Reducer; 6—Central shaft; 7—Tilt Screw; 8—Motor; 9—Base
1—刮刀架; 2—刮刀; 3—圆盘; 4—伞齿轮; 5—减速机; 6—中心轴; 7—调倾角螺杆; 8—电动机; 9—底座

量轻, 电能消耗少, 操作方便。其缺点是单机产量低。

Roasting Equipment
焙烧设备

At present, there are mainly three kinds of roasting equipment for pellet: shaft furnace, belt roaster and grate—rotary kiln.

目前, 球团矿的焙烧设备主要有竖炉、带式焙烧机和链算机—回转窑三种。

- Shaft Furnace
- 竖炉

Shaft furnace is one of the earliest pelletizing equipment. Its structure is shown in Figure 1-8. In the middle of the shaft furnace is the roasting chamber, two sides are the combustion chamber, and the lower part is the discharge roller and the sealing gate. The cross section of the roasting chamber and the combustion chamber are mostly rectangular. The specification of the shaft furnace is expressed by the area of the furnace mouth (the area at the charging line). The furnace mouth area of shaft furnace is mostly $4 \sim 8m^2$, the general width is not more than 1.8m, the length is $3 \sim 3.25$ times of the width, and the distance from furnace mouth to discharge roller is $7.5 \sim 10m$. The cross-sectional area of foreign shaft furnaces is $2.13 \sim 6.40m^2$, and the height is 13.7m.

竖炉是最早被采用的一种球团焙烧设备, 其结构如图 1-8 所示。竖炉中间是焙烧室, 两侧是燃烧室, 下部是卸料辊和密封闸门, 焙烧室和燃烧室的横截面多为矩形。竖炉的规格以炉口面积 (料线处的面积) 表示。竖炉的炉口面积多为 $4 \sim 8m^2$, 一般宽度不超过 1.8m, 长度为宽度的 $3 \sim 3.25$ 倍, 从炉口到卸料辊的距离为 $7.5 \sim 10m$。国外竖炉的横截面积为 $2.13 \sim 6.40m^2$, 高为 13.7m。

- Belt Roaster
- 带式焙烧机

The belt roaster is a kind of equipment with early history, large flexibility and wide range of

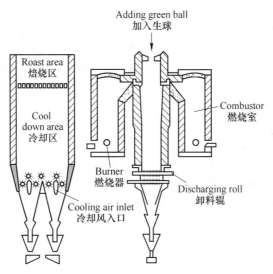

Figure 1-8　Shaft furnace
图 1-8　竖炉

use. At present, it has become the largest roasting equipment in pellet production. The structure of the belt roaster is similar to that of the belt sintering machine. But with multi roller distributor, the suction system is more complex than that of the sintering machine, and the heat transfer mode is also different. Figure 1-9 shows a schematic diagram of $162m^2$ belt roaster in an iron and steel plant. The roasting of the belt roaster depends on the external heat supply, which is divided into drying section, preheating section, roasting section, soaking section and cooling section along the length of the roaster. The length and thermal system of each section will not be asked according to different raw material conditions. The working faces of each section are covered with hoods, and they form an air circulation system through pipes, fans, etc., which makes full use of the heat energy in the roasting process. The temperatures of each section are: the drying section is not higher than 800℃, the preheating section is not higher than 1100℃, and the roasting section is about 1250℃. At present, nearly 60% of the pellets in the world are produced by belt roaster.

带式焙烧机是一种历史悠久、灵活性大、使用范围广的细粒造块设备，目前它已成为球团矿生产中产量最大的焙烧设备。带式焙烧机的构造与带式烧结机相似，但其采用多辊布料器，抽风系统比烧结机复杂，传热方式也不同。如图 1-9 所示是某钢铁厂 $162m^2$ 带式焙烧机示意图。带式焙烧机焙烧全靠外部供热，沿焙烧机长度分为干燥段、预热段、焙烧段、均热段和冷却段。每段的长度和热工制度随着原料条件的不同而不同。各段工作面均用机罩覆盖，它们之间通过管道、风机等组成一个气流循环系统，使焙烧过程中的热能得到充分利用。各段温度为：干燥段不高于 800℃，预热段不超过 1100℃，焙烧段为 1250℃ 左右。目前世界上近 60% 的球团矿采用带式焙烧机生产。

- Grate Kiln
- 链箅机—回转窑

Once the grate kiln pelletizing process came out, it got the attention of steel enterprises all

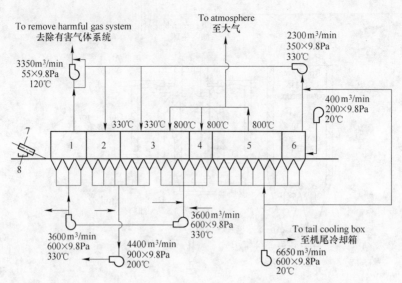

Figure 1-9　Diagram of 162m² belt roaster in an iron and steel plant

图 1-9　某钢铁厂 162m² 带式焙烧机示意图

1，2—Drying section; 3—Preheating roasting section; 4—Soaking section; 5—Cooling section Ⅰ; 6—Cooling section Ⅱ;
7—Belt feeder; 8—Edge and bottom material feeder

1，2—干燥段；3—预热焙烧段；4—均热段；5—冷却一段；6—冷却二段；7—带式给料机；8—铺边、铺底料给料机

over the world and developed rapidly. At present, it has become an important method of roasting pelletizing. Grate rotary kiln is a kind of combined unit, which is composed of grate, rotary kiln, cooler and auxiliary equipment. Therefore, the characteristics of this pelletizing process are that the drying, preheating, roasting and cooling of raw pellets are carried out on three different equipment respectively. The raw pellets are dried, dehydrated and preheated on the grate, then baked in the rotary kiln, and finally cooled on the cooler. The structural diagram of grate kiln is shown in Figure 1-10.

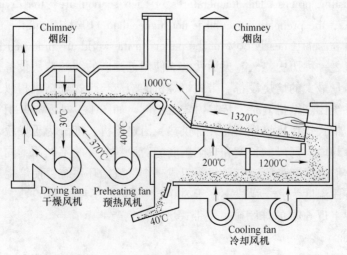

Figure 1-10　Structure diagram of chain computer rotary kiln

图 1-10　链箅机—回转窑的结构示意图

链箅机—回转窑球团工艺一经问世就得到世界各国钢铁企业的重视，得到迅速发展，目前已成为焙烧球团的一种重要方法。链箅机—回转窑是一种联合机组，由链箅机、回转窑、冷却机和附属设备联合组成。因此，这种球团工艺的特点是生球的干燥、预热、焙烧和冷却分别在三台不同的设备中进行。生球首先在链箅机上干燥、脱水、预热，而后进入回转窑内焙烧，最后在冷却机上完成冷却。链箅机—回转窑的结构示意图如图 1-10 所示。

Task 1.2 Blast Furnace Ironmaking
任务 1.2 高炉炼铁

1.2.1 Overview of Blast Furnace Ironmaking
1.2.1 高炉炼铁概述

1.2.1.1 Characteristics of Blast Furnace Ironmaking
1.2.1.1 高炉炼铁的特点

Blast furnace ironmaking is an important part of modern iron and steel complex. The task of blast furnace ironmaking is to separate the iron and oxygen of iron oxide from the ore by reduction method, and to separate the reduced metal iron from gangue by high temperature melting separation method. Then getting qualified pig iron after flow separation and carburization. The general production process of blast furnace ironmaking is shown in Figure 1-11.

高炉炼铁在现代钢铁联合企业中是极其重要的一环。高炉炼铁的任务是用还原法将矿石中铁氧化物的铁与氧分离，用高温熔化分离法将已还原的金属铁与脉石分离，然后经过脱硫、渗碳，最后得到合格生铁。高炉炼铁的一般生产工艺流程如图 1-11 所示。

The smelting process is carried out in the body of blast furnace, and the whole process is completed in the mutual contact process from top to bottom of charge and from bottom to top of gas. The charge is put into the furnace from the top of the furnace in a certain order, and then it is blown into the internal hot blast furnace from the tuyere to heat to the hot blast of $1000 \sim 1300$℃. The coke in the charge reacts with the oxygen in the blast in front of the tuyere to produce high temperature and reducing gas. During the rising process of the furnace, the charge that slowly drops is heated, and the iron oxide in the iron ore is reduced to metallic iron. When the ore rises to a certain temperature, it will soften. In the process of melting and dropping, the last reduced substance in the ore forms slag to separate slag and iron. The molten iron and slag have been gathered in the hearth and many reactions have been made. Finally, the composition and temperature of the molten iron are adjusted to the end point, and the slag and pig iron are discharged periodically. The temperature of the rising high-temperature gas flow decreases continuously due to the energy transmitted to the charge, and finally the blast furnace gas is formed and discharged from the top of the furnace. The whole process depends on the combustion of coke in front of the tuyere, a series of heat and mass transfer between the rising gas flow and the falling charge, as well as physical and chemical changes such as drying, evaporation,

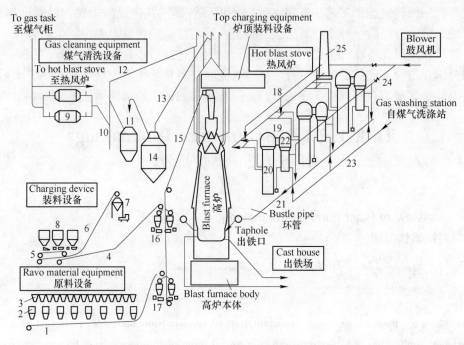

Figure 1-11 Production process of blast furnace ironmaking
图 1-11 高炉炼铁生产工艺流程

1—Ore conveying belt conveyor; 2—Weighing funnel; 3—Ore storage tank; 4—Coke conveying belt conveyor; 5—Feeder;
6—Coke conveying belt conveyor; 7—Coke bin; 8—Coke storage tank; 9—Electrostatic precipitator; 10—Regulating valve;
11—Venturi tube precipitator; 12—Net gas discharge pipe; 13—Downcomer; 14—Gravity precipitator;
15—Feeding belt conveyor; 16—Coke weighing funnel; 17—Ore weighing funnel; 18—Cold air pipe; 19—Flue;
20—Regenerator; 21—Hot air main pipe; 22—Combustion chamber; 23—Gas main pipe; 24—Mixed air pipe; 25—Chimney
1—矿石输送皮带机；2—称量漏斗；3—储矿槽；4—焦炭输送皮带机；5—给料机；6—粉焦输送皮带机；
7—粉焦仓；8—储焦槽；9—电除尘器；10—调节阀；11—文氏管除尘器；12—净煤气放散管；13—下降管；
14—重力除尘器；15—上料皮带机；16—焦炭称量漏斗；17—矿石称量漏斗；18—冷风管；19—烟道；
20—蓄热室；21—热风主管；22—燃烧室；23—煤气主管；24—混风管；25—烟囱

volatilization, decomposition, reduction, soft melting, slagging, carburization, desulfurization, etc. Therefore, the blast furnace is essentially a reactor with two reverse flow movements: one is the decrease of charge and the other is the rise of gas.

冶炼过程在高炉本体内进行，全过程是在炉料自上而下、煤气自下而上的相互接触过程中完成。炉料按一定顺序从炉顶装入炉内，从风口鼓入内热风炉加热到 1000~1300℃ 的热风，炉料中焦炭在风口前与鼓风中的氧发生燃烧反应，产生高温和还原性气体。在炉内上升过程中加热缓慢下降的炉料，并将铁矿石中的铁氧化物还原为金属铁，矿石升到一定温度后软化、熔融滴落，矿石中未被还原的物质形成熔渣，从而实现渣铁分离。已熔化的渣铁聚集在炉缸内，发生诸多反应，最后调整铁液的成分和温度达到终点，定期从炉内排放炉渣和生铁。上升的高温煤气流由于将能量传给炉料，使得温度不断下降，最终形成高炉煤气，从炉顶导出管排出。整个过程取决于风口前焦炭的燃烧，上升煤气流与下降炉料之间进行的一系列传热、传质，以及干燥、蒸发、挥发、分解、还原、软熔、造渣、渗碳、脱硫等物理化学变化。因此，高炉实质是一个炉料下降、煤气上升两个逆向流运动的反应器。

1.2.1.2 Products of Blast Furnace Ironmaking
1.2.1.2 高炉炼铁的产品

The main product of blast furnace ironmaking is pig iron, and the by-product is slag, blast furnace gas and furnace dust.

高炉炼铁的主要产品是生铁，副产品是炉渣、高炉煤气以及其带出的炉尘。

Pig iron can be divided into steel-making pig iron and casting pig iron. Cast iron is mainly used to produce pressure resistant castings, accounting for about 10% of the output of pig iron. Blast furnace can also be used to produce special pig iron, such as ferromanganese, brick iron, etc. At present, The national standards of steel-making pig iron and cast pig iron are shown in Table 1-5.

生铁可分为炼钢生铁和铸造生铁。炼钢生铁供转炉、电炉炼钢使用，约占生铁产量的90%；铸造生铁主要用于生产耐压铸件，约占生铁产量的10%。高炉也可用来生产特殊生铁，如锰铁、砖铁等。目前，炼钢生铁、铸造生铁的国家标准见表1-5。

Table 1-5　National standard for pig iron（GB/T 717—1998）
表1-5　生铁的国家标准（GB/T 717—1998）

Iron species 铁种			Steelmaking pig iron 炼钢生铁			Cast pig iron 铸造生铁					
Iron number 铁号	Brand name 牌号		Pig iron for steelmaking 04 炼04	Pig iron for steelmaking 08 炼08	Pig iron for steelmaking 10 炼10	Casting pig 34 铸34	Casting pig 30 铸30	Casting pig 26 铸26	Casting pig 22 铸22	Casting pig 18 铸18	Casting pig 14 铸14
	Code name 代号		L04	L08	L10	Z34	Z30	Z26	Z22	Z18	Z14
Components (by mass) /% 化学成分 (质量分数) /%	Si		≤0.45	>0.45~0.85	>0.85~1.25	>3.20~3.60	>2.80~3.20	>2.40~2.80	>2.00~2.40	>1.60~2.00	>1.25~1.60
	Mn	One group（一组）	≤0.40			≤0.05					
		Two group（二组）	>0.40~1.00			>0.05~0.90					
		Three group（三组）	>1.00~2.00			>0.90~1.30					
	P	One group（一组）	>0.100~0.150			≤0.06					
		Two group（二组）	>0.150~0.250			>0.06~0.10					
		Three group（三组）	>0.250~0.400			>0.10~0.20					
		Four group（四组）	Special grade≤0.100			>0.20~0.40					
		Five group（五组）	—			>0.40~0.90					

Continued Table 1-5

Iron species 铁种			Steelmaking pig iron 炼钢生铁	Cast pig iron 铸造生铁	
Components (by mass) /% 化学成分 (质量分数) /%	S	Special grade (特等)	≤0.02	—	—
		One category (一类)	>0.020~0.030	≤0.03	≤0.04
		Two category (二类)	>0.030~0.050	≤0.04	≤0.05
		Three classes of initials (三类)	>0.050~0.070	≤0.05	≤0.06
	C	—	≥3.50	>3.30	

Due to the difference of iron ore grade, coke ratio and ash content of coke, the amount of slag per ton of pig iron produced by blast furnace smelting is quite different. Generally, the amount of slag per ton of pig iron is between 0.2t and 0.5t. The amount of slag per ton of pig iron in blast furnace with poor raw material conditions and low technical level even exceeds 0.6t. The blast furnace slag containing CaO, SiO_2, MgO, Al_2O_3, etc. is usually granulated into water slag through rapid cooling, which is used to manufacture cement and building materials. It can also be steam blown into slag cotton for sound insulation and heat preservation materials.

出于铁矿石品位、焦比及焦炭灰分的不同，高炉冶炼每吨生铁产生的渣量差异很大，一般每吨生铁的渣量在 0.2~0.5t 之间。原料条件差、技术水平低的高炉每吨生铁的渣量甚至超过 0.6t。高炉渣中含有 CaO、SiO_2、MgO、Al_2O_3 等，一般通过急冷粒化成水渣，用于制造水泥和建筑材料；也可用蒸汽吹成渣棉，作隔声、保温材料。

Blast furnace gas with chemical composition of CO, CO_2, N_2, H_2 and CH_4 can be produced by smelting each ton of pig iron, among which $[w(CO) = 20\% \sim 25\%]$, $[w(CO_2) = 15\% \sim 25\%]$, $[w(H_2) = 1\% \sim 3\%]$ and a small amount of CH_4 are combustible gases. The calorific value of blast furnace gas after dedusting treatment is $3350 \sim 4200 kJ/m^3$, which is a good gas fuel. It is mainly used as hot blast furnace fuel, and also for power, coking, sintering, steelmaking, steel rolling and other departments.

冶炼每吨生铁可产生 1600~3000m^3 的高炉煤气，化学成分为 CO、CO_2、N_2、H_2 及 CH_4 等，其中 $[w(CO)=20\%\sim25\%]$、$[w(CO_2)=15\%\sim25\%]$、$[w(H_2)=1\%\sim3\%]$ 及少量 CH_4 为可燃性气体。经除尘处理后的高炉煤气发热值为 3350~4200kJ/m^3，是良好的气体燃料，主要作为热风炉燃料，也可供动力、炼焦、烧结、炼钢、轧钢等部门使用。

1.2.1.3 Main Technical and Economic Indexes of Blast Furnace Production
1.2.1.3 高炉生产主要技术经济指标

The technical and economic indicators of blast furnace production are used to measure the following aspects of blast furnace production technology:

高炉生产技术经济指标是用来衡量高炉生产技术以下几项：

(1) Effective volume utilization coefficient of blast furnace η_u. The effective volume utilization coefficient of blast furnace refers to the tonnage of pig iron produced in a day and night by the effective volume of $1m^3$ blast furnace, which is given by:

(1) 高炉有效容积利用系数 η_u。高炉有效容积利用系数是指 $1m^3$ 高炉有效容积一昼夜生产的生铁吨数，其计算公式为：

$$\eta_u = P/V_u \tag{1-9}$$

where η_u——the effective volume utilization coefficient in $t/(m^3 \cdot d)$;
  P——the iron production per day and night in t/d;
  V_u——the effective volume of blast furnace in m^3.

式中 η_u——有效容积利用系数，$t/(m^3 \cdot d)$；
  P——高炉每昼夜产铁量，t/d；
  V_u——高炉有效容积，m^3。

The effective volume utilization coefficient of blast furnace is an important index of blast furnace smelting. The larger η_u is, the higher the blast furnace productivity is. At present, the effective volume utilization coefficient of blast furnace is generally $2.00 \sim 2.50 t/(m^3 \cdot d)$, and some advanced blast furnaces can reach $3.5 t/(m^3 \cdot d)$.

高炉有效容积利用系数是高炉冶炼的一个重要指标，η_u 越大，高炉生产率越高。目前高炉的有效容积利用系数一般为 $2.00 \sim 2.50 t/(m^3 \cdot d)$，一些先进的高炉可达到 $3.5 t/(m^3 \cdot d)$。

(2) Utilization coefficient of blast furnace hearth area η_A. The utilization coefficient of blast furnace hearth area refers to the tonnage of pig iron produced in a day and night for the effective area of $1m^2$ blast furnace hearth, given by:

(2) 高炉炉缸面积利用系数 η_A。高炉炉缸面积利用系数是指 $1 m^2$ 高炉炉缸有效面积一昼夜生产的生铁吨数，其计算公式为：

$$\eta_A = P/S_A \tag{1-10}$$

Where η_A——the utilization coefficient of hearth area in $t/(m^2 \cdot d)$;
  S_A——the effective area of blast furnace hearth in m^2.

式中 η_A——炉缸面积利用系数，$t/(m^2 \cdot d)$；
  S_A——高炉炉缸有效面积，m^2。

The utilization coefficient of blast furnace hearth area is an important reference index to measure blast furnace efficiency. The utilization coefficient of domestic $1000 \sim 5000 m^3$ blast furnace hearth area is generally $65 \sim 70 t/(m^2 \cdot d)$.

高炉炉缸面积利用系数是衡量高炉效率的一个重要参考指标，国内 $1000 \sim 5000 m^3$ 高炉的炉缸面积利用系数一般为 $65 \sim 70 t/(m^2 \cdot d)$。

(3) Coke ratio K (kg/t). Coke ratio refers to the coke consumption per ton of pig iron smelting, given by:

(3) 焦比 K (kg/t)。焦比是指冶炼每吨生铁所消耗的焦炭量，其计算公式为：

$$K = Q_k/P \tag{1-11}$$

Where K——the coke ratio in kg/t;
　　Q_k——the dry coke consumption per day and night of blast furnace in kg/d.

式中　K——焦比，kg/t；
　　Q_k——高炉每昼夜消耗的干焦量，kg/d。

The coke ratio of blast furnace is generally 400~500kg/t, which can be effectively reduced by injecting fuel.

高炉的焦比一般为400~500kg/t，喷吹燃料可以有效降低焦比。

(4) Coal ratio Y. Coal ratio refers to the pulverized coal consumption per ton of pig iron smelting, given by:

(4) 煤比 Y。煤比是指冶炼每吨生铁所消耗的煤粉量，其计算公式为：

$$Y = Q_y/P \tag{1-12}$$

Where Y——the coal ratio in kg/t;
　　Q_y——the pulverized coal consumption per day and night of blast furnace in kg/d.

式中　Y——煤比，kg/t；
　　Q_y——高炉每昼夜消耗的煤粉量，kg/d。

(5) Smelting strength I. Smelting strength refers to the average coke consumption per day and night of effective volume of 1m³ blast furnace, given by:

(5) 冶炼强度 I。冶炼强度是指1m³高炉有效容积每昼夜平均消耗的焦炭量，其计算公式为：

$$I = Q_y/V_u \tag{1-13}$$

Where I——the smelting strength in t/(m³·d);
　　V_u——the effective volume of blast furnace in m³.

式中　I——冶炼强度，t/(m³·d)；
　　V_u——高炉有效容积，m³。

The relationship among the effective volume utilization coefficient η_u, coke ratio K and smelting strength I is as follows:

高炉有效容积利用系数 η_u、焦比 K 和冶炼强度 I 三者的关系如下：

$$\eta_u = I/K \tag{1-14}$$

(6) Qualified rate of pig iron. Pig iron with chemical composition meeting the national standard is called qualified pig iron. And the percentage of qualified pig iron output in total pig iron output is called qualified pig iron rate.

(6) 生铁合格率。化学成分符合国家标准的生铁称为合格生铁。合格生铁产量占生铁总产量的百分数称为生铁合格率。

(7) Pig iron cost. Pig iron cost refers to the sum of all the costs of raw materials, fuel, materials, water and electricity, labor, etc. for the production of 1t qualified pig iron.

(7) 生铁成本。生铁成本是指每生产1t合格生铁所消耗的所有原料、燃料、材料、水电、人工等一切费用的总和。

(8) Rest rate. The air shut-off rate refers to the percentage of blast furnace air shut-off time in the specified operation time (i.e. calendar time minus planned major and medium maintenance

time). It reflects the level of maintenance and operation of blast furnace equipment. The shut-off rate of advanced blast furnace is less than 1%.

（8）休风率。休风率是指高炉休风时间占规定作业时间（即日历时间减去计划大、中修时间）的百分数，它反映高炉设备维护和高炉操作水平的高低。先进高炉的休风率在1%以下。

（9）Blast furnace life. The first generation life of blast furnace refers to the smelting time from ignition and opening to shutdown and overhaul, or the smelting time between two adjacent overhauls of blast furnace. The first generation life of large blast furnace is 10~15 years.

（9）高炉寿命。高炉一代寿命是指从点火开炉到停炉大修之间的冶炼时间，或者指高炉相邻两次大修之间的冶炼时间。大型高炉一代寿命为10~15年。

1.2.2　Basic Principles of Blast Furnace Ironmaking
1.2.2　高炉炼铁基本原理

1.2.2.1　Charge Shapes and Main Reactions in Each Area of Blast Furnace
1.2.2.1　高炉内各区域的炉料形态及主要反应

Blast furnace is a closed and continuous countercurrent reactor. How to carry out the internal reaction can not be directly observed. According to the anatomical study of blast furnace at home and abroad, the change of charge shape in blast furnace is shown in Figure 1-12, which can be divided into five areas (or five belts), as follows:

高炉是一个密闭的、连续的逆流反应器，其内部的反应如何进行不能直接观察。通过国内外高炉解剖研究可知，高炉内炉料形态变化如图1-12所示，其可以分为五个区域（或称五个带）。

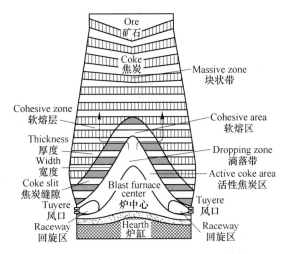

Figure 1-12　Schematic diagram of charge shape change in blast furnace
图1-12　高炉内炉料形态变化示意图

（1）Massive zone. In this area, the burden obviously keeps the layered state (ore layer and coke layer) during charging, and there is no liquid slag iron. With the decrease of burden, the layer tends to be horizontal and the thickness becomes thinner.

（1）块状带。在该区域中，炉料明显保持装料时的分层状态（矿石层和焦炭层），没有液态渣铁。随着炉料下降，其层状逐渐趋于水平，且厚度逐渐变薄。

（2）Soft melt zone. The zone from the beginning of ore softening to complete melting is called the soft melting zone. It is composed of many solid coke layers and semi molten ore layers bonded together. Coke alternates with ore with distinct layers. The gas mainly passes through the coke layer like a window, because the ore is in the form of soft melting and the permeability is very poor. So it is called "Coke Window". The upper edge of the soft melt part is the softening line (i.e. solid phase line), and the lower edge of the soft melt zone is the melting line (i.e. liquid phase line), as shown in Figure 1-13.

（2）软熔带。矿石从开始软化到完全熔化的区间称为软熔带。软熔带是由许多固态焦炭层和黏结在一起的半熔矿石层组成。焦炭与矿石相间，层次分明。由于矿石呈软熔状，透气性极差，煤气主要从焦炭层通过，像窗口一样，因此称其为"焦窗"。软熔件的上沿是软化线（即固相线），软熔带的下沿是熔化线（即液相线），如图1-13所示。

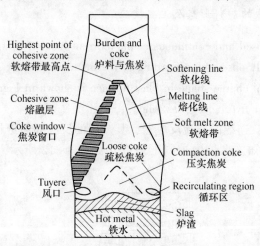

Figure 1-13　Schematic diagram of soft melt zone
图1-13　软熔带示意图

（3）Drip belt. Below the soft melting zone, the melted slag iron passes through the solid coke gap and drips like a raindrop, so it is called the dropping zone. The area where the coke in the dropping belt basically stable for a long time is called central dead zone (dead zone). The area where coke looseness decreases is called active coke area.

（3）滴落带。软熔带以下，已熔化的渣铁穿过固体焦炭空隙，像雨滴一样滴落，故称为滴落带。焦炭在滴落带长时间处于基本稳定状态的区域称为中心呆滞区（死料区）。焦炭松动下降的区域称为活动性焦炭区。

（4）Tuyere burning belt. The area in front of the tuyere where the coke makes gyratory motion under the action of the blast kinetic energy is called tuyere combustion zone, which means coke gyratory zone. The center of the gyratory zone is in a half empty state. The combustion of coke in this area is not only the main producing place of heat and gas reducing agent in blast furnace, but also the only oxidizing area in blast furnace.

（4）风口燃烧带。风口前在鼓风动能作用下焦炭作回旋运动的区域称为风口燃烧带，亦称焦炭回旋区，这个回旋区中心呈半空状态。该区域内焦炭燃烧，是高炉内热量和气体还原剂的主要产生地，也是高炉内唯一存在的氧化性区域。

（5）Slag iron belt. The area below the tuyere (i.e. hearth) is mainly composed of liquid slag iron and coke immersed in it. In this area, when the iron drop passes through the slag layer and the slag iron interface, the necessary slag iron reaction is finally completed to obtain qualified pig iron. And the pig iron is discharged from the slag port and iron port out of the furnace intermittently.

（5）渣铁带。风口以下（即炉缸）区域，主要由液态渣铁以及浸入其中的焦炭组成。在这一区域内，铁滴穿过渣层时以及在渣—铁界面最终完成必要的渣铁反应，得到合格生铁，并间断地从渣口、铁口排出炉外。

The main reactions and characteristics of each area in the blast furnace are shown in Table 1-6.

高炉内各区域的主要反应及特征见表 1-6。

Table 1-6 Main reactions and characteristics of each zone in blast furnace

表 1-6 高炉内各区域的主要反应及特征

Zones 区域	Functions 功能		
	Opposite motion 相向运动	Heat exchange 热交换	Reaction 反应
Massive zone 块状带	The solid (ore, coke) falls under the action of gravity, and the gas rises under the action of forced blast 固体（矿、焦）在重力作用下下降，煤气在强制鼓风作用下上升	The rising gas preheats and dries the solid charge 上升的煤气对固体炉料进行预热和干燥	Indirect reduction of ore, evaporation and decomposition of water in charge 矿石间接还原，炉料中水分蒸发、分解
Soft melt zone 软熔带	Affect gas flow distribution 影响煤气流分布	Heat transfer and melting of softened semi melted layer by rising gas 上升煤气对软化半熔层进行传热熔化	Direct reduction and carburization of ore. The gasification reaction of coke is $CO_2+C=2CO$ 矿石直接还原和渗碳，焦炭的气化反应为 $CO_2+C=2CO$
Dropping zone 滴落带	The solid (coke) and liquid (molten iron slag) decrease, and the gas rises to supply coke to the raceway 固体（焦炭）、液体（铁水熔渣）下降，煤气上升向回旋区供给焦炭	Rising gas makes molten iron, slag and coke warm up, and the molten iron, slag and coke dropped are subject to heat exchange 上升煤气使铁水、熔渣、焦炭升温，滴下的铁水、熔渣和焦炭进行热交换	Direct reduction carburization of ore, gasification reaction of coke $CO_2+C=2CO$ 非铁元素的还原，脱硫、渗碳，焦炭的气化反应 $CO_2+C=2CO$

Continued Table 1-6

Zones 区域	Functions 功能		
	Opposite motion 相向运动	Heat exchange 热交换	Reaction 反应
Tuyere burning belt 风口燃烧带	The blast whirled the coke 鼓风使焦炭做回旋运动	Reaction exothermic, make gas temperature rise 反应放热,使煤气温度上升	Oxygen and steam in the blast make coke burn 鼓风中的氧和蒸汽使焦炭燃烧
Slag iron belt 渣铁带	The molten iron and slag are stored. During tapping, the molten iron and slag move in a circular flow, while the coke immersed in the slag iron moves in a slow floating motion with the slag tapping. Some of the coke is squeezed into the tuyere combustion zone for gasification 铁水、炉渣存放,出铁时,铁水和炉渣做环流运动,而浸入渣铁中的焦炭则随出渣出铁而做缓慢的沉浮运动,部分被挤入风口燃烧带气化	Heat exchange between molten iron, slag and slow moving coke 铁水、熔渣和缓慢运动的焦炭进行热交换	Final slag iron reaction 最终的渣铁反应

1.2.2.2 Reduction Process and Formation of Pig Iron
1.2.2.2 还原过程与生铁的形成

The purpose of blast furnace ironmaking is to reduce iron and some useful elements in iron ore, So reduction reaction is the most basic chemical reaction in blast furnace.

高炉炼铁的目的是将铁矿石中的铁和一些有用元素还原出来,所以还原反应是高炉内最基本的化学反应。

Basic Theory of Reduction Reaction
还原反应的基本理论

The affinity between metal and oxygen is very strong. Except that some metals can be decomposed from their oxides, and almost all metals can not be separated from oxides by simple heating method. We must rely on some reducing agent to capture oxygen of oxides and make them become metal elements. The smelting process of blast furnace is basically the reduction process of iron oxide. In addition to iron reduction, there are also a small amount of silicon, manganese, phosphorus and other elements reduction in the blast furnace. From the top of the blast furnace to the bottom of the hearth (tuyere area), the reduction reaction almost runs through the whole blast furnace.

金属与氧的亲和力很强,除个别的金属能从其氧化物中分解出来外,几乎所有金属都不能靠简单加热的方法从氧化物中分离出来，必须依靠某种还原剂夺取氧化物的氧，使之变成金属元素。高炉冶炼过程基本上就是铁氧化物的还原过程。除铁的还原外，高炉内还有少量硅、锰、磷等元素的还原。炉料从高炉顶部装入后开始直至到达下部炉缸（除风口区域），还原反应几乎贯穿整个高炉冶炼的始终。

The general formula of reduction reaction of metal oxides can be expressed as follows:

金属氧化物的还原反应通式可表示为：

$$MeO + B = Me + BO \tag{1-15}$$

Where　MeO——the reduced metal oxide;

　　　　Me——the metal obtained by reduction;

　　　　B——the reducing agent, which can be gas or solid, or metal or nonmetal;

　　　　BO——the product obtained by oxidation after the reductant taking oxygen from the metal oxide.

式中　MeO——被还原的金属氧化物；

　　　Mc——还原得到的金属；

　　　B——还原剂，可以是气体或固体，也可以是金属或非金属；

　　　BO——还原剂夺取金属氧化物中的氧后被氧化得到的产物。

It can be seen from reaction (1-15) that the loss of O in MeO is reduced to Me, and the gain of O in B is oxidized to BO. Which substance can be used as reducing agent to capture oxygen in metal oxide can be determined by the chemical affinity of substance and oxygen. All substances with higher affinity to oxygen than to metal elements can be used as reducers of the metal oxides. Obviously, the greater the affinity between reductant and oxygen, the stronger the ability to capture oxygen, or the stronger the reduction ability. For the reduced metal oxide, the stronger the affinity between the metal elements and oxygen is, the more difficult it is to reduce. The affinity between a substance and oxygen can be measured by the decomposition pressure of the oxide. The higher the decomposition pressure of the oxide (p_{O_2}), the smaller the affinity between the substance and oxygen, the more unstable the oxide is, and the easier it is to decompose. On the contrary, it is the opposite.

从式（1-15）可以看出，MeO 失去 O 被还原成 Me，B 得到 O 而被氧化成 BO。哪种物质可以充当还原剂来夺取金属氧化物中的氧，可以通过物质与氧的化学亲和力的大小来判断。凡是与氧的亲和力比与金属元素的亲和力大的物质，都可以作为该金属氧化物的还原剂。很明显，还原剂与氧的亲和力越大，夺取氧的能力越强，或者说还原能力越强。而对被还原的金属氧化物来说，其金属元素与氧的亲和力越强，该氧化物越难还原。某物质与氧亲和力的大小又可用该物质氧化物的分解压来衡量，氧化物分解压（p_{O_2}）越大，说明该物质与氧的亲和力越小，氧化物越不稳定，越易分解，反之则相反。

At present, the sequence (from easy to difficult) of reduction of various metal elements in blast furnace smelting is: Cu, Pb, Ni, Co, Fe, Cr, Mn, V, Si, Ti, Al, Mg, Ca. From the thermodynamic point of view, the elements behind iron can be used as reductants of iron oxides. However, according to the specific conditions of blast furnace production, as reducing agent in blast furnace production, it is the fixed carbon in coke and the CO produced after combustion of coke, as well as the H_2 produced by the decomposition of blowing water and ejecta. Because they are rich in reserves, easy to obtain and the lowest in price.

目前，高炉冶炼常遇到的各种金属元素还原的难易顺序（由易到难）为：Cu，Pb，Ni，Co，Fe，Cr，Mn，V，Si，Ti，Al，Mg，Ca。从热力学角度来讲，在彼此顺序排列的各元素中，排在铁后面的各元素均可作为铁氧化物的还原剂。但是，根据高炉生产的特定条件，在高炉生产中作为还原剂的是焦炭中的固定碳和焦炭燃烧后产生的 CO，以及鼓风水分和喷吹物分解产生的 H_2。这是因为它们储量丰富，易于获取，价格最低廉。

It can also be concluded from the above that under the condition of blast furnace smelting, Cu, Pb, Ni, Co, and Fe are elements that are easy to be reduced completely, Cr, Mn, V, Si, and Ti are elements that can only be partially reduced, and Al, Mg, and Ca are elements that cannot be reduced.

由上还可得出，在高炉冶炼条件下，Cu，Pb，Ni，Co，Fe 为易被全部还原的元素；Cr，Mn，V，Si，Ti 为只能被部分还原的元素；Al，Mg，Ca 为不能被还原的元素。

Reduction Sequence of Iron Oxides
铁氧化物的还原顺序

The main existing forms of iron oxides in blast furnace raw materials are Fe_2O_3, Fe_3O_4, Fe_2SiO_4, FeS_2, etc. But they are all reduced to metal Fe by FeO.

高炉原料中铁氧化物的存在形态主要有 Fe_2O_3、Fe_3O_4、Fe_2SiO_4、FeS_2 等，但最后都是经 FeO 形态被还原成金属 Fe。

The reduction sequence of various iron oxides is the same as that of decomposition. When the temperature is higher than 570℃, it is $Fe_2O_3 \rightarrow Fe_3O_4 \rightarrow FeO \rightarrow Fe$. At this time, the oxygen loss of each stage is:

各种铁氧化物的还原顺序与分解顺序相同。当温度高于 570℃ 时为 $Fe_2O_3 \rightarrow Fe_3O_4 \rightarrow FeO \rightarrow Fe$，此时各阶段的失氧量为：

$$3Fe_2O_3 \xrightarrow{1/9} 2Fe_3O_4 \xrightarrow{2/9} 6FeO \xrightarrow{6/9} 6Fe \qquad (1-16)$$

From equation (1-16), it can be seen that the amount of oxygen loss in the first stage ($Fe_2O_3 \rightarrow Fe_3O_4$) is small, so reduction is easy. The more oxygen loss in the later stage, the more difficult reduction. More than half of the oxygen is captured in the final stage (the reduction from FeO to Fe). So the reduction of FeO in iron oxide is of the most important significance.

由式（1-16）可见，第一阶段（$Fe_2O_3 \rightarrow Fe_3O_4$）失氧数量少，因此还原容易，越到后面失氧量越多，还原越困难。有一半以上的氧是在最后阶段（从 FeO 还原到 Fe 的过程）中被夺取的，所以铁氧化物中 FeO 的还原具有最重要的意义。

When the temperature is lower than 570℃, due to the instability of FeO, it will immediately decompose as follows:

当温度低于 570℃ 时，由于 FeO 不稳定，会发生如下分解：

$$4FeO = Fe_3O_4 + Fe \qquad (1-17)$$

So the reduction sequence is:

所以此时的还原顺序为：
$$Fe_2O_3 \longrightarrow Fe_3O_4 \longrightarrow Fe$$

Reduction of Iron Oxide with CO

用 CO 还原铁氧化物

After the ore enters the blast furnace, the oxygen in the iron oxide is captured by the CO in the gas of the middle and upper part of the blast furnace with the heating temperature less than 1000℃. This reduction process is not directly using carbon in coke as reducing agent, so it is called indirect reduction.

矿石进入高炉后，在加热温度未超过 1000℃ 的高炉中上部，铁氧化物中的氧被煤气中 CO 夺取产生 CO_2。这种还原过程不是直接用焦炭中的碳作还原剂，故称为间接还原。

When the temperature is below 570℃, the reduction reaction is divided into the following two steps：

当温度低于 570℃ 时，还原反应分为以下两步：

$$3Fe_2O_3 + CO \Longrightarrow 2Fe_3O_4 + CO_2 + 27130 kJ/mol \tag{1-18}$$

$$Fe_3O_4 + 4CO \Longrightarrow 3Fe + 4CO_2 + 17160 kJ/mol \tag{1-19}$$

When the temperature is higher than 570℃, the reduction reaction is divided into the following three steps：

当温度高于 570℃ 时，还原反应分为以下三步：

$$3Fe_2O_3 + CO \Longrightarrow 2Fe_3O_4 + CO_2 + 27130 kJ/mol \tag{1-20}$$

$$Fe_3O_4 + CO \Longrightarrow 3FeO + CO_2 - 20888 kJ/mol \tag{1-21}$$

$$FeO + CO \Longrightarrow Fe + CO_2 + 13600 kJ/mol \tag{1-22}$$

As K_p of different temperatures and iron oxides are different, when the above reactions reach equilibrium, the relationship between temperature and gas phase composition is shown in Figure 1-14.

由于不同温度和不同铁氧化物的 K_p 是不同的，所以上述各反应达到平衡时，其温度与气相组成的关系如图 1-14 所示。

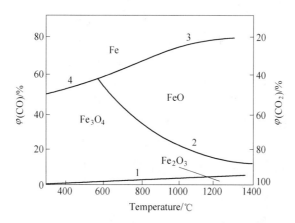

Figure 1-14 Relationship between equilibrium gas composition and temperature of iron oxide reduction by CO

图 1-14 用 CO 还原铁氧化物的平衡气相成分与温度的关系

In Figure 1-14, curve 1 shows the relationship between the equilibrium gas composition and temperature of reaction $3Fe_2O_3+CO = 2Fe_3O_4+CO_2$; curve 2 shows the relationship between the equilibrium gas composition and temperature of reaction $Fe_3O_4+CO = 3FeO+CO_2$; curve 3 shows the relationship between the equilibrium gas composition and temperature of reaction $Fe_3O_4 + 4CO = 3Fe+4CO_2$; and curve 4 shows the relationship between the equilibrium gas composition and temperature of reaction $Fe_3O_4+4CO = 3Fe+4CO_2$. It shows that in order to carry out the reduction of iron oxide, in addition to Co participating in the reaction, it also needs excessive CO to maintain the balance of chemical reaction. Otherwise, the reduction reaction can not be carried out, and even the reduced substances may be oxidized by CO_2. At the same time, the concentration of CO needed to maintain the chemical equilibrium is different under different temperatures. The four curves in Figure 1-14 divide the stable existence regions of Fe_2O_3, Fe_3O_4, FeO and Fe respectively, that is, the substance ciron oxide or mefallic iron can only exist stably if the temperature and gas phase composition conditions are within the range of their stable regions. Otherwise the reverse reaction of reduction reaction will occur.

如图1-14所示，曲线1为反应$3Fe_2O_3+CO = 2Fe_3O_4+CO_2$的平衡气相成分与温度的关系线；曲线2为反应$Fe_3O_4+CO = 3FeO+CO_2$的平衡气相成分与温度的关系线；曲线3为反应$FeO+CO = Fe+CO_2$的平衡气相成分与温度的关系线；曲线4为反应$Fe_3O_4+4CO = 3Fe+4CO_2$的平衡气相成分与温度的关系线。由此表明，要使铁氧化物还原反应进行，除了需要参加反应的CO外，还需要过量的CO来维持化学反应的平衡。否则，还原反应不但不能进行，甚至可能出现已还原的物质被CO_2氧化的情况。同时，不同温度下各种铁氧化物还原反应维持化学平衡所需的CO浓度是不一样的。图1-14中的四条曲线划分出了Fe_2O_3、Fe_3O_4、FeO、Fe各自的稳定存在区域，即只有温度和气相组成条件处于各自稳定区域的范围内，该物质（铁氧化物或金属铁）才能稳定存在，否则将发生还原反应的逆反应。

Reduction of Iron Oxide with H_2

用H_2还原铁氧化物

In the blast furnace without fuel injection, the content of H_2 in the gas is only 1.8%~2.5% (by volume), which is mainly produced by the high temperature decomposition of water in the blast in front of the tuyere. In the blast furnace with fuel injection, the content of H_2 in the gas increases significantly, up to 5%~8% (by volume). The affinity between hydrogen and oxygen is very strong, so hydrogen is also a reducing agent in blast furnace smelting. The reduction of iron oxide with H_2 is the same as that with CO, which can also be called indirect reduction.

在不喷吹燃料的高炉上，煤气中的H_2含量（体积分数）只有1.8%~2.5%，它主要由鼓风中的水分在风口前高温分解产生。在喷吹燃料的高炉内，煤气中的H_2含量显著增加，可达5%~8%（体积分数）。氢与氧的亲和力很强，所以氢也是高炉冶炼中的还原剂。用H_2还原铁氧化物过程与用CO一样，也可称为间接还原。

When the temperature is below 570℃, the reduction reaction is divided into the following two steps:

当温度低于570℃时，还原反应分为以下两步：

$$3Fe_2O_3+H_2 = 2Fe_3O_4+H_2O+21800kJ/mol \quad (1-23)$$

$$Fe_3O_4+4H_2 \Longrightarrow 3Fe+4H_2O-146650kJ/mol \qquad (1-24)$$

When the temperature is higher than 570℃, the reduction reaction is divided into the following three steps:

当温度高于570℃时，还原反应分为以下三步：

$$3Fe_2O_3+H_2 \Longrightarrow 2Fe_3O_4+H_2O+21800kJ/mol \qquad (1-25)$$

$$Fe_3O_4+H_2 \Longrightarrow 3FeO+H_2O-63570kJ/mol \qquad (1-26)$$

$$FeO+H_2 \Longrightarrow Fe+H_2O-27700kJ/mol \qquad (1-27)$$

When chemical equilibrium is established, $K_p = P_{H_2O}/P_{H_2} = \varphi(H_2O)/\varphi(H_2)$, and the relationship between equilibrium gas phase and temperature is shown in Figure 1-15.

反应建立化学平衡时，$K_p = P_{H_2O}/P_{H_2} = \varphi(H_2O)/\varphi(H_2)$，其平衡气相与温度的关系如图1-15所示。

There are similarities and differences in the reduction of iron oxides with H_2 and CO. For comparison, Figure 1-14 and Figure 1-15 are superposed to get Figure 1-16. Compared with the reduction of CO, H_2 has the following characteristics:

用H_2与CO还原铁氧化物的过程有相同点也有不同点。为便于比较，将图1-14与图1-15叠加得到图1-16。

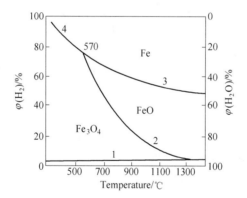

Figure 1-15　Relationship between equilibrium gas composition and temperature of iron oxide reduction by H_2

图1-15　用H_2还原铁氧化物的平衡气相成分与温度的关系

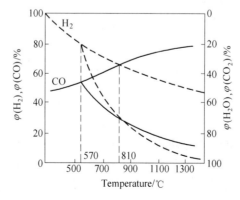

Figure 1-16　Comparison of gas phase equilibrium components between Fe-O-C and Fe-O-H systems

图1-16　Fe-O-C与Fe-O-H系气相平衡成分比较

H_2 与 CO 的还原相比有以下特点：

(1) Like co-reduction, it belongs to indirect reduction. The volume of gas phase (H_2 and H_2O) has no change before and after the reaction. For the other words, the reaction is not affected by pressure.

(1) 与 CO 还原一样，均属于间接还原。反应前后气相体积（H_2 与 H_2O）没有变化，即反应不受压力影响。

(2) In addition to the reduction of Fe_2O_3, the reduction of Fe_3O_4 and FeO is reversible, and there is a fixed equilibrium gas component at a certain temperature. In order to make the reduction of iron oxide complete, an excessive reductant is needed.

(2) 除 Fe_2O_3 的还原外，Fe_3O_4、FeO 的还原均为可逆反应。在一定温度下有固定的平衡气相成分，为了使铁氧化物还原彻底，都需要过量的还原剂。

(3) The reaction is endothermic. With the increase of temperature, the reduction ability of H_2 increases. When the temperature is lower than 810℃, the reduction capacity of CO is stronger than that of H_2; When the temperature is higher than 810℃, the reduction capacity of H_2 is stronger than that of CO.

(3) 反应为吸热过程，随着温度升高，H_2 的还原能力增强。温度低于 810℃ 时，CO 的还原能力比 H_2 强；温度高于 810℃ 时，H_2 的还原能力比 CO 强。

(4) Under the condition of blast furnace smelting, H_2 can also accelerate the reduction of CO and C. the reactions are as follows:

(4) 在高炉冶炼条件下，用 H_2 还原铁氧化物时还可促进 CO 和 C 还原反应的加速进行，其反应如下：

$$FeO + H_2 =\!=\!= Fe + H_2O \quad (1-28)$$
$$H_2O + C =\!=\!= CO + H_2 \quad (1-29)$$

The total reaction can be written as follows:

$$FeO + C =\!=\!= Fe + CO \quad (1-30)$$

The results show that H_2 only plays the role of oxygen transport, and does not consume itself, which can promote the reduction of CO and C.

反应结果表明，H_2 只起传输氧的作用，本身不消耗，可促进 CO 和 C 的还原。

Reduction of Iron Oxide with Solid Carbon
用固体碳还原铁氧化物

When iron oxide is reduced by solid carbon, the gaseous product is Co, which is called direct reduction, such as $FeO + C =\!=\!= Fe + CO$. In the blast furnace, iron ore is first reduced indirectly in the slow movement from top to bottom, because the reduction capacity of blast furnace gas is not fully utilized in the lower part of the blast furnace. And the rising gas flow still has considerable reduction capacity, which can participate in indirect reduction. Therefore, the ore has been reduced to a certain extent before reaching the high temperature zone, and the remaining iron oxide mainly exists in the form of FeO. Because the contact area between ore and coke before softening and melting is very small and the reaction speed is very slow, the direct reduction of solid carbon in blast furnace is completed by two steps.

用固体碳还原铁氧化物,生成的气相产物是 CO,这种还原称为直接还原,如 FeO+C⸺Fe+CO。在高炉内,铁矿石在自上而下的缓慢运动中先进行间接还原,这是由于高炉煤气的还原能力在高炉下部并未得到充分利用,上升的煤气流仍具有相当的还原能力,可参加间接还原。因此,矿石在到达高温区之前已受到一定程度的还原,残存下来的铁氧化物主要以 FeO 形式存在。由于矿石在软化和熔化之前与焦炭的接触面积很小,反应速度很慢,所以高炉内固体碳参加的直接还原是通过两步来完成的。

First, the reactions of indirect restore are as follows:
第一步,间接还原反应为:

$$Fe_3O_4 + CO \Longrightarrow 3FeO + CO_2 \quad (1-31)$$

$$FeO + CO \Longrightarrow Fe + CO_2 \quad (1-32)$$

In the second step, the product CO_2 reacts with solid carbon to produce carbon gasification, written as:
第二步,产物 CO_2 与固体碳发生碳的气化反应为:

$$CO_2 + C \Longrightarrow 2CO \quad (1-33)$$

The final reactions of the above two steps are:
以上两步反应的最终结果是:

$$FeO + CO \Longrightarrow Fe + CO_2 \quad (1-34)$$

$$CO_2 + C \Longrightarrow 2CO \quad (1-35)$$

The total reation can be written as follows:

$$FeO + C \Longrightarrow Fe + CO - 152190 kJ/mol \quad (1-36)$$

Therefore, the reduction of iron oxides by solid carbon is controlled by the gasification of carbon. According to the test, the general metallurgical coke starts gasification reaction at 800℃, and it is intense at 1100℃. In the region above 1100℃, the concentration of CO is almost 100% and the concentration of CO_2 is almost zero. Direct reduction and indirect reduction in blast furnace are divided into regions, as shown in Figure 1-17. The area with temperature lower than 800℃ (shown in area Ⅰ in Figure 1-17) does not have carbon gasification reaction with no direct reduction, so it is called indirect reduction area; the area with temperature between 800℃ and 1100℃ (shown in area Ⅱ in Figure 1-17) has both indirect reduction and direct reduction; and the area with temperature higher than 1100℃ (shown in area Ⅲ in Figure 1-17) does not have CO_2 in the gas phase, so it can be considered that there is no indirect reduction, called direct restore area.

因此,固体碳还原铁氧化物受碳的气化反应所控制。据测定,一般冶金焦炭在 800℃ 时开始气化反应,到 1100℃ 时激烈进行。在 1100℃ 以上的区域气相中 CO 浓度几乎达到 100%,CO_2 浓度几乎为零。高炉内直接还原和间接还原是划分了区域的,如图 1-17 所示。温度低于 800℃ 的区域(见图 1-17 中区域Ⅰ)内不存在碳的气化反应,也就不存在直接还原,故称为间接还原区域;温度在 800~1100℃ 的区域(见图 1-17 中区域Ⅱ)内间接还原和直接还原都存在;温度高于 1100℃ 的区域(见图 1-17 中区域Ⅲ)内气相中不存在 CO_2,也可认为不存在间接还原,所以称为直接还原区域。

In addition to the above two-step direct reduction, there are also the following ways of reduction in the lower high temperature zone:

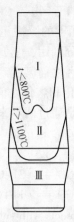

Figure 1-17 Distribution of reduction zone of iron in blast furnace
图 1-17 高炉内铁的还原区分布示意图

高炉内的直接还原除了上述的两步式直接还原外，在下部高温区还存在以下方式的还原：

$$(FeO)+C_{焦} = [Fe]+CO(g) \tag{1-37}$$
$$(FeO)+[Fe_3C] = 4[Fe]+CO(g) \tag{1-38}$$

In general, only the content of Fe with 0.2%~0.5% (by mass) enters into the slag. In case of abnormal furnace conditions, there are more FeO in the slag, resulting in the increase of direct reduction. Moreover, due to a large number of endothermic reactions, the temperature will fluctuate violently.

一般情况下只有 0.2%~0.5%（质量分数）的 Fe 进入炉渣中。如遇到炉况失常，渣中 FeO 较多，则造成直接还原增加，而且由于发生大量吸热反应，会引起温度剧烈波动。

Compared with indirect reduction, indirect reduction uses gas as reducing agent, which is a reversible reaction. The reducing agent can not be fully utilized, and a certain amount of reducing agent is needed. However, the reaction itself is mostly exothermic reaction, with little heat consumption, while direct reduction is just the opposite. Therefore, direct reduction or indirect reduction is not good in blast furnace. Only when direct reduction and indirect reduction are in proper proportion, can fuel consumption be reduced and the best effect be achieved. The ideal situation is that the direct reduction degree is 0.2~0.3, while the direct reduction degree is 0.4~0.5 or higher in the actual operation of gas blast furnace. So the goal of blast furnace workers is to reduce direct reduction and develop indirect reduction. All measures that can reduce direct reduction are conducive to reducing coke consumption.

直接还原与间接还原相比，间接还原以气体作还原剂，是可逆反应，还原剂不能全部利用，需要有一定过量的还原剂，但反应本身多为放热反应，热量消耗不大。而直接还原刚好相反。因此，高炉内全部为直接还原或全部为间接还原都不好，只有直接还原和间接还原在适宜的比例范围内，才能降低燃料消耗，取得最佳效果。理想的情况是直接还原度在 0.2~0.3。但在煤气高炉实际操作中，直接还原度在 0.4~0.5 之间，甚至更高，因此需要降低直接还原，发展间接还原。凡是能降低直接还原的措施都有利于降低焦炭消耗。

Reduction of Non-ferrous Elements
非铁元素的还原

In addition to Fe, Si, Mn, P and other elements are reduced. According to the decompression of each oxide, Cu, As, Co and Ni are almost reduced in the blast furnace; Mn, Si, V, and Ti are difficult to be reduced, and only part of them are reduced into pig iron.

高炉内除铁元素外,还有硅、锰、磷等其他元素的还原。根据各氧化物分解压的大小可知,铜、砷、钴、镍在高炉内几乎全部还原;锰、硅、钒、钛等较难还原,只有部分还原进入生铁。

- Reduction of Mn
- 锰的还原

Mn is a common metal in blast furnace smelting. Mn in blast furnace is mainly brought in by manganese ore. generally, and iron ore also contains a small amount of Mn. The reduction of manganese oxide in blast furnace is also carried out step by step from high price to low price:

Mn 是高炉冶炼中常遇到的金属。高炉中 Mn 主要由锰矿石带入,一般铁矿石中也都含有少量 Mn。高炉内锰氧化物的还原也是从高价到低价逐级进行的,其顺序为:

$$MnO_2 \rightarrow Mn_2O_3 \rightarrow Mn_3O_4 \rightarrow MnO \rightarrow Mn \tag{1-39}$$

It is easy for gaseous reducing agents (CO, H_2) to reduce MnO_2 to low-cost MnO. But MnO can only be reduced to Mn by direct reduction. The initial reduction temperature is between 1000℃ and 1200℃, and the reactions are as follows:

气体还原剂(CO、H_2)能够很容易地把 MnO_2 还原成低价 MnO,但 MnO 只能由直接还原方式还原为 Mn,其开始还原温度在 1000~1200℃ 之间,反应式如下:

$$MnO+CO = Mn+CO_2 - 121500 kJ/mol \tag{1-40}$$

$$CO_2+C = 2CO - 165800 kJ/mol \tag{1-41}$$

The total reaction can be written as follows:

$$MnO+C = Mn+CO - 287300 kJ/mol \tag{1-42}$$

Compared with iron reduction, the heat consumption of 1kg of Mn reduction is twice that of 1kg of Fe, which is more difficult to reduce than iron. So high temperature is the primary condition for manganese reduction.

与铁的还原相比,还原 1kg Mn 的耗热量是还原 1kg Fe 的两倍,其比铁更难还原,所以高温是锰还原的首要条件。

Since Mn has entered the liquid slag before reduction, it can quickly combine with SiO_2 in the slag to form $MnSiO_3$ at 1100~1200℃, which is more difficult to reduce than free MnO. When CaO content in slag is high, MnO can be replaced to make reduction easier, as follows:

由于 Mn 在还原之前已进入液态炉渣,在 1100~1200℃ 时能迅速与炉渣中的 SiO_2 结合成 $MnSiO_3$,此时要比自由的 MnO 更难还原。当渣中 CaO 含量高时,可将 MnO 置换出来,使还原变得容易些,反应式如下:

$$MnSiO_3+CaO = CaSiO_3+MnO+58990 kJ/mol \tag{1-43}$$

$$MnO+C = Mn+CO - 287300 kJ/mol \tag{1-44}$$

The total reaction can be written as follows:

$$MnSiO_3+CaO+C \Longrightarrow Mn+CaSiO_3+CO-228310kJ/mol \qquad (1-45)$$

During the smelting of common pig iron, the content of Mn with 40%~60% (by mass) is reduced into pig iron, the content of Mn with 5%~10% is volatilized into gas, and the rest of Mn enters into slag.

在冶炼普通生铁时，有40%~60%的Mn还原进入生铁，5%~10%的Mn挥发进入煤气，其余的Mn进入炉渣。

- Reduction of Si
- Si 的还原

Different kinds of iron have different requirements for the content of Si. In general, the content of Si in steel-making pig iron should be less than 1% (by mass). At present, it in blast furnace has been reduced to 0.2%~0.3% (by mass), even 0.1% or lower. The content of Si in cast iron should be in the range of 1.25%~4.0% (by mass).

不同的铁种对硅含量有不同的要求。一般炼钢生铁的硅含量（质量分数）应小于1%，目前高炉冶炼低硅炼钢生铁时，其硅含量（质量分数）已降低到0.2%~0.3%，甚至达0.1%或更低。铸造生铁则要求硅含量（质量分数）在1.25%~4.0%范围内。

Si in pig iron mainly comes from gangue of ore and SiO_2 in ash of coke. SiO_2 is a relatively stable compound, so the reduction of Si is more difficult than that of Fe and Mn. SiO_2 can only be directly reduced by solid carbon in high temperature liquid state, and the reaction is as follows:

生铁中的硅主要来自矿石的脉石和焦炭灰分中的SiO_2，SiO_2是比较稳定的化合物，所以 Si 的还原比 Fe 和 Mn 都要困难。SiO_2只能在高温液态下依靠固体碳直接还原，其反应如下：

$$SiO_2+2C \Longrightarrow Si+2CO-627980kJ/mol \qquad (1-46)$$

The heat consumption of reduction of Si with 1kg is 8 times of that of 1kg of Fe. So higher reduction temperature, greater heat consumption and more difficult reduction are required. In blast furnace smelting, only a little Si is reduced into pig iron, and most of it enters into slag with SiO_2.

还原 1kg Si 的耗热相当于还原 1kg Fe 的 8 倍，因此要求还原温度更高，热消耗更大，还原更困难。高炉冶炼中只有少部分硅还原进入生铁，大多数以SiO_2进入炉渣。

- Reduction of P
- 磷的还原

P in the furnace charge mainly exists in the form of calcium phosphate $[(CaO)_3 \cdot P_2O_5$, also called apatite], and sometimes in the form of iron phosphate $[(FeO)_3 \cdot P_2O_5 \cdot 8H_2O$, also called blue iron ore]. For example calcium phosphate is a very stable compound, which firstly enters the slag in the blast furnace and is replaced by SiO_2 in the slag to produce P_2O_5. At 1100~1300℃, C is used as reducing agent to reduce C. The reduction reaction is as follows:

炉料中的磷主要以磷酸钙$[(CaO)_3 \cdot P_2O_5$，又称磷灰石]的形态存在，有时也以磷酸铁$[(FeO)_3 \cdot P_2O_5 \cdot 8H_2O$，又称蓝铁矿]的形态存在。以磷酸钙为例，它是很稳定的化合物，在高炉内首先进入炉渣，被炉渣中的SiO_2置换出自出态P_2O_5。在 1100~1300℃时用 C 作还原剂还原 P，其还原反应为：

$$2Ca_3(PO_4)_2 + 3SiO_2 = 3Ca_2SiO_4 + 2P_2O_5 - 917340 kJ/mol \quad (1-47)$$
$$2P_2O_5 + 10C = 4P + 10CO - 1921290 kJ/mol \quad (1-48)$$

The total reaction can be written as follows:
$$2Ca_3(PO_4)_2 + 3SiO_2 + 10C = 3Ca_2SiO_4 + 4P + 10CO - 2838630 kJ/mol \quad (1-49)$$

The heat consumption of reducing 1kg of P is 8 times of that of 1kg of Fe. So the heat consumption of reducing phosphorus is large.

还原 1kg P 的耗热量相当于还原 1kg Fe 的 8 倍，所以磷的还原耗热大。

Due to various conditions favorable for phosphorus reduction in blast furnace, it can be said that all phosphorus can be reduced into pig iron when smelting common pig iron. Because phosphorus is harmful to steel, the content of phosphorus in pig iron should be controlled, which can only be achieved by controlling the amount of phosphorus brought in by raw materials.

由于高炉内的各种条件有利于 P 还原，所以在冶炼普通生铁时，P 能全部还原进入生铁。由于 P 对钢材有害，应控制生铁中 P 含量，因此只有通过控制原料带入的磷量来实现。

- Reduction of Pb, Zn and As
- Pb、Zn、As 的还原

Some iron ores contain Pb、Zn、As and other elements, which are easy to be reduced under blast furnace smelting conditions.

一些铁矿石含有 Pb、Zn、As 等元素，这些元素在高炉冶炼条件下易被还原。

The reduction of Pb is insoluble in molten iron, and it is easy to deposit at the bottom of the furnace because its density is higher than pig iron. It also infiltrate into the brick seam, and damage the bottom of the furnace. Some Pb volatilizing from the blast furnace and rising, will be oxidized when encountering CO_2 and H_2O, and it will be reduced when it falls with the burden, so as to circulate in the blast furnace.

还原出来的 Pb 不溶于铁水，其密度大于生铁且易沉积于炉底，能够渗入砖缝，破坏炉底。部分 Pb 自高炉内挥发上升，遇到 CO_2 和 H_2O 时将被氧化，随炉料一起下降时又被还原，从而在高炉内循环。

The reduction of Zn volatilizes in the furnace and oxidizes into ZnO. It also expands in volume, destroys the furnace lining and forms furnace lump.

还原出来的 Zn 在炉内挥发，氧化成 ZnO，体积膨胀，从而破坏炉衬，形成炉瘤。

The combination of reduced arsenic and iron will affect the properties of steel and reduce the weldability of steel.

还原出来的 As 与铁化合，会影响钢铁性能，降低钢的焊接性能。

Formation and Carburization of Pig Iron
生铁的形成与渗碳过程

The formation process of pig iron mainly includes carburization and the entry of reduced elements into pig iron. Finally, qualified pig iron containing Fe, C, Si, Mn, P, S and other elements is obtained.

生铁的形成过程主要包括渗碳和已还原的元素进入生铁中，最终得到含 Fe、C、Si、

Mn、P、S 等元素的合格生铁。

In the upper part of the blast furnace, some iron ores have been gradually reduced to metallic iron. The newly reduced iron is like porous sponge, which is called sponge iron. During the decline of sponge iron, C, Si, Mn, P, S, etc. infiltrate into it. With the increase of temperature, the sponge iron finally becomes liquid pig iron, which is deposited in the hearth and discharged regularly to obtain molten iron.

在高炉上部就已有部分铁矿石逐渐被还原成金属铁。刚还原出来的铁呈多孔海绵状，称为海绵铁。海绵铁在下降过程中，C、Si、Mn、P、S 等渗入其中，伴随着温度升高，最后变成液态生铁沉积于炉缸中，定期排出得到铁水。

1.2.2.3　Slag and Desulfurization
1.2.2.3　炉渣与脱硫

In blast furnace production, not only metal iron is reduced from iron ore, but also reduced iron, unreduced oxides and other impurities can be melted into liquid state and separated from each other. Finally, iron and slag flow out of the furnace smoothly in the form of molten iron and slag liquid. The quantity and performance of slag have a direct impact on the running of blast furnace, the output and quality of pig iron and coke ratio. So it has a decisive influence on blast furnace production. If you want to make good iron, you must make good slag.

高炉生产不仅能从铁矿石中还原出金属铁，而且还原出的铁、未还原的氧化物和其他杂质都能熔化成液态并相互分开，最后以铁水和渣液的形态顺利流出炉外。炉渣的数量和性能直接影响高炉的顺行、生铁的产量和质量以及焦比，所以其对高炉生产有决定性的影响。要想炼好铁，必须造好渣。

Composition, Function and Requirements of Slag
炉渣的成分、作用与要求

- Composition of Slag
- 炉渣的成分

Generally, blast furnace slag is mainly composed of SiO_2, Al_2O_3, CaO, MgO and other oxides. In addition, it contains a small amount of other oxides and sulfides. The approximate range of its composition is shown in Table 1-7.

一般高炉渣主要由 SiO_2、Al_2O_3、CaO、MgO 等氧化物组成，此外还含有少量其他氧化物和硫化物，其成分的大致范围见表 1-7。

Table 1-7　Composition range of blast furnace slag
表 1-7　高炉渣成分范围

Formation 组成	SiO_2	Al_2O_3	CaO	MgO	MnO	FeO	GaS	K_2O+Na_2O
Components (mass fraction)/% 化学成分（质量分数）/%	30~40	8~18	35~50	<12	<3	<1	<2.5	0.5~1.5

The compositions and quantities mainly depend on the composition of raw materials and pig iron varieties smelted in blast furnace. When smelting special iron ore, the blast furnace slag also contains other components. For example, when smelting fluorine-containing iron ore, the content of CaF_2 in the slag is about 18% (by mass); and when smelting vanadium titanium magnetite, the content of TiO_2 in slag contains 20%~25% (by mass).

这些成分及数量主要取决于原料的成分和高炉冶炼的生铁品种。冶炼特殊铁矿石时的高炉渣还会含有其他成分。例如，在冶炼含氟铁矿石时，渣中 CaF_2 含量（质量分数）为18%左右；在冶炼钒钛磁铁矿时，渣中含有20%~25%（质量分数）的 TiO_2。

Various components in slag can be divided into basic oxide and acid oxide. The basicity of slag is usually expressed by the ratio of the mass fraction of alkali oxide and acid oxide in slag, which is expressed by R.

炉渣中的各种成分可分为碱性氧化物和酸性氧化物两大类。通常以炉渣中碱件氧化物与酸性氧化物的含量（质量分数）之比表示炉渣碱度，用 R 表示。

Functions and Requirements of Slag
炉渣的作用与要求

Blast furnace slag should have the characteristics of low melting point, low density and insoluble in molten iron, so that it can be effectively separated from iron to obtain pure pig iron, which is the basic role of blast furnace slag. The blast furnace slag shall meet the following requirements:

高炉渣应具有熔点低、密度小、不溶于铁水的特点，使其能够与铁有效分离，从而获得纯净的生铁的基本作用。高炉渣应满足以下要求：

(1) It should have proper chemical composition and good physical properties, and it can be able to melt into liquid and separate from metal in the blast furnace. It can also flow out of the furnace smoothly.

(1) 具有合适的化学成分和良好的物理件质，在高炉内能够熔融成液体，并与金属分离，能够顺利流出炉外。

(2) It should have sufficient desulfurization capacity to ensure high quality pig iron.

(2) 具有充分的脱硫能力，保证炼出优质生铁。

(3) It should be beneficial to the smooth operation of the furnace and to the good technical and economic indexes of the furnace.

(3) 有利于炉况顺行，能够使高炉获得良好的技术经济指标。

(4) Its composition should be beneficial to the reduction of some elements and inhibit the reduction of others (i.e. selective reduction), which can adjust the composition of pig iron.

(4) 其成分有利于一些元素的还原，抑制另一些元素的还原（即选择还原），具有调整生铁成分的作用。

(5) It should be beneficial to protect the lining and prolong the service life of blast furnace.

(5) 有利于保护炉衬，延长高炉寿命。

Slag Forming Process in Blast Furnace
高炉内的成渣过程

In the relative movement of gas and charge, gas transfers heat to charge with the temperature of charge rising continuously after heating. The solid softens to molten drop, and finally becomes liquid pig iron and slag. The formation of blast furnace slag has a long process from the beginning to the end. It can be divided into three steps:

在煤气与炉料的相对运动中，煤气将热量传递给炉料，炉料受热后温度不断升高，由固体经软化到熔滴，最后变成液态生铁和炉渣。高炉渣从开始形成到最后排出经历了相当长的过程，其步骤可分为：

(1) The formation of primary slag: The formation of primary slag includes four stages: solid phase reaction, softening, melting and dropping.

(1) 初渣的生成：初渣的生成包括固相反应、软化、熔融和滴落四个阶段。

The selective reactions are reacted between solid oxides (such as FeO and SiO_2, MnO and SiO_2, CaO and SiO_2), and the formation of new low melting point compounds are the beginning of slag making process.

固体氧化物之间（如 FeO 与 SiO_2、MnO 与 SiO_2、CaO 与 SiO_2 之间）发生选择性的反应，生成新的低熔点化合物。固相反应是造渣过程的开始。

The resulting low melting point compounds soften and melt with the increase of temperature. The molten slag is the primary slag. Generally, the content of FeO in the primary slag is high.

生成的低熔点化合物随温度升高发生软化、熔融。洒落下来的熔融炉渣就是初渣，一般初渣中的 FeO 含量较高。

(2) Change of intermediate slag. The composition of primary slag is very different and FeO content is high. During the decline process, with the reduction of FeO, MnO and SiO_2 and the increase of temperature, its performance will fluctuate, which has a great impact on the smelting process of blast furnace.

(2) 中间渣的变化。初渣的成分差异很大，且 FeO 含量较高，下降过程中伴随 FeO、MnO 和 SiO_2 的还原和温度的升高，其性能会发生波动，对高炉冶炼过程影响很大。

(3) Formation of final slag. After passing through the tuyere area, the composition and performance of the intermediate slag tend to be stable again, and it is deposited in the hearth. Desulfurization reaction also occurs, and the composition is further homogenized. Generally speaking, blast furnace slag is the final slag, which has a very important influence on the control of pig iron composition and the assurance of pig iron quality.

(3) 终渣的形成。中间渣经过风口区域，其成分和性能再次变化后趋于稳定，沉积于炉缸，发生脱硫反应，成分进一步均匀化。一般所说的高炉渣就是终渣，终渣对控制生铁成分、保证生铁质量有非常重要的影响。

Slag Desulphurization
炉渣脱硫

S is a harmful element in pig iron. It is an important task for blast furnace process to obtain molten iron with qualified content of S.

硫是生铁中的有害元素，保证获得硫含量合格的铁水是高炉冶炼中的重要任务。

- **Changes of S in Blast Furnace and Factors Determining the content of S in Pig Iron**
- **S 在高炉中的变化及决定生铁硫含量的因素**

S in the blast furnace comes from coke, injection fuel and ore. The total amount of S brought in by the furnace charge during the smelting of each ton of pig iron is called sulfur load, which is generally 4~8kg/t. The amount of S brought in by coke is the most, accounting for 60%~80%, and the amount of S brought in by ore generally does not exceed 1/3 of the total sulfur.

高炉内的 S 来自焦炭、喷吹燃料和矿石。冶炼每吨生铁时由炉料带入的总硫量称为硫负荷，一般为 4~8kg/t。炉料中焦炭带入的硫量最多，占 60%~80%，而矿石带入的硫量一般不超过总硫量的 1/3。

As the amount of sulfur volatilized with the gas not changing much under certain smelting conditions, it is necessary to reduce the sulfur content of pig iron. Firstly, try to control the total flow brought in by the furnace charge. Secondly, try to improve the desulfurization capacity of the slag and increase the amount of sulfur taken away by the slag.

由于随煤气挥发的硫量在一定冶炼条件下变化不大，因此要降低生铁的 S 含量。一是尽量控制炉料带入的总流量；二是尽可能提高炉渣的脱硫能力，增加炉渣带走的硫量。

Desulfurization Capacity of Slag
炉渣的脱硫能力

Under certain smelting conditions, the desulfurization of pig iron is mainly realized by improving the desulfurization capacity of slag.

在一定冶炼条件下，生铁的脱硫主要通过提高炉渣的脱硫能力来实现。

The basic oxides of CaO, MgO, MnO and other basic oxides are the main desulfurizers in the slag, among which CaO is the strongest desulfurizer. The desulfuration reaction between slag and iron in blast furnace begins immediately after the formation of primary slag, and is carried out more frequently in the bosh or dropping zone, and finally completed in the hearth. There are two kinds of desulfuration in hearth: One is the desulfuration in slag when molten iron passes through slag layer; the other is the desulfuration on slag—iron interface. The desulfurization reaction is divided into the following three steps:

炉渣中起脱硫作用的主要是碱性氧化物 CaO、MgO、MnO 等，其中 CaO 是最强的脱硫剂。高炉内渣—铁之间的脱硫反应在初渣生成后开始，在炉腹或滴落带中较多地进行，在炉缸中完成。炉缸中的脱硫存在两种情况，一是当铁水穿过渣层时在渣中脱硫，二是在渣—铁界面上进行。脱硫反应分为以下三步：

(1) The sulfur diffusion from pig iron to slag（生铁中 S 向渣中扩散）：

$$[FeS] = (FeS) \tag{1-50}$$

(2) The reaction with CaO in slag（与渣中 CaO 发生反应）：

$$(FeS)+(CaO) = (CaS)+(FeO) \tag{1-51}$$

(3) FeO is reduced by C（生成的 FeO 被 C 还原）：

$$(FeO)+C = [Fe]+CO(g) \tag{1-52}$$

The total desulfulfurization reaction can be written as follows:
脱硫总反应可写成：

$$[FeS]+(CaO)+C = (CaS)+[Fe]+CO-149140kJ/mol \tag{1-53}$$

The ways to improve the desulfurization capacity of slag are as follows:
提高炉渣脱硫能力的途径如下:

(1) The basicity of slag can be improved so that the sulfur in pig iron can be converted into CaS or MgS and transferred into slag stably.

(1) 提高炉渣碱度，有利于将生铁中的 S 转变为 CaS 或 MgS 而稳定转入炉渣。

(2) Raise the hearth (slag iron) temperature. The desulfurization reaction is endothermic, and the increase of temperature is beneficial to its operation. At the same time, the high temperature can improve the fluidity of slag and the transfer speed of sulfur in slag.

(2) 提高炉缸（渣铁）温度。脱硫反应是吸热反应，提高温度有利于其进行。同时，高温可提高炉渣的流动性，增加硫在渣中的传递速度。

(3) The strong reducing atmosphere can make FeO in slag reduced continuously, which is propitious to the desulfurization.

(3) 提供强烈的还原气氛，可使渣中的 FeO 不断被还原，有利于反应向脱硫方向进行。

1.2.2.4 Combustion of Fuel and Changes of gas in Blast Furnace
1.2.2.4 燃料的燃烧及煤气在高炉内的变化

The main fuel for blast furnace smelting is coke, followed by pulverized coal. Most of the carbon in coke is burned in front of the tuyere, except that a small part of carbon takes part in direct reduction and dissolves in pig iron (carburizing). The fuel injected from the tuyere also burns when it meets the hot air in front of the tuyere.

高炉冶炼的燃料主要是焦炭，其次是煤粉。焦炭中的碳除少部分参与直接还原反应和溶解于生铁（渗碳）外，大部分在风口前燃烧。从风口喷吹的燃料也是在风口前与鼓入的热风相遇进行燃烧。

The combustion of the fuel in front of the tuyere is one of the most important reactions in the blast furnace, which plays a very important role in the smelting process of the blast furnace, as follows:

风口前燃料的燃烧是高炉内最重要的反应之一，它对高炉冶炼过程有着十分重要的作用，具体如下:

(1) Fuel combustion produces reducing gas CO and H_2, and releases a lot of heat, which meets the needs of heating, decomposition, reduction, slagging and other processes of blast furnace burden. It is the source of heat energy and chemical energy for blast furnace smelting.

(1) 燃料燃烧产生还原性气体 CO 和 H_2，并放出大量热，满足高炉对炉料的加热、分解、还原、造渣等过程的需要，是高炉冶炼热能和化学能的来源。

(2) The combustion reaction makes the solid carbon gasify continuously and forms a free space in the hearth, which creates a precondition for the upper change to descend continuously. Whether the fuel combustion before the tuyere is uniform and effective has great influences on the charge and gas movement. Without fuel combustion, the movement of blast furnace charge and gas can not be carried out.

(2) 燃烧反应使固体碳不断气化，在炉缸内形成自由空间，为上部炉料不断下降创造了先决条件。风口前燃料的燃烧是否均匀有效,对炉料和煤气运动具有重大影响。没有燃料燃烧,高炉炉料和煤气的运动也就无法进行。

In addition to the combustion of fuel in the hearth, the uncompleted reactions such as direct reduction, carburization and desulfurization shall be concentrated in the hearth and finally completed, forming molten iron and slag and discharging from the hearth. Therefore, hearth reaction is not only the starting point of blast furnace smelting process, but also the end point of blast furnace smelting process. The quality of hearth work plays a decisive role in blast furnace smelting.

炉缸内除了燃料的燃烧外,直接还原、渗碳、脱硫等尚未完成的反应都要集中在炉缸内最后完成,最终形成铁水和炉渣,从炉内排出。因此,炉缸反应既是高炉冶炼过程的起点,也是高炉冶炼过程的终点。炉缸工作的好坏对高炉冶炼起决定性的作用。

Combustion of Fuel
燃料的燃烧

- Combustion Reactions
- 燃烧反应

The combustion reactions in the hearth of blast furnace are different from the general combustion process. They are carried out in the environment full of coke, under the condition of a certain amount of air and excess coke.

高炉炉缸内的燃烧反应不同于一般的燃烧过程,它是在充满焦炭的环境中进行的,即在空气量一定而焦炭过剩的条件下进行。

At the tuyere, there is enough oxygen. At first, the reactions of complete combustion and incomplete combustion exist at the same time. The products are CO and CO_2. The reactions are as follows:

在风口的氧气比较充足,最初完全燃烧和不完全燃烧反应同时存在,产物为 CO 和 CO_2,反应式为:

(1) The reaction of complete combustion (equivalent to 1kg of C heat releases 33390kJ) is written as:

(1) 完全燃烧(相当于1kg C 放热33390kJ)的反应式为:

$$C+O_2 =\!=\!= CO_2+400660 kJ/mol \qquad (1-54)$$

(2) The reaction of incomplete combustion (equivalent to 1kg of C heat releases 9790kJ) is written as:

(2) 不完全燃烧(相当于1kgC 放热9790kJ)的反应式为:

$$C+\frac{1}{2}O_2 =\!=\!= CO+117490 kJ/mol \qquad (1-55)$$

In the distance from the tuyere, due to the lack of oxygen and the presence of a large number of coke, and the high temperature in the hearth, the CO_2 produced in the place with sufficient oxygen will also react with the solid carbon for carbon gasification, expressed as:

在离风口较远处,由于氧的缺乏和大量焦炭的存在,且炉缸内温度很高,氧充足的地方产生的 CO_2 也会与固体碳进行碳的气化反应,其反应式为

$$CO_2+C =\!=\!= 2CO-165800 kJ/mol \qquad (1-56)$$

The composition of dry air is $\varphi(O_2):\varphi(N_2)=21:79$, while N_2 does not participate in the reaction. If there is no water, the combustion reaction products in the hearth are CO and N_2. And the total reaction can be expressed as:

干空气的成分为 $\varphi(O_2):\varphi(N_2)=21:79$，而 N_2 不参加反应。如果没有水分存在，则炉缸中的燃烧反应产物为 CO 和 N_2，总的反应式可表示为：

$$2C+O_2+\frac{79}{21}N_2 = 2CO+\frac{79}{21}N_2 \tag{1-57}$$

There is a certain amount of water in the blast, which reacts with carbon at high temperature as follows:

鼓风中含有一定量的水分，水分在高温下与碳发生以下反应：

$$H_2O+C = CO+H_2-124450 kJ/mol \tag{1-58}$$

The combustion of pulverized coal injected into blast furnace in front of tuyere is similar to that of coke, except that the hydrocarbon in the volatile of pulverized coal will decompose to produce H_2. Therefore, under the actual production conditions, the final product of fuel combustion in front of tuyere is composed of CO, H_2 and N_2.

高炉喷吹煤粉在风口前的燃烧与焦炭类似，不同之处在于煤粉挥发分中的碳氢化合物会分解产生 H_2。所以在实际生产条件下，风口前燃料燃烧的最终产物由 CO、H_2 和 N_2 组成。

- Tuyere Raceway and Combustion Zone
- 风口回旋区与燃烧带

In modern blast furnace smelting, the air blown from the tuyere is injected into the blast furnace at a speed of more than 100m/s to form a nearly spherical cavity before sealing, which is called tuyere raceway, as shown in Figure 1-18.

在现代高炉冶炼中，从风口鼓入的风以 100m/s 以上的速度喷射入高炉，使封口前形成一个近似球形的空腔，称为风口回旋区，如图 1-18 所示。

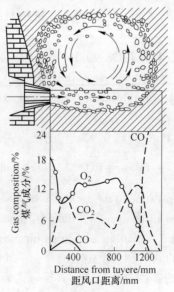

Figure 1-18　Schematic diagram of tuyere raceway
图 1-18　风口回旋区示意图

The range of combustion zone is basically the same as that of tuyere raceway. But tuyere raceway refers to the area where coke makes mechanical movement under the action of blast kinetic energy. While combustion zone refers to the area of combustion reaction, which is determined according to gas composition. The combustion zone is slightly larger than the tuyere raceway. The distribution of combustion zone on the hearth section is shown in Figure 1-19.

燃烧带与风口回旋区的范围基本一致,但风口回旋区是指在鼓风动能的作用下焦炭做机械运动的区域,而燃烧带是指燃烧反应的区域,其是根据煤气成分来确定的。燃烧带比风口回旋区略大。炉缸截面上燃烧带的分布如图1-19所示。

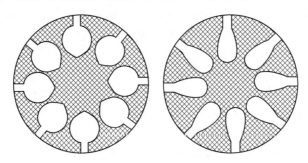

Figure 1-19 Distribution of combustion zone on hearth section
图1-19 炉缸截面上燃烧带的分布

The influences of combustion zone on the smelting process of blast furnace inclucle, the size and distribution of combustion zone, the homogenization of hearth work and the smooth operation of blast furnace smelting.

燃烧带对高炉冶炼过程的影响包括:燃烧带的大小和分布,对炉料和煤气的运动与分布,以及炉缸工作的均匀化和高炉冶炼的顺利进行。

Combustion zone is the origin of blast furnace gas, which determines the distribution of gas in the hearth, and affects the distribution of gas in the process of rising in the blast furnace. If the combustion zone extends to the center of the hearth, the central gas will develop and the central temperature of the hearth will be high. On the contrary, if the combustion zone is reduced to the edge of the hearth, at this time, the marginal gas flow will develop, the central temperature of the hearth will decrease, and the central heat of the hearth will be insufficient, which is adverse to the chemical reaction. It is usually desirable that the combustion zone extend more towards the center of the hearth.

燃烧带是高炉煤气的发源地,其决定着煤气在炉缸内的分布,同时在很大程度上决定和影响煤气在高炉内上升过程中的分布。若燃烧带伸向炉缸中心,则中心煤气发展,炉缸中心温度高;相反,若燃烧带缩小至炉缸边缘,此时边缘煤气流发展,炉缸中心温度降低,炉缸中心热量不足,对化学反应不利。通常希望燃烧带较多地伸向炉缸中心。

The combustion of fuel in the combustion zone makes room for the decline of charge, which is the main factor to promote the decline of charge. The charge above the combustion belt is always looser than other places, and the discharge is faster. Properly expanding the combustion zone (including the radius direction and the circumference direction of the hearth) can reduce the sluggish area of the burden falling and expand the area of the active area of the hearth, which is conducive to the smooth operation of the blast furnace.

燃料在燃烧带燃烧为炉料的下降腾出了空间，它是促进炉料下降的主要因素。燃烧带上方的炉料总是比其他地方松动，而且下料快。适当扩大燃烧带（包括炉缸半径方向和圆周方向）可以缩小炉料下降的呆滞区域，扩大炉缸活跃区域的面积，有利于高炉顺行。

There are many factors affecting the size of combustion zone, mainly depending on:

影响燃烧带大小的因素。影响燃烧带大小的因素很多，主要取决于：

(1) Blowing function: The size of the blowing function determines the size of the combustion zone. Blast kinetic energy refers to the ability of blast to overcome the resistance of material layer in front of tuyere and penetrate into the center of hearth. It is the main reason that coke circulates in front of tuyere to form raceway. All the factors that affect the kinetic energy of the blast will affect the size of the combustion zone, such as blast volume, blast temperature, blast pressure, tuyere diameter, etc. For example, when the air volume is fixed in production, the methods to reduce the central air flow and develop the edge air flow is carried out: expanding the tuyere diameter, reducing the kinetic energy of the blast, shortening the combustion zone along the hearth radius direction and increasing along the circumference direction.

(1) 鼓风功能。鼓风功能的大小决定了燃烧带的大小。鼓风动能是指鼓风克服风口前料层的阻力，向炉缸中心穿透的能力，它是风口前焦炭做循环运动形成回旋区的主要原因。凡是影响鼓风动能的因素都将影响燃烧带的大小，如鼓风量、鼓风温度、鼓风压力、风口直径等，例如生产中在风量一定的条件下，扩大风口直径，鼓风动能减小，燃烧带沿炉缸半径方向缩短而沿圆周方向增大。这就是减少中心气流、发展边缘气流的手段。

(2) Combustion reaction rate: The combustion reaction rate has a certain influence on the size of combustion zone. If the combustion speed is fast, the combustion time is short, the combustion space is small, and the combustion zone is reduced; On the contrary, the combustion zone is increased. At present, the influence of combustion speed on combustion zone is limited in blast furnace.

(2) 燃烧反应速度。燃烧反应速度对燃烧带的大小有一定影响。燃烧速度快，则燃烧时间短，燃烧进行的空间就小，燃烧带缩小；相反，燃烧带增大。目前高炉上，燃烧速度对燃烧带的影响有限。

(3) Charge distribution: The charge distribution also has some influences on the combustion zone. The charge is loose, the air permeability is good, the resistance to gas is small, the blast penetration is strong, and the combustion zone is increased.

(3) 炉料分布。炉料分布对燃烧带也有一定的影响。炉料疏松，透气性好，对煤气的阻力小，鼓风穿透能力强，燃烧带增大。

Changes of Gas in Blast Furnace
煤气在高炉内的变化

The gas and heat, produced by the fuel combustion in front of the tuyere, transport a series of heat and material with the falling charge in the process of rising. The volume, composition, temperature and pressure of gas have changed a lot.

风口前燃料燃烧产生的煤气和热量，在上升过程中与下降的炉料进行一系列热量与物质的传递和输送。煤气的体积、成分、温度和压力等都发生重大变化。

- Changes of Gas Volume and Composition
- 煤气体积和成分的变化

The volume and composition change of gas in the rising process is shown in Figure 1-20.
煤气在上升过程中体积和成分变化如图1-20所示。

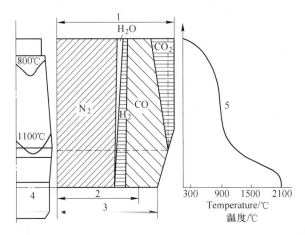

Figure 1-20 Changes of volume and component temperature along the height of blast furnace during gas rising
图1-20 煤气上升过程中体积、成分温度沿高炉高度的变化
1—Top gas volume; 2—Air volume; 3—Hearth combustion with gas volume; 4—Center line of tuyere; 5—Gas temperature
1—炉顶煤气量；2—风量；3—炉缸燃烧带煤气量；4—风口中心线；5—煤气温度

The total volume of gas has increased from bottom to top. Generally, the amount of gas in hearth (by volume) is about 1.21 times of the blast volume, and the amount of gas in top is 1.35~1.37 times of the blast volume. When injecting fuel, the gas quantity of hearth is about 1.3 times of the blast quantity, and the gas quantity of top is about 1.4~1.45 times of the blast quantity. The increase of gas volume is mainly due to the direct reduction of Fe, Si, Mn, P and other elements in the ore to generate a part of CO. CO_2 decomposed from carbonate in the high-temperature zone reacts with C to generate a double volume of CO. CO_2 decomposed from the middle temperature zone also directly increases the gas volume.

煤气总的体积自下而上有所增大。通常炉缸煤气量（体积分数）约为鼓风量的1.21倍，炉顶煤气量为鼓风量的1.35~1.37倍。喷吹燃料时，炉缸煤气量约为鼓风量的1.3倍，炉顶煤气量为鼓风量的1.4~1.45倍。煤气体积的增加主要是由于矿石中Fe、Si、Mn、P等元素的直接还原生成一部分CO，碳酸盐在高温区分解出的CO_2与C作用生成两倍体积的CO，而中温区分解出的CO_2也直接增加了煤气体积。

The changes of volume and compositions of gas in the rising process are as follows:
煤气在上升过程中体积和成分的变化情况如下：

(1) CO. In the high temperature region, the volume of CO increases gradually, which is due to the direct reduction of Fe, Si, Mn, P and other elements to produce CO. In the middle temperature region, CO takes part in indirect reduction and consumes a part, so the amount of CO increases firstly and then decreases.

(1) CO。在高温区，CO的体积逐渐增大，这是由于Fe、Si、Mn、P等元素直接还原产生CO；在中温区，CO参加间接还原又消耗一部分，所以，CO的量是先增加后降低。

(2) CO_2. In the high temperature zone, there is no indirect reduction and CO_2 does not exist. In the middle temperature zone, indirect reduction produces CO_2, while carbonate decomposition releases CO_2, and the amount of CO_2 increases gradually.

(2) CO_2。在高温区,没有间接还原,CO_2 不存在;在中温区,间接还原产生 CO_2,同时碳酸盐分解放出 CO_2,CO_2 的量逐渐增加。

(3) H_2. During the rising process, $1/3 \sim 1/2$ of H_2 brought in by blowing water, coke volatiles, injection fuel, etc., participates in indirect reduction and becomes H_2O.

(3) H_2。鼓风水分、焦炭挥发分、喷吹燃料等带入的 H_2,在上升过程中有 $1/3 \sim 1/2$ 参加间接还原,变成 H_2O。

(4) N_2. A large amount of N_2 is brought in by blast, and a small amount is organic N_2 in coke. N_2 does not take part in any chemical reaction, so its absolute amount remains unchanged.

(4) N_2。大量的 N_2 由鼓风带入,少量是焦炭中的有机 N_2。N_2 不参加任何化学反应,故其绝对量不变。

(5) CH_4. CH_4 is produced by a small amount of C and H_2 in the high temperature zone. And CH_4 is added into the volatile of coke in the process of gas rising, but the amount is very small.

(5) CH_4。在高温区有少量的 C 与 H_2 生成 CH_4,煤气上升过程中又有焦炭挥发分中的 CH_4 加入,但数量均很少。

In general, the total amount of CO and CO_2 in the top gas is relatively stable, ranging from 38% to 42%. The range of gas composition to the top of the blast furnace is shown in Table 1-8.

一般情况下,炉顶煤气中 CO 与 CO_2 的总量比较稳定,为 38%~42%。最后到达高炉炉顶的煤气成分范围见表 1-8。

Table 1-8 Composition range of blast furnace top gas
表 1-8 高炉炉顶煤气成分范围

Formation 组成 Components (volume fraction)/% 化学成分(体积分数)/%	CO_2	CO	N_2	H_2	CH_4
	15~22	20~25	55~57	about (约) 2.0	about (约) 0.3

- Changes of Gas Temperature
- 煤气温度的变化

The temperature distribution of the gas in the hearth is shown in Figure 1-21. The highest temperature is about 1000mm from the front of the tuyere, which is also the highest temperature in the blast furnace.

煤气在炉缸内的温度分布如图 1-21 所示。其温度最高点在距风口前沿 1000mm 左右的地方,也是高炉内的最高温度。

In the process of gas rising, the temperature of gas is higher than that of charge, and heat is transferred to charge, which causes heat exchange. And the temperature is gradually reduced. At the same time, the temperature of decreased charge is gradually increased. Due to the different chemical reactions of the charge in different areas, the heating rate of the charge along the height direction of the blast furnace is different from that of the gas, as shown in Figure 1-22. In the upper area, gas cooling is relatively slow, and the heating rate of charge is relatively fast; in the

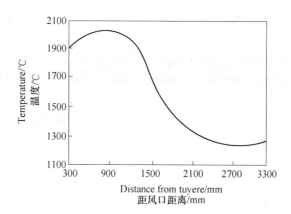

Figure 1-21　Changes of gas temperature in hearth along radius direction

图 1-21　沿半径方向炉缸内煤气温度的变化

lower area, gas cooling is relatively fast, and the heating rate of charge is relatively slow; and in the middle area, the temperature difference between gas and charge is small, the heat exchange is less, and the range of gas cooling and charge heating is very small.

　　煤气在上升过程中，其温度高于炉料的温度，将热量传递给炉料，发生热交换，温度逐渐降低。与此同时，下降的炉料温度逐渐升高。由于不同区域的炉料发生的化学反应不同，沿高炉高度方向上炉料的升温速度与煤气的降温速度不同，如图 1-22 所示。在上部区域，煤气降温比较慢，炉料升温速度比较快；在下部区域，煤气降温比较快，炉料升温速度比较慢；而在中部区域，煤气与炉料温差小，热交换少，煤气降温和炉料升温幅度都很小。

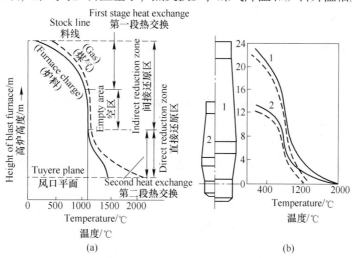

Figure 1-22　Schematic diagram of heat exchange processes in blast furnace

图 1-22　高炉内热交换过程示意图

(a) Division of heat exchange processes in blast furnace; (b) Changes of temperature of charge and gas along blast furnace height in large and small blast furnaces

(a) 高炉内热交换过程分区；(b) 大、小高炉内炉料和煤气的温度沿高炉高度的变化

1—Large blast furnace; 2—Small blast furnace

1—大高炉；2—小高炉

- Changes of Gas Pressure
- 煤气压力的变化

When the gas rises from the hearth and passes through the soft melting zone and block zone to the top of the furnace, its own pressure decreases, and the pressure in the lower parts of the blast furnace decreases faster than that in the upper parts of the blast furnace during the rising process, which is mainly caused by the increase of resistance of the lower charge to the gas passing after softening and melting, as shown in Figure 1-23.

煤气从炉缸上升,穿过软熔带、块状带到达炉顶,其本身压力降低,且上升过程中在高炉下部比在高炉上部压力降低要快。这主要是由于下部炉料软化熔融后对煤气通过的阻力增大所致,如图1-23所示。

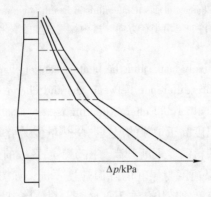

Figure 1-23 Distribution diagram of Blast furnace gas with different smelting strength static pressure of Δp

图1-23 不同冶炼强度下高炉煤气静压力 Δp 分布示意图

1.2.2.5 Movement of Charge
1.2.2.5 炉料的运动

In the smelting process of blast furnace, the moving state of the charge in the furnace is a slow moving bed of solid bulk material. The uniform and rhythmic underground drop of the charge is an important sign of the blast furnace's going forward.

在高炉冶炼过程中,炉料在炉内的运动状态是一个固体散料的缓慢移动床,炉料均匀而有节奏地下降是高炉顺行的重要标志。

The conditions for the lowering of the furnace charge are as follows: firstly, there should be space for the lowering; and secondly, there should be a force for the lowering. Both are indispensable.

炉料下降的条件为:一是要有下降的空间;二是要有下降的力。两者缺一不可。

Space Condition of Charge Lowering
炉料下降的空间条件

The basic condition of charge falling is that there is a free space for charge falling in the blast furnace. There are four factors to form the burden falling space in blast furnace:

炉料下降的基本条件是在高炉内不断产生供炉料下降的自由空间。高炉内形成炉料下降空间的因素有以下四个方面：

(1) The coke in front of tuyere is burned, and the solid coke is converted into gas.

(1) 风口前焦炭燃烧，固体焦炭转化为气体。

(2) Above the tuyere area, the volume of coke decreases due to the direct reduction of fixed carbon consumed by coke.

(2) 风口区以上，由于直接还原消耗焦炭的固定碳，焦炭体积减小。

(3) The ore is rearranged, compacted and melted into liquid phase in the process of falling, so that the volume is reduced.

(3) 矿石在下降过程中重新排列，压紧并熔化成液相，从而使体积缩小。

(4) Slag and iron are continuously discharged from the hearth.

(4) 炉缸不断放出渣和铁。

Mechanical Condition of Charge Falling
炉料下降的力学条件

The burden has the space to drop, but also must have the force to drop. The lowering of charge depends on its own gravity. But at the same time, it is subject to the friction resistance between charge and charge, the friction resistance between charge and furnace wall, and the resistance of rising gas to the lowering of charge. The formula of the mechanical condition is as follows：

炉料不仅具有下降的空间，还须具备下降的力。炉料下降依靠自身重力，但同时又受到炉料与炉料之间的摩擦阻力，炉料与炉墙之间的摩擦阻力，以及上升煤气对炉料下降产生阻力的影响。其计算公式为：

$$p = (W_{charge} - p_{wall\ friction} - p_{material\ friction}) - \Delta p = W_{effective} - \Delta p$$
$$p = (W_{炉料} - p_{墙摩} - p_{料摩}) - \Delta p = W_{有效} - \Delta p \tag{1-59}$$

Where p——the force determining the lowering of charge；

W_{charge}——the total gravity of charge in the furnace；

$p_{wall\ friction}$——the friction resistance between charge and wall；

$p_{material\ friction}$——the friction resistance between particles when the material blocks move with each other；

Δp——the resistance of rising gas to charge (supporting force or buoyancy)；

$W_{effective}$——the effective gravity of charge, which is given by：

$$W_{effective} = W_{charge} - p_{wall\ friction} - p_{charge\ friction} \tag{1-60}$$

式中 p——决定炉料下降的力；

$W_{炉料}$——炉料在炉内的总重力；

$p_{墙摩}$——炉料与炉墙之间的摩擦阻力；

$p_{料摩}$——料块相互运动时颗粒之间的摩擦阻力；

Δp——上升煤气对炉料的阻力（支撑力或浮力）；

$W_{有效}$——炉料的有效重力，其公式为：

$$W_{有效} = W_{炉料} - p_{墙摩} - p_{料摩} \tag{1-60}$$

Obviously, the mechanical condition of the charge falling is $p > 0$, i.e. $W_{effective} > \Delta p$. The larger the p value (or $W_{effective}$) is, the smaller the Δp is, and the more favorable for the charge moving forward. When $W_{effective}$ is close to or equal to Δp, the burden is difficult to move or suspend.

显然，炉料下降的力学条件是 $p>0$，即 $W_{有效}>\Delta p$，其中，p 值越大（或 $W_{有效}$ 越大），Δp 越小，越有利于炉料顺行。当 $W_{有效}$ 接近或等于 Δp 时，炉料难行或悬料。

If $\Delta p>0$, because of the supporting force of the rising gas greater than the effective gravity of the charge, the charge cannot be lowered, and there is suspension or pipeline stroke.

若 $\Delta p>0$，由于上升煤气的支撑力大于炉料的有效重力，炉料不能下降，因此会出现悬料或者管道行程。

It is worth that $p>0$ is the mechanical condition of whether the charge lowered. The larger the value is, the more favorable the charge will be lowered. However, the size of p has little effect on the rate of the decrease. The main factor affecting the cutting speed is the quantity of coke combustion per unit time. For the other words the cutting speed is directly proportional to the amount of blast.

值得注意的是，$p>0$ 是炉料能否下降的力学条件，其值越大，越有利于炉料下降，但 p 值的大小对炉料下降的快慢影响并不大。影响下料速度的因素主要是单位时间内焦炭燃烧的数量，即下料速度与鼓风量成正比。

1.2.3 Equipment of Blast Furnace Ironmaking
1.2.3 高炉炼铁设备

The blast furnace ironmaking equipment consists of a complete set of composite continuous equipment system, as shown in Figure 1-24. In addition to the individual blast furnace, the main equipment also includes the charging system at the back of the furnace, the charging system at the top of the furnace, the air supply system, the gas dedusting system, the slag iron treatment system, the injection system, etc.

高炉炼铁设备由一整套复合连续设备系统构成，如图 1-24 所示。其主体设备除了高炉个体以外，还包括炉后供料和炉顶装料系统、送风系统、煤气除尘系统、渣铁处理系统、喷吹系统等。

1.2.3.1 Blast Furnace Body
1.2.3.1 高炉本体

Blast furnace body is the main equipment for smelting pig iron, including furnace base, furnace lining, cooling equipment, furnace shell, pillar and furnace top frame. Among them, the furnace base is reinforced concrete and heat-resistant concrete structure, the furnace lining is made of refractory materials, and the rest of the equipment are all structural parts. The lower part of the blast furnace is provided with a tuyere, an iron port and a slag port. And the upper part is provided with a charging inlet and a gas outlet.

高炉本体是冶炼生铁的主体设备，包括炉基、炉衬、冷却设备、炉壳、支柱及炉顶框架等。其中，炉基为钢筋混凝土和耐热混凝土结构，炉衬由耐火材料砌筑而成，其余设备均为合属结构件。在高炉的下部设置有风口、铁口和渣口，上部设置有炉料装入口和煤气导出口。

Internal Type of Blast Furnace
高炉内型

Blast furnace is a kind of blast shaft furnace for producing liquid pig iron. Its working space is

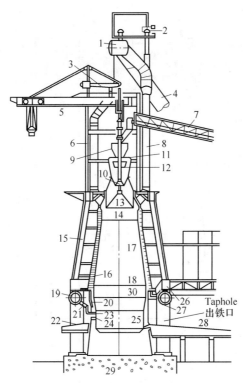

Figure 1-24 General drawing of BF body equipment
图 1-24 高炉炉体设备总图

1—Header; 2—Top gas vent valve; 3—Bell balance bar; 4—Downcomer; 5—Top crane; 6—Top frame; 7—Belt loader; 8—Riser; 9—Fixed hopper; 10—Small bell; 11—Seal valve; 12—Rotary chute; 13—Big bell; 14—Throat; 15—Shaft support; 16—Cooling water tank; 17—Furnace body; 18—Waist; 19—Shroud; 20—Cooling wall; 21—Air supply branch pipe (elbow); 22—Tuyere platform; 23—Tuyere; 24—Slag outlet; 25—Hearth; 26—Intermediate beam; 27—Support beam; 28—Cast yard; 29—Blast furnace foundation; 30—Bosh

1—集合管；2—炉顶煤气放散阀；3—料钟平衡杆；4—下降管；5—炉顶起重机；6—炉顶框架；7—带式上料机；8—上升管；9—固定料斗；10—小料钟；11—密封阀；12—旋转溜槽；13—大料钟；14—炉喉；15—炉身支柱；16—冷却水箱；17—炉身；18—炉腰；19—围管；20—冷却壁；21—送风支管（弯管）；22—风口平台；23—风口；24—出渣口；25—炉缸；26—中间梁；27—支承梁；28—出铁场；29—高炉基础；30—炉腹

made of refractories. The internal shape of blast furnace refers to the internal section shape of blast furnace working space. It is very important to obtain good technical and economic indexes and prolong the life of blast furnace with reasonable internal mold. Modern blast furnace internal mold consists of five sections: hearth, bosh, waist, body and throat. Among them, the hearth, waist and throat of the furnace are cylindrical, the bosh of the furnace is inverted cone shaped, and the furnace body is truncated cone shaped. The meaning of the dimensions and symbols of blast furnace internal mol is shown in Figure 1-25.

高炉是一种生产液态生铁的鼓风竖炉，其工作空间用耐火材料砌筑而成。高炉内型是指高炉工作空间的内部剖面形状。合理的高炉内型对获得良好的技术经济指标和延长高炉寿命具有重要的意义。现代高炉内型由炉缸、炉腹、炉腰、炉身和炉喉五段组成。其中，炉缸、炉腰和炉喉呈圆筒形，炉腹呈倒锥台形，炉身呈截锥台形。高炉内型尺寸及各符号所表示的意义如图 1-25 所示。

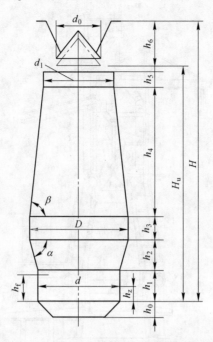

Figure 1-25 Representation method of BF internal dimensions
图 1-25 高炉内型尺寸表示方法

H—Full height, mm; H_u—Effective height, mm; h_1—Hearth height, mm; h_2—Bosh height, mm; h_3—Furnace waist height, mm; h_4—Shaft height, mm; h_5—Throat height, mm; h_6—Bottom height from top flange to falling position of large bell, mm; h_f—Height from center line of iron port to center line of air port, mm; h_z—Height from center line of iron port to center line of slag port, mm; h_0—Height from bottom of dead iron layer to center line of iron port, mm; d—Height of furnace basin diameter, mm; D—Furnace waist diameter, mm; d_1—Furnace throat diameter, mm; d_0—Large bell diameter, mm; α—Furnace belly angle, (°); β—Furnace body angle, (°)

H—全高,mm; H_u—有效高度,mm; h_1—炉缸高度,mm; h_2—炉腹高度,mm; h_3—炉腰高度,mm; h_4—炉身高度,mm; h_5—炉喉高度,mm; h_6—炉顶法兰盘至大料钟下降位置的底面高度,mm; h_f—铁口中心线至风口中心线的高度,mm; h_z—铁口中心线至渣口中心线的高度,mm; h_0—死铁层最底面至铁口中心线的高度,mm; d—炉缸直径,mm; D—炉腰直径,mm; d_1—炉喉直径,mm; d_0—大料钟直径,mm; α—炉腹角,(°); β—炉身角,(°)

The blast furnace size is indicated by effective volume. The effective volume of blast furnace, V_u, is the sum of the five volumes of hearth, bosh, waist, body and throat. At present, the most effective volume of blast furnace in the world is 6183m³.

高炉大小用有效容积表示。高炉有效容积 V_u 为炉缸、炉腹、炉腰、炉身和炉喉五段容积之和。目前,世界上高炉有效容积最大的是 6183m³。

Blast Furnace Lining

高炉炉衬

Blast furnace lining is made of refractory material which can resist high temperature and chemical erosion. The main function of furnace lining is to form working space, reduce heat loss and protect metal structure from thermal stress and chemical erosion. Prolonging lining life is an important task of blast furnace design and operation.

高炉炉衬是用能够抵抗高温和化学侵蚀作用的耐火材料砌筑而成的。炉衬的主要作用是构成工作空间、减少散热损失以及保护金属结构件免遭热应力和化学侵蚀作用。延长炉衬寿命是高炉设计的重要任务，也是高炉操作的重要任务。

Blast furnace lining is generally made of ceramic materials (clay and high aluminum) and carbon materials (carbon brick and carbon rammed graphite, etc.). The erosion and damage of furnace lining are closely related to smelting conditions, and the damage mechanism of each part is not the same. It is important to study the damage mechanism of furnace lining with reasonable selection of refractory materials and design of furnace lining structure. In summary, the damage mechanism of furnace lining mainly includes the following four aspects:

高炉炉衬一般以陶瓷材料（黏土质和高铝质）和碳质材料（炭砖和炭捣石墨等）砌筑。炉衬的侵蚀和破损与冶炼条件密切相关，各部位的破损机理并不相同，研究炉衬的破损机理与合理选择耐火材料及设计炉衬结构有重要关系。归纳起来，炉衬的破损机理主要有以下四个方面：

(1) Infiltration and erosion of high temperature slag iron;

(1) 高温渣铁的渗透和侵蚀；

(2) High temperature and thermal shock damage;

(2) 高温和热震破损；

(3) Friction and erosion of the charge and gas flow, and destruction of the carbon deposition of the gas;

(3) 炉料和煤气流的摩擦冲刷及煤气碳素沉积的破坏作用；

(4) Destruction of alkali metals and other harmful elements.

(4) 碱金属及其他有害元素的破坏作用。

Blast Furnace Cooling Equipment

高炉冷却设备

The furnace lining must be cooled. And the cooling medium is usually water, steam water mixture and air. The common characteristics of these cooling media are large heat transfer capacity, convenient transportation, safety and reliability, easy access and low cost.

高炉炉衬必须冷却，冷却介质通常为水、汽水混合物及空气。这些冷却介质的共同特点是传热能力大、输送方便、安全可靠、易于获取及成本低等。

Due to different working conditions, the cooling effects of each part of the blast furnace are not same. Generally speaking, the cooling of the blast furnace has the following functions:

高炉各部位由于工作条件不同，冷却的作用也不完全相同。总体来说，高炉冷却有以下几方面的作用：

(1) It can reduce the temperature of refractory brick lining, make it maintain enough strength and reasonable working space of blast furnace.

(1) 降低耐火砖衬温度，使其能保持足够的强度，维持高炉合理的工作空间；

(2) It can make the surface of furnace lining form protective slag skin, and rely on the slag

skin to protect or replace the furnace lining to maintain a reasonable operation furnace type.

（2）使炉衬表面形成保护性渣皮，并依靠渣皮保护或代替炉衬工作，维持合理的操作炉型；

(3) It can protect the furnace shell and metal components from damage under heat load.

（3）保护炉壳及金属构件，使其不致在热负荷作用下遭到损坏；

(4) It does not affect the air tightness and strength of the furnace shell.

（4）不影响炉壳的气密性和强度。

Cooling forms include water spray cooling outside the furnace and cooler cooling. The main coolers of blast furnace include cooling plate (shown in Figure 1-26), cooling water tank (shown in Figure 1-27), cooling wall (shown in Figure 1-28), tuyere and slag inlet water jacket (shown in Figure 1-29), air cooling or water cooling pipe (shown in Figure 1-30), etc. The working principle of the cooler is to take the heat from the furnace lining or components away from the cooling medium, so that the furnace lining or components can be cooled.

冷却的形式有炉外喷水冷却和冷却器冷却。高炉的主要冷却器有冷却板（见图1-26）、冷却水箱（见图1-27）、冷却壁（见图1-28）、风口和渣口水套（见图1-29）以及风冷或水冷管（见图1-30）等。冷却器的工作原理是将自炉衬或构件传来的热量由冷却介质带走，使炉衬或构件得以冷却。

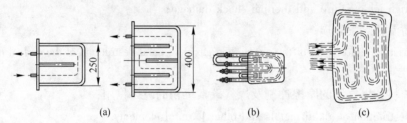

Figure 1-26 Cooling plate

图 1-26 冷却板

(a) Cast copper cooling plate; (b) Embedded cooling plate; (c) Cast iron cooling plate

(a) 铸铜冷却板；(b) 埋入式冷却板；(c) 铸铁冷却板

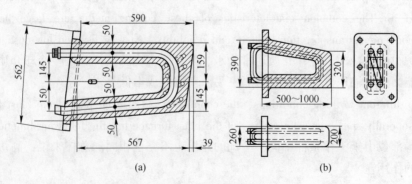

Figure 1-27 Cooling water tank

图 1-27 冷却水箱

(a) Beam type water tank; (b) Flat water tank

(a) 支梁式水箱；(b) 扁水箱

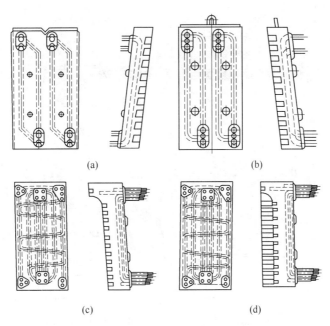

Figure 1-28 Brick lined cooling stave of blast furnace

图 1-28 高炉镶砖冷却壁

(a) First generation; (b) Second generation; (c) Third generation; (d) Fourth generation

(a) 第1代; (b) 第2代; (c) 第3代; (d) 第4代

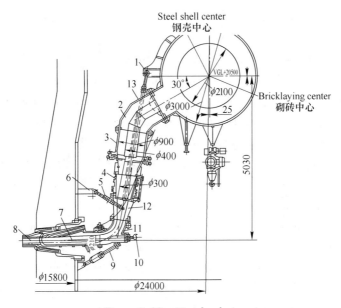

Figure 1-29 Air inlet device

图 1-29 风口装置

1—Beam; 2—A-1 pipe; 3—A-2 pipe; 4—Expansion pipe; 5—Tension screw; 6—Ring beam; 7—Blowpipe;
8—Air outlet; 9—Fastening device; 10—Peephole; 11—Elbow; 12—Reducer; 13—Hanger

1—横梁; 2—A-1管; 3—A-2管; 4—伸缩管; 5—拉紧螺丝; 6—环梁; 7—直吹管; 8—风口;
9—紧固装置; 10—窥视孔; 11—弯管; 12—异径管; 13—吊挂装置

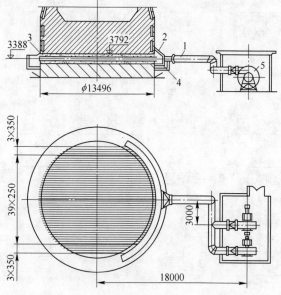

Figure 1-30　Layout of air-cooled bottom of 2000m³ blast furnace
图 1-30　2000m³ 高炉风冷炉底布置图
1—Air inlet pipe; 2—Air inlet box; 3—Dust guard; 4—Air cooling pipe; 5—Blower
1—进风管; 2—进风箱; 3—防尘板; 4—风冷管; 5—鼓风机

Blast Furnace Foundation
高炉基础

The blast furnace foundation bears the gravity transmitted by the blast furnace body, pillar and other related auxiliary facilities, and transmits these gravity to the stratum evenly. The foundation of blast furnace must be stable, and large uneven sinking is not allowed to avoid large changes in the relative position of blast furnace and its surrounding equipment, so as to destroy the connection between them causing dangerous deformation.

高炉基础承受着高炉炉体、支柱及其他有关附属设施所传递的重力，并将这些重力均匀地传递给地层。高炉基础必须稳定，不允许发生较大的不均匀下沉，以免高炉与其周围设备的相对位置发生大的变化，从而破坏它们之间的联系，并使之发生危险的变形。

The blast furnace foundation is generally composed of the foundation buried in the underground part and the foundation pier exposed on the ground. The function of foundation pier is to insulate heat and adjust the elevation of iron port, which is used to resist the temperature of 900~1000℃. And it is also made of heat-resistant concrete. Its shape is cylindrical, and its diameter and size are suitable for the furnace bottom. The blast furnace foundation is required to be wrapped in the furnace shell. The main function of the foundation is to transmit the load from above to the stratum. Its bottom area is large, so as to reduce the pressure on the foundation per unit area. The base is made of ordinary reinforced concrete. In order to reduce the thermal stress, it is better to make it into a circle. However, considering the convenience of construction, it is generally a regular polygon.

高炉基础一般由埋在地下部分的基座和露在地面的基墩组成。基墩的作用是隔热和调节铁口标高，用来抵抗 900~1000℃ 的温度，由耐热混凝土制成。高炉基础的形状为圆柱形，直径尺寸与炉底相适应，并要求能包于炉壳之内。基座的主要作用是将上面传来的载荷传递给地层，其底面积较大，能够减小单位面积的地基所承受的压力。基座用普通钢筋混凝土制成，为减少热应力作用，最好将其制作成圆形。但考虑施工方便，一般都为正多边形。

Blast Furnace Steel Structure
高炉钢结构

Blast furnace steel structure refers to the external structure of blast furnace body. In large and medium-sized blast furnace, the parts of steel structure are furnace shell, pillar, waist support ring (waist support ring), furnace top frame, inclined bridge, various pipes, platforms, bridges and ladders. The requirements for steel structure are: simple and durable, safe and reliable, convenient operation, easy maintenance and material saving.

高炉钢结构是指高炉本体的外部结构。在大中型高炉上采用钢结构的部位有炉壳、支柱、炉腰托圈（炉腰支圈）、炉顶框架、斜桥、各种管道、平台、过桥以及走梯等。对钢结构的要求是：简单耐用，安全可靠，操作便利，容易维修和节省材料。

The early blast furnace wall is very thick. It is not only the refractory lining but also the structure supporting the blast furnace and its equipment. The structural form of blast furnace mainly depends on the way of load transfer from top and shaft to foundation, lining thickness and cooling mode of each part of furnace body. There are basically four types of blast furnaces, as shown in Figure 1-31.

早期的高炉炉墙很厚，它既是耐火炉衬又是支撑高炉及其设备的结构。高炉的结构形式主要取决于炉顶和炉身载荷传递到基础的方式，以及炉体各部位的内衬厚度和冷却方式。高炉基本上有四种结构形式，如图 1-31 所示。

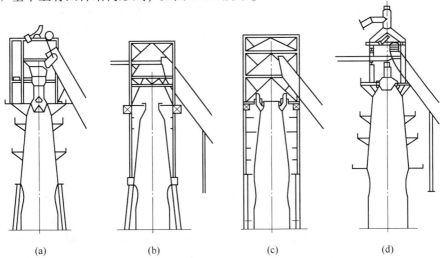

Figure 1-31　Blast furnace structure
图 1-31　高炉的结构形式
(a) Hearth pillar type; (b) Hearth and shaft pillar type; (c) Furnace frame type; (d) Self-supporting type
(a) 炉缸支柱式；(b) 炉缸、炉身支柱式；(c) 炉体框架式；(d) 自力式

The main function of the furnace shell is to bear the load, fix the cooling equipment and use the water spray outside the furnace to cool the furnace lining, so as to ensure the overall soundness of the furnace lining and to make the furnace body have a certain degree of air tightness. In addition to the huge gravity, the furnace shell is also under the action of thermal stress and internal gas pressure. And sometimes it is also resistant to the impact of sudden accidents such as gas explosion, material collapse and sitting. Therefore, the furnace shell is required to have sufficient strength.

炉壳的主要作用是承受载荷,固定冷却设备和利用炉外喷水来冷却炉衬,以保证高炉炉衬的整体坚固性和使炉体具有一定的气密程度。炉壳除承受巨大的重力外,还受热应力和内部煤气压力的作用,有时还要抵抗煤气爆炸、崩料、坐料等突发事故的冲击。因此要求炉壳具有足够的强度。

The pillar can be divided into three types: hearth pillar, furnace body pillar and furnace body frame, as shown in Figure 1-31(a)~(c).

支柱可分为炉缸支柱、炉身支柱和炉体框架三种,如图1-31(a)~(c)所示。

In order to facilitate the overhaul and maintenance of the furnace top equipment, a furnace top platform is arranged on the horizontal surface of the furnace top flange. The roof platform has a roof frame, which is used to support the balance bar of big and small bell, install the girder and receiving funnel, etc.

为了便于炉顶设备的检修和维护,在炉顶法兰水平面上设有炉顶平台。炉顶平台上有炉顶框架,用来支撑大小料钟的平衡杆、安装大梁和受料漏斗等。

1.2.3.2 Post-furnace Feed and Top Charge System
1.2.3.2 炉后供料和炉顶装料系统

The task of post-furnace feed and top charge system is to ensure continuous and balanced supply of raw materials for blast furnace smelting, and to load the change into the blast furnace and to make its reasonbly distribution.

炉后供料和炉顶装料系统的任务是保证连续、均衡地供应高炉冶炼所需原料,将炉料装入高炉并使之分布合理。

Modern large blast furnaces need ten thousands of tons of raw materials and fuels every day and night. The supply of raw materials and fuels is guaranteed by the charging system at the back of the furnace and the charging system at the top of the furnace. The charging system at the back of the furnace and the top of the furnace, including the charging equipment and the feeding belt conveyor, as well as various unloading, screening, weighing and transportation equipment under the trough, shall meet the following requirements: it has large production capacity, continuous feeding, adapting to the requirements of the intensified production of the blast furnace and the change of the variety of raw materials; it also has good anti-wear performance, high mechanical strength, and high temperature continuous operation for a long time under dusty conditions; the sealing structure of the furnace top must be tight and reliable, and the sealing material can work normally for a long time at 250℃; and the structure is simple, easy to operate and maintain; and the manual operation should be abolished to fully realize mechanized and automatic feeding.

现代大型高炉每昼夜连续需要原燃料上万吨。原燃料的供应由高炉炉后供料和炉顶装料系统来保证。炉后供料和炉顶装料系统包括装料设备和上料胶带运输机，以及槽下各种卸料、筛分、称量、运输设备所组成的系统。该系统应当满足下列要求：生产能力大，能连续供料，能适应高炉强化生产的供料要求和原料品种变化后的要求；抗磨性能好，机械强度高，并能在高温、多粉尘条件下长时间地连续工作；炉顶密封结构必须严密、可靠，密封材料能在250℃温度下长时期正常工作；结构简单，操作方便，易于维护；应废除人工操作，全面实现机械化和自动化供料。

Rear Furnace Feeding System
炉后供料系统

After furnace feeding refers to the process of transporting raw materials from the blast furnace workshop to the top of the blast furnace. The feeding system behind the furnace mainly includes:

炉后供料是指将原料从高炉车间运送到高炉炉顶的过程。炉后供料系统主要包括：

(1) Ore storage tank and coke storage tank. The ore storage tank and coke storage tank behind the blast furnace are used to receive and store the furnace charge, to buffer the production imbalance between the sintering plant and the coking plant and the blast furnace, as well as the impact of the accident or maintenance of the conveyor belt conveyor. In addition, a certain number of miscellaneous ore tanks should be set up to store fluxes and furnace washing materials.

（1）储矿槽与储焦槽。高炉炉后储矿槽和储焦槽是用来接受和储存炉料的，并用以缓冲烧结厂和焦化厂与高炉间的生产不平衡，以及运料胶带运输机发生事故或检修时所带来的影响。此外，还应设置一定数目的杂矿槽，以储存熔剂和洗炉料等。

(2) Screening under the tank. The screening under the tank is the last screening of the furnace charge before entering the furnace. Its purpose is to further screen the powder in the furnace charge to improve the air permeability of the charge column in the furnace. Sometimes the sieve also acts as a feed.

（2）槽下筛分。槽下筛分是炉料在入炉前的最后一次筛分，其目的是进一步筛除炉料中的粉末，从而改善炉内料柱透气性。有时筛子还可以起到给料的作用。

(3) Weighing. Weighing is divided into two ways: weighing car and weighing funnel. Weighing vehicle is a kind of electric transport vehicle with weighing and handling mechanism. Weighing funnel can be used to weigh sinter, raw ore, pellet and coke.

（3）称量。称量分为称量车称量和称量漏斗称量两种方式。称量车是一种带有称量和装卸机构的电动运输车辆。称量漏斗可以用来称量烧结矿、生矿、球团矿和焦炭等。

(4) Transportation under the tank. The belt conveyor is generally used for under trough transportation. It is the best scheme to realize the automatic operation under the blast furnace trough that the feeding of the belt conveyor is matched with the weighing of the weighing funnel.

（4）槽下运输。槽下运输普遍采用胶带运输机供料。胶带运输机供料与称量漏斗称量相配合，是高炉槽下实现自动化操作的最佳方案。

(5) Charging car type feeder. The charging car type is to use the charging car to walk on the inclined bridge and then send the charge to the top of the blast furnace. The charging car type feeder system is mainly composed of discharging car, inclined bridge and winch of charging car, as shown in Figure 1-32. Some of the winch rooms of the charging car are arranged above the inclined bridge, and some are arranged below the inclined bridge. Considering the influence of many factors, most of the newly-built blast furnaces arrange the winch rooms under the inclined bridge.

(5) 料车式上料机。料车式上料是利用料车在斜桥上行走，将炉料送到高炉炉顶。料车式上料机系统主要由出料车、斜桥和料车卷扬机等几部分组成，如图1-32所示。料车卷扬机室有的布置在斜桥上方，有的布置在斜桥下方。考虑到多种因素的影响，大多数新建高炉都把卷扬机室布置在斜桥的下方。

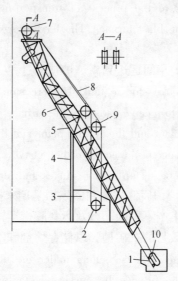

Figure 1-32　Feeder system
图 1-32　料车式上料机系统

1—Dumper pit; 2—Dumper winch; 3—Winch room; 4—Pillar; 5—Track; 6—Inclined bridge;
7, 9—Rope pulley; 8—Steel rope; 10—Dumper

1—料车坑；2—料车卷扬机；3—卷扬机室；4—支柱；5—轨道；6—斜桥；7, 9—绳轮；8—钢绳；10—料车

(6) Belt feeder. Due to the large-scale and automation of blast furnace, belt feeder system has become a mainstream configuration, which is mainly composed of belt, drive drum, drive motor and drive device. The working diagram of the belt feeder is shown in Figure 1-33.

(6) 胶带式上料机。由于高炉的大型化和自动化，胶带式上料机系统已经成为一种主流配置，它主要由胶带、驱动卷筒、驱动电动机及传动装置等组成。胶带式上料机的工作示意图如图1-33所示。

Top Charging System
炉顶装料系统

The main task of the top charging system is to load the charge into the blast furnace and make

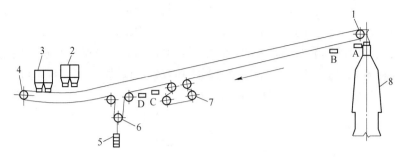

Figure 1-33　Schematic diagram of belt feeder
图 1-33　胶带式上料机的工作示意图

1—Top head wheel; 2—Ore funnel; 3—Coke funnel; 4—Tail wheel; 5—Counterweight; 6—Belt tensioning device;
7—Belt drive device; 8—Blast furnace; A—Raw material arrival detection; B—Top charging preparation detection;
C—Ore end detection; D—Coke end detection

1—炉顶头轮；2—矿石漏斗；3—焦炭漏斗；4—尾轮；5—配重；6—胶带张紧装置；7—胶带传动装置；8—高炉；
A—原料到达炉顶检测；B—炉顶装料准备检测；C—矿石终点检测；D—焦炭终点检测

its distribution reasonable. The equipment mainly includes charging, distributing, probing and pressure equalizing parts. The charging system mainly includes bell top, bell valve top and bell less top. Bell type furnace top mainly includes receiving hopper, rotary distributor, large and small bells, large and small hoppers, balance bar mechanism of large and small bell, electric hoist or hydraulic drive device for large and small bells, material detection device and its hoist, etc. The bell valve type furnace top also has the storage tank and the seal valve. The bell less furnace top is not equipped with a bell, and the rotary chute is used for distribution. Other main equipment is basically the same as the bell valve type furnace top.

炉顶装料系统的主要任务是将炉料装入高炉，并使其合理分布。该系统设备主要包括装料、布料、探料及均压几部分。装料系统的类型主要有钟式炉顶、钟阀式炉顶和无料钟炉顶。钟式炉顶主要包括受料漏斗、旋转布料器、大小料钟、大小料斗、大小料钟平衡杆机构、大小料钟电动卷扬机（或液压驱动装置）、探料装置及其卷扬机等。钟阀式炉顶还有储料罐及密封阀门。无料钟炉顶不设置料钟，并采用旋转溜槽布料，其他主要设备与钟阀式炉顶大体相同。

- Equipment of Bell Type and Bell Valve Type Top Charging
- 钟式与钟阀式炉顶装料设备

The bell roof is divided into two bell type, three bell type and four bell type. The purpose of increasing the number of charging clocks is to strengthen the sealing of top gas. But it will make the structure of top charging equipment more complicated. The double bell top structure is widely used in blast furnaces. Bell valve roof is developed on the basis of double bell roof, and its main purpose is to strengthen the sealing of top gas. According to the number of storage tanks, bell valve top can be divided into two types: double bell double valve type and double bell four valve type. At present, these two types of top are used in blast furnaces. Figure 1-34 shows the equipment of bell top charging, and Figure 1-35 shows the equipment of double bell and double valve top charging.

钟式炉顶分为双钟式、三钟式和四钟式。增加料钟个数的目的是为了加强炉顶煤气的密封，但会使炉顶装料设备的结构更加复杂化。高炉普遍采用双钟式炉顶结构。钟阀式炉顶是在双钟式炉顶的基础上发展起来的，其主要目的也是为了加强炉顶煤气的密封。钟阀式炉顶按照储料罐个数的不同又分为双钟双阀式和双钟四阀式两种，目前这两种炉顶在高炉上均有采用。如图1-34所示为钟式炉顶装料设备，如图1-35所示为双钟双阀式炉顶装料设备。

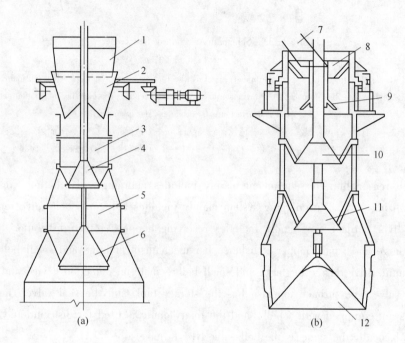

Figure 1-34　Bell top charging equipment
图1-34　钟式炉顶装料设备
(a) Double bell top with quick distributor; (b) Three bell top
(a) 带有快速布料器的双钟炉顶；(b) 三钟炉顶

1—Fixed receiving hopper; 2—Quick distributor; 3—Small hopper; 4, 10—Small material clock; 5—Large material hopper;
6, 12—Large material clock; 7—Receiving funnel; 8—Rotary chute; 9—Charge distributor;
11—Medium material clock

1—固定受料漏斗；2—快速布料器；3—小料斗；4，10—小料钟；5—大料斗；6，12—大料钟；7—受料漏斗；
8—旋转溜槽；9—炉料分布器；11—中料钟

- Equipment of Bell Less Top Charging
- 无料钟炉顶装料设备

In 1970s, Paul Wurth (PW) Company in Luxembourg launched PW type bell less top charging equipment, as shown in Figure 1-36. The reason why the bell less top charging equipment has developed rapidly since it was invented is that it not only has many means of distribution, flexible distribution and increased means for the upper part of the blast furnace, but also provides guarantee for the high-pressure operation of the top of the blast furnace and the improvement of the high-pressure operation rate. It can also effectively control the gas flow

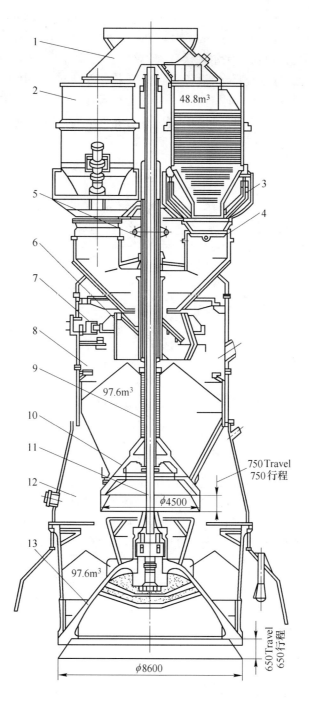

Figure 1-35 Double bell and double valve top charging equipment
图 1-35 双钟双阀式炉顶装料设备

1—Belt chute; 2—Storage hopper; 3—Gate; 4—Disc valve; 5—Spreader drive; 6—Spreader;
7—Stop roller; 8—Small hopper; 9—Small bell rod; 10—Small bell; 11—Big bell pole; 12—Big hopper; 13—Big bell
1—皮带溜槽;2—储料斗;3—闸门;4—盘式阀;5—布料器传动装置;6—布料器;7—挡辊;8—小料斗;
9—小钟杆;10—小料钟;11—大钟杆;12—大料斗;13—大料钟

distribution in the furnace and create conditions for the smooth operation of the blast furnace. Bell less top charging equipment is used in most of the newly built 1000m³ or above blast furnaces.

20世纪70年代，卢森堡保尔·沃特（PW）公司推出了PW型无料钟炉顶装料设备，如图1-36所示。无料钟炉顶装料设备自问世以来之所以发展迅速，是因为它不仅布料手段多、布料灵活、为高炉上部调剂增加了手段，而且还为高炉炉顶实现高压操作、提高高压作业率提供了保证，有效地控制炉内煤气流分布，为高炉顺行创造了条件。绝大部分新建1000m³级以上的高炉均采用无料钟炉顶装料设备。

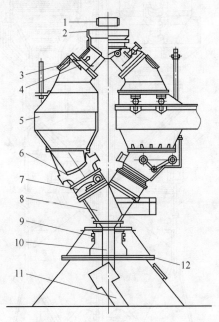

Figure 1-36 Top charging equipment of parallel tank type bell less furnace
图1-36 并罐式无料钟炉顶装料设备

1—Belt conveyor；2—Receiving funnel；3—Discharging net valve；4—Upper sealing valve；5—Material skip；
6—Material flow regulating valve；7—Lower sealing valve；8—Fork pipe；9—Central pipe；10—Distributor；
11—Rotary slug；12—Steel ring

1—胶带机；2—受料漏斗；3—排料网阀；4—上密封阀；5—料跳；6—料流调节阀；7—下密封阀；8—叉形管；
9—中心喉管；10—布料器；11—旋转溜槽；12—钢圈

1.2.3.3 Air Supply System
1.2.3.3 送风系统

The task of the air supply system is to supply the hot air needed for blast furnace smelting in a timely, continuous, stable and reliable manner. Its main equipment includes blast furnace blower, hot blast furnace, waste gas waste heat collection device, hot air pipe, cold air pipe, and control valves on the cold and hot air pipes.

送风系统的任务是及时、连续、稳定、可靠地供给高炉冶炼所需热风，其主要设备包括高炉鼓风机、热风炉、废气余热回收装置、热风管道、冷风管道以及冷、热风管道上的控制阀门等。

Blower

鼓风机

Blast furnace blower is the most important power equipment in blast furnace smelting. It not only provides the oxygen needed for the blast furnace smelting directly, but also provides the necessary power for the gas flow in the furnace to overcome the resistance movement of the material column. Blast furnace blower is the heart of blast furnace.

高炉鼓风机是高炉冶炼最重要的动力设备。它不仅可以直接为高炉冶炼提供所需要的氧气，而且还可以为炉内煤气流克服料柱阻力运动提供必需的动力。高炉鼓风机是高炉的心脏。

There are three types of blast furnace blowers commonly used: centrifugal type, axial flow type (shown in Figure 1-37) and constant volume type.

常用高炉鼓风机的类型有离心式、轴流式（见图 1-37）和定容式三种。

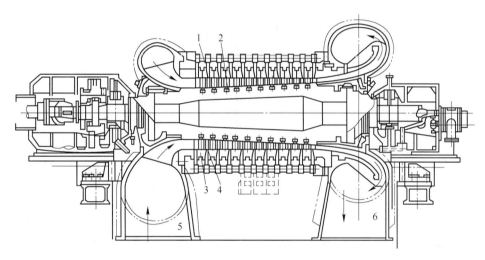

Figure 1-37   Axial flow blower

图 1-37   轴流式鼓风机

1—Casing; 2—Rotor; 3—Working blade; 4—Guide blade; 5—Suction port; 6—Exhaust port

1—机壳；2—转子；3—工作叶片；4—导流叶片；5—吸气口；6—排气口

Hot Blast Stove

热风炉

Hot blast furnace is the heating equipment of blast furnace hot blast, which is essentially a heat exchanger. In modern blast furnace, regenerative hot blast furnace is widely used. In order to ensure continuous air supply to blast furnace, three or four hot blast furnaces are usually equipped for each blast furnace. The size of the hot blast furnace and the size of each part depend on the required blast temperature and air volume.

热风炉是高炉热风的加热设备，其实质是一个热交换器。现代高炉普遍采用蓄热式热风炉。由于燃烧和送风交替进行，为保证向高炉连续送风，通常每座高炉配置三座或四座热风炉。热风炉的大小及各部位尺寸取决于高炉所需的风温及风量。

According to the different layout of combustion chamber and regenerator, the hot blast furnace can be divided into three basic structural forms, namely internal combustion hot blast furnace (shown in Figure 1-38), external combustion hot blast furnace (shown in Figure 1-39) and top combustion hot blast furnace (shown in Figure 1-40). The brief introductions of the working principle of combustion hot blast furnace are as follows:

根据燃烧室和蓄热室布置形式的不同, 热风炉分为三种基本结构形式, 即内燃式热风炉 (见图 1-38)、外燃式热风炉 (见图 1-39) 和顶燃式热风炉 (见图 1-40)。其工作原理以内燃式热风炉为例, 如下所示:

(1) Combustion chamber and regenerator are built in the same furnace shell, and there is no partition between them.

(1) 燃烧室和蓄热室砌在同一炉壳内, 它们之间没有隔墙。

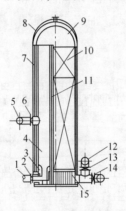

Figure 1-38 Internal combustion hot blast furnace
图 1-38 内燃式热风炉

1—Gas pipe; 2—Gas valve; 3—Burner; 4—Combustion chamber; 5—Hot air pipe; 6—Hot air valve; 7—Big wall; 8—Furnace shell; 9—Vault; 10—Regenerator; 11—Partition wall; 12—Cold air pipe; 13—Cold air valve; 14—Flue valve; 15—Furnace grate and pillar

1—煤气管道; 2—煤气阀; 3—燃烧器; 4—燃烧室; 5—热风管道; 6—热风阀; 7—大墙; 8—炉壳; 9—拱顶; 10—蓄热室; 11—隔墙; 12—冷风管道; 13—冷风阀; 14—烟道阀; 15—炉箅子和支柱

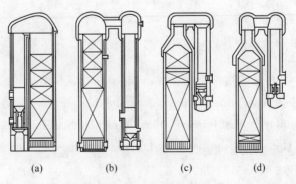

(a) (b) (c) (d)

Figure 1-39 Structure diagram of external combustion hot blast furnace
图 1-39 外燃式热风炉结构示意图

(a) Dide style; (b) Cowbay style; (c) Horse piano style; (d) Nippon Steel style
(a) 地得式; (b) 考贝式; (c) 马琴式; (d) 新日铁式

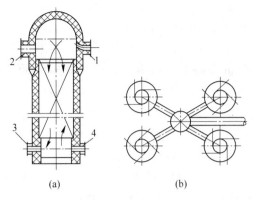

Figure 1-40　Top burning hot blast furnace
图 1-40　顶燃式热风炉
(a) Structure diagram; (b) Layout plan
(a) 结构示意图; (b) 平面布置图
1—Burner; 2—Hot air outlet; 3—Flue gas outlet; 4—Cold air inlet
1—燃烧器; 2—热风出口; 3—烟气出口; 4—冷风入口

(2) The gas and air are sent to the burner through the valve to the pipeline and burned in the combustion chamber. The hot flue gas moves upward, changes the direction through the vault, and passes through the regenerator downward. then it enters the flue, and is discharged into the atmosphere through the chimney.

(2) 煤气和空气由管道经阀门送入燃烧器并在燃烧室内燃烧，燃烧的热烟气向上运动，经拱顶改变方向，向下穿过蓄热室，然后进入烟道，经烟囱排入大气。

(3) When the hot flue gas passes through the regenerator, the lattice bricks in the regenerator are heated. After the lattice brick is heated and stored with a certain amount of heat, the hot blast furnace stops burning and turns to air supply. Air supply means that the cold air enters the regenerator through the cold air valve from the lower cold air pipeline.

(3) 在热烟气穿过蓄热室时，将蓄热室内的格子砖加热。格子砖被加热并蓄存一定热量后，热风炉停止燃烧，转入送风。送风是指使冷风从下部冷风管道经冷风阀进入蓄热室。

(4) Air is heated by lattice brick, and enters combustion chamber through vault. then it is sent to blast furnace through hot air outlet, hot air valve and hot air main pipe.

(4) 空气通过格子砖被加热，经拱顶进入燃烧室，再经热风出口、热风阀、热风总管送至高炉。

1.2.3.4　Gas Dedusting System
1.2.3.4　煤气除尘系统

The task of gas dedusting system is to dedust and cool down the blast furnace gas, so as to meet the user's requirements for gas quality.

煤气除尘系统的任务是对高炉煤气进行除尘降温处理，以满足用户对煤气质量的要求。

Blast furnace smelting produces a lot of gas. Blast furnace gas contains combustible gas components such as CO, H_2 and CH_4. Its calorific value is generally $3350 \sim 4200 kJ/m^3$, which can be used as fuel for hot blast furnace, sintering ignition and boiler. However, the blast furnace gas without dust removal contains $10 \sim 30 g/m^3$ (up to $60 \sim 100 g/m^3$) of dust. If it is directly used, it will not only block the pipeline during transportation, but also erode and destroy the refractory brick lining of hot blast furnace and burner. Therefore, blast furnace gas can only be used as fuel after dedusting.

高炉冶炼产生大量煤气。高炉煤气中含有 CO、H_2 和 CH_4 等可燃气体成分,其发热值一般为 $3350 \sim 4200 kJ/m^3$,可作为热风炉、烧结点火和锅炉的燃料。但是,未经除尘的高炉煤气中含有 $10 \sim 30 g/m^3$(高的可达 $60 \sim 100 g/m^3$)的灰尘,若直接使用,不仅会在运送时堵塞管道,而且会使热风炉和燃烧器等的耐火砖衬被侵蚀破坏。因此,高炉煤气必须除尘后才能作为燃料使用。

The blast furnace gas becomes clean gas after dedusting. In order to improve the heating value of the purified gas, facilitate the transportation and ensure the combustion safety of users, it is generally required that the dust content of the purified gas is less than $10 mg/m^3$, the temperature is less than 35℃, the mechanical water content is less than $30 g/m^3$, and the pressure is greater than 8000Pa. The problem of generating electricity and recovering energy by using residual pressure of gas should also be considered for the net gas on the top of high pressure furnace.

高炉煤气除尘后变为净煤气。为了提高净煤气的发热值、方便输送及保证用户燃烧安全,一般要求净煤气含尘量小于 $10 mg/m^3$,温度低于 35℃,机械水含量小于 $30 g/m^3$,压力大于 8000Pa。高压炉顶的净煤气还应考虑利用煤气余压发电和回收能源问题。

At present, there are mainly two dedusting processes for blast furnace gas: wet process and dry process. The equipment of wet dedusting system includes gravity deduster, washing tower, venturi, dehydrator, electrostatic precipitator, high-pressure valve group, etc. And the turbine is also included when there is gas residual pressure power generation. The equipment of gas dedusting system with dry dedusting mainly includes gravity deduster, bag box or plate type electric deduster. These different forms of dust removal equipment are used to facilitate the removal of dust of different particle sizes in the gas.

目前,高炉煤气除尘工艺主要有湿法和干法两种。湿法除尘系统的设备包括重力除尘器、洗涤塔、文氏管、脱水器、电除尘器、高压阀组等,有煤气余压发电的还包括透平机。采用干法除尘的煤气除尘系统,其设备主要包括重力除尘器、布袋箱体或板式电除尘器。采用这些不同形式的除尘设备,有利于清除煤气中不同粒级的灰尘。

1.2.3.5　Slag Iron Treatment System
1.2.3.5　渣铁处理系统

The task of the slag iron treatment system is to treat the slag and iron discharged from the blast furnace in time to ensure the normal operation of production. Its main equipment includes the taphole opener, the taphole mud gun, the molten iron tank car, the taphole stopper, the slag granulation device, the slag pool and the slag filtration device.

渣铁处理系统的任务是及时处理高炉排出的渣和铁,从而保证生产的正常进行。其主要设备包括开铁口机、堵铁口泥炮、铁水罐车、堵渣口机、炉渣粒化装置、水渣池及水渣过滤装置等。

A tuyere platform and a tapping field are arranged below the horizontal surface of the blast furnace damper and the tap hole. Slag discharge ditch is arranged on the air outlet platform, and iron ditch and slag discharge ditch are arranged on the iron field. The cast-in-place is also provided with a traveling crane and a smoke dust-proof device. There are air changer under the hot air shroud or on the air outlet platform. At present, the general process of blast furnace slag iron treatment is shown in Figure 1-41.

在高炉风门和出铁口水平面以下设置有风口平台和出铁场。在风口平台上布置有出渣沟，在出铁场上布置有铁水沟和放渣沟。在出铁场还设置有行车和烟气防尘装置。在热风围管下或风口平台上设有换风口机等。目前高炉渣铁处理的一般流程如图1-41所示。

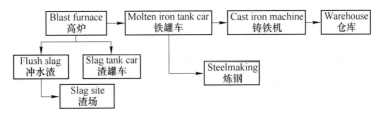

Figure 1-41　Flow chart of blast furnace slag iron treatment system
图1-41　高炉渣铁处理系统流程图

Tuyere Platform and Casting Yard
风口平台和出铁场

In the lower part of the blast furnace, the working platform set below the tuyere plane around the blast furnace hearth is the air gate platform. The operator shall observe the furnace condition through the tuyere, replace the tuyere, discharge the slag, maintain the slag port and slag ditch, check the cooling equipment and operate some valves, etc. For the convenience of operation, the air outlet platform is generally 1150~1250mm lower than the center line of the air outlet. And it shall be kept flat except for the upper slag ditch, and only the discharge slope shall be reserved.

在高炉下部，沿高炉炉缸周围风口平面以下设置的工作平台为风门平台。操作人员要通过风口观察炉况，更换风口、放渣、维护渣口和渣沟、检查冷却设备以及操纵一些阀门等。为了操作方便，风口平台一般比风口中心线低1150~1250mm，除上渣沟部位外应保持平坦，只留泄水坡度。

The casting yard is a working platform in front of the furnace for arranging iron trench and slag chute, installing equipment in front of the furnace, discharging slag and tapping operation. Due to the different elevation of the taphole and slag mouth, the taphole is generally about 1500mm lower than the tuyere. The area of the taphole depends on the layout of the slag iron ditch and the needs of the operation in front of the furnace. The length of the large and medium-sized blast furnace is 40~60m. The width is 15~25m, and the height is required to ensure that the lower edge of any slag iron flow nozzle is not less than 5m, so that the slag iron tank car can pass through. Above the iron field is arranged an iron ditch and slag ditch. The area of the main iron ditch should be kept flat, and the rest parts can be made into a slope from the center to both sides and from the iron mouth to the end consistent with the strike of the slag iron ditch. Generally, there is only one tapping site for medium and small blast furnaces, and two or three tapping sites for large blast furnaces.

出铁场是布置铁沟和下渣沟、安装炉前设备、进行放渣和出铁操作的炉前工作平台。由于铁口和渣口标高不同,出铁场一般比风口低约1500mm。出铁场的面积取决于渣铁沟的布置和炉前操作的需要,大中型高炉长度为40~60m,宽度为15~25m,高度则要求能保证任何一个渣铁流嘴下沿不低于5m,以便渣铁罐车通过。出铁场上面布置有出铁沟和下渣沟。在主铁沟区域应保持平坦,其余部分由中心向两侧和由出铁口向端部随渣铁沟走向一致的坡度。中小型高炉一般只有一个出铁场,大型高炉有2个或3个出铁场。

Molten Iron Treatment

铁水处理

Most of the molten iron produced by blast furnace is sent to steelmaking plant for steelmaking, and a little part is used to cast iron block. Hot metal is transported by hot metal tank car.

高炉生产的铁水绝大部分送往炼钢厂进行炼钢,小部分用于铸成铁块。铁水采用铁水罐车进行运输。

Slag Disposal

炉渣处理

The treatment methods of blast furnace slag depend on the choices of the utilization way. At present, water quenching treatment is widely used, followed by dry slag utilization. In addition, a small amount of slag is used to produce slag cotton and other purposes.

高炉炉渣的处理方法取决于对其利用途径的选择。目前广泛采用的是水淬处理,其次是干渣块利用,此外还有少量炉渣用于生产渣棉及其他用途。

1.2.3.6 Injection system

1.2.3.6 喷吹系统

The main task of the injection system is to evenly and stably inject pulverized coal into the blast furnace, so as to promote the energy saving and consumption reduction of blast furnace production.

喷吹系统的主要任务是均匀、稳定地向高炉喷吹煤粉,促进高炉生产的节能降耗。

Blast furnace fuel injection is a technology of injecting fuel into the hearth through the tuyere when adopting the direction of high air temperature and oxygen enriched blast. Its development has strengthened the competition between blast furnace ironmaking process and new non blast furnace ironmaking process, eased the pressure of ironmaking production limited by resources, investment, cost, energy, environment, transportation and other aspects. the fuel has become the core of ironmaking system process structure optimization and energy structure change. The fuel injected by blast furnace includes natural gas, coke oven gas, heavy oil, coke powder, coal, etc. At present, the main blast furnace is coal injection.

高炉喷吹燃料是在采用高风温和富氧鼓风的同时，通过风口向炉缸喷吹燃料的技术。它的发展增强了高炉炼铁工艺与新型非高炉炼铁工艺竞争的力量，缓解了炼铁生产受到资源、投资、成本、能源、环境、运输等多方面限制的压力，已成为炼铁系统工艺结构优化、能源结构变化的核心。高炉喷吹的燃料有天然气、焦炉煤气、重油、焦粉、煤等。目前高炉主要以喷煤为主。

The blast furnace coal injection system is composed of raw coal storage and transportation, coal preparation, coal transportation, injection, dry gas preparation and power gas supply systems. The process flow is shown in Figure 1-42.

高炉喷煤系统由原煤储运、煤料制备、煤料输送、喷吹、干燥气体制备和动力供气等系统组成，工艺流程如图1-42所示。

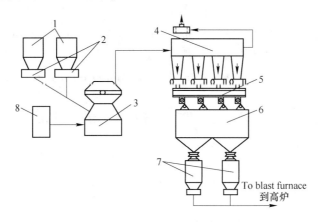

Figure 1-42　Process flow of blast furnace coal injection system
图 1-42　高炉喷煤系统工艺流程

1—Raw coal bin；2—Belt scale；3—Coal mill；4—Air box bag type powder collector；5—Scraper；6—Coal powder bin；
7—Injection tank；8—Flue gas furnace
1—原煤仓；2—皮带秤；3—磨煤机；4—气箱式布袋收粉器；5—刮板机；6—煤粉仓；7—喷吹罐；8—烟气炉

Raw coal storage and transportation system is to transport raw coal to the coal storage yard for storage, control drying, mixing, etc. And then it is sent to the raw coal bunker with a belt conveyor.

原煤储运系统是将原煤运至储煤场进行存放、控干、混匀等，然后用皮带机将其送入原煤仓内。

The pulverized coal preparation system (pulverizing system) is to make the raw coal into dry pulverized coal through the pulverizer, and then separate the pulverized coal from the dry gas and store it in the pulverized coal bunker.

煤粉制备系统（制粉系统）是将原煤经过磨煤机制成干燥煤粉后，再将煤粉从干燥气中分离出来存入煤粉仓内。

The pulverized coal conveying and injection system is to inject pulverized coal into the blast furnace through the conveying pipeline and spray gun by pressurizing in the injection tank.

煤粉输送和喷吹系统是通过在喷吹罐内加压，将煤粉经输送管道和喷枪喷入高炉。

The dry gas preparation system is to send the blast furnace gas into the combustion furnace for combustion, and the generated hot flue gas is sent to the coal powder preparation system as the dry gas.

干燥气体制备系统,是将高炉煤气等送入燃烧炉内进行燃烧,生成的热烟气送入煤粉制备系统作为干燥气。

Power supply system refers to the compressed air, oxygen, nitrogen and steam required for the whole coal injection system.

动力供气系统是指供给整个喷煤系统所需的压缩空气、氧气、氮气及蒸汽等。

Exercises
思考题

(1) What are the functions of the blast furnace smelting fuel, and what kinds are there?

(1) 高炉冶炼用燃料的作用是什么,有哪些种类?

(2) What kinds of iron ore are commonly used in blast furnace smelting, and what metallurgical characteristics do they have?

(2) 高炉冶炼中常用的铁矿石种类有哪几种,各有什么冶金学特征?

(3) What is the purpose of iron ore powder?

(3) 铁矿粉造块的目的是什么?

(4) What are the products of blast furnace smelting products?

(4) 高炉冶炼的产品有哪些?

(5) What happens to all parts of the blast furnace?

(5) 高炉内各部分都发生什么样的反应?

(6) What are the equipment of blast furnace smelting, and what role does each have?

(6) 高炉冶炼设备有哪些,各有什么作用?

Project 2　Metallurgy of Steel
项目2　钢冶金

　　Steelmaking is an engineering science that studies the heating and melting of blast furnace molten iron (pig iron), direct reduction iron (DRI), hot pressed block iron (HBI) or scrap iron (iron), removing harmful impurity elements from molten iron through chemical reaction, adding alloy and pouring into semi-finished products (billet or ingot). With the development of steelmaking technology for more than 200 years, its technical level and automation level have been greatly improved. At present, steelmaking is developing in the direction of high efficiency, high cleanliness and high quality billet. This chapter briefly introduces the basic theory of steelmaking, converter steelmaking, electric furnace steelmaking, off furnace refining and continuous cast steel.

　　炼钢学是研究将高炉铁水（生铁）、直接还原铁（DRI）、热压块铁（HBI）或废钢（铁）加热、熔化，通过化学反应去除铁液中的有害杂质元素，配加合金并浇注成半成品（铸坯或钢锭）的工程科学。炼钢技术经过200多年的发展，其技术水平、自动化程度得到很大提高，当前炼钢正向高效、高洁净度、高质量铸坯方向发展。本章简要介绍有关炼钢的基本理论、转炉炼钢、电炉炼钢、炉外精炼及连续铸钢方面的知识。

Task 2.1　Basic Theory of Steelmaking
任务2.1　炼钢基础理论

　　The steelmaking process is essentially a comprehensive process of many very complex high-temperature physicochemical transformations, which involves a variety of components in different aggregation states, such as solid state (furnace charge, auxiliary material and furnace lining), liquid state (liquid metal and slag) and gas state (furnace gas, air or oxygen blown into the metal, etc.).

　　炼钢过程实质上是许多非常复杂的高温物理化学转变的综合过程。它涉及多种以不同聚集状态存在的组元，如固态（炉料、辅助材料及炉衬等）、液态（液体金属及炉渣）和气态（炉气、吹入金属内的空气或氧气等）。

　　In the process of metallurgical reaction, the direction and limit of reaction change under certain conditions and the final product problem obtained are calculated by using the thermodynamics of metallurgical reaction, and the concentration, temperature and pressure are selected as the calculation parameters. The mechanism, rate of reaction process and its relationship with various factors are studied by using the kinetics of metallurgical reaction, and the measures to strengthen the metallurgical process are determined.

在冶金反应过程中，可以应用冶金反应热力学计算一定条件下反应变化的方向和限度，以及将得到的最终产物问题，一般选择浓度、温度和压力等作为计算参数，同时也可以应用冶金反应动力学研究反应过程的机理、速率以及其与各种因素的关系，确定强化冶金过程的措施。

2.1.1 Physical Properties of Liquid Steel
2.1.1 钢液的物理性质

2.1.1.1 Density of Liquid Steel
2.1.1.1 钢液的密度

The density of liquid steel refers to the mass per unit volume of liquid steel in kg/m^3. The main factors affecting the density of liquid steel are temperature and chemical composition of liquid steel. Generally speaking, the increase of temperature and the decrease of density of molten steel are due to the increase of atomic distance. When the density of bulk pure iron is $7880kg/m^3$, the density of liquid at 1550℃ is $7040kg/m^3$, and the change of steel is similar to that of pure iron.

钢液的密度是指单位体积钢液所具有的质量，单位通常采用 kg/m^3。影响钢液密度的因素主要有温度和钢液的化学成分，温度升高，钢液密度降低，其原因是因为原子间距增大。固体纯铁的密度为 $7880kg/m^3$，1550℃时液态纯铁的密度为 $7040kg/m^3$。其中，钢的变化与纯铁类似。

The influences of various metal and non-metal elements on the density of steel are different, and the influences of carbon are more complex.

各种金属和非金属元素对钢密度的影响不同。其中，碳对钢密度的影响较大且比较复杂。

2.1.1.2 Melting Point of Steel
2.1.1.2 钢的熔点

The melting point of steel refers to the temperature at which the solid begins to precipitate when the steel is completely transformed into a liquid state (or when condensed). And it is an important parameter to determine smelting and pouring temperature. The melting point of pure iron is about 1538℃. When an element dissolves, the force between pure iron atoms are weakened, and the melting point of iron decreased.

钢的熔点是指钢完全转变成液体状态时（或冷凝时）开始析出固体的温度，是确定冶炼和浇注温度的重要参数。纯铁的熔点约为1538℃。当某元素溶入后，纯铁原子之间的作用力减弱，铁的熔点降低。

2.1.1.3 Viscosity of Liquid Steel
2.1.1.3 钢液的黏度

Viscosity is an important property of molten steel, which has a great influence on the determination of smelting temperature parameters, the floatation of non-metallic inclusions, the removal of gas and the solidification and crystallization of steel. Internal friction will be produced between various layers of liquid moving at different speeds. Usually, the internal friction coefficient or viscosity coefficient is called viscosity. Generally, the symbol u is used to represent the dynamic viscosity in Pa·s (N·s/m² or 1P = 0.1Pa·s); and the symbol v is used to represent the kinematic viscosity in m²/s.

黏度是钢液的一个重要性质,它对冶炼温度参数的制定,非金属夹杂物的上浮和气体的去除,以及钢的凝固结晶都有很大影响。各种以不同速度运动的液体各层之间会产生内摩擦力,通常将内摩擦系数或黏度系数称为黏度。一般用符号 u 表示动力黏度,单位为 Pa·s (N·s/m² 或 P (泊)), 1P = 0.1Pa·s); 用符号 v 表示运动黏度,单位为 m²/s。

The viscosity of molten steel is much smaller than that of normal slag, and its value is 0.002~0.003Pa·s at 1600℃. The viscosity of pure iron liquid at 1600℃ is 0.0005Pa·s.

钢液的黏度比正常熔渣的黏度要小得多,1600℃时其值为 0.002~0.003Pa·s。纯铁液在1600℃时的黏度为 0.0005Pa·s。

The main factors affecting the viscosity of molten steel are temperature and composition. As the temperature increases, the viscosity decreases. The carbon in molten steel has a great influence on the viscosity, because the carbon content makes the density and melting point of steel change cause the viscosity change. At the same temperature, the fluidity of high carbon steel liquid steel is better than that of low carbon steel liquid steel.

影响钢液黏度的因素主要是温度和成分。温度升高,黏度降低。钢液中的碳对黏度的影响非常大,这主要是因为碳含量使钢的密度和熔点发生变化,从而引起黏度的变化。同一温度下,高碳钢钢液的流动性比低碳钢钢液的好。

The influences of non-metallic inclusions on the viscosity of molten steel are as follows:

非金属夹杂物对钢液黏度的影响有:

(1) When the content of non-metallic inclusions in molten steel increases, the viscosity of molten steel increases and the fluidity becomes worse.

(1) 钢液中非金属夹杂物含量增加,则钢液黏度增加,流动性变差。

(2) The deoxidizing products in molten steel also have a great influence on the fluidity. When the molten steel is deoxidized with Si, Al or Cr respectively, the inclusion content is high and the viscosity increases at the initial stage due to the formation of deoxidizing products. However, the viscosity decreases with the increasing of inclusions, or the formation of low melting point inclusions.

(2) 钢液中的脱氧产物对流动性的影响也很大,当钢液分别用 Si、Al 或 Cr 脱氧时,初期由于脱氧产物生成,夹杂物含量高,黏度增大。但随着夹杂物不断上浮,或形成低熔点夹杂物,黏度又会下降。

Therefore, if deoxidation is poor, the fluidity of molten steel is generally poor.

因此,如果脱氧不良,钢液的流动性一般不好。

2.1.1.4 Surface Tension of Liquid Steel

2.1.1.4 钢液的表面张力

The force that causes the surface of liquid steel to shrink is called the surface tension of liquid steel in N/m. In fact, the surface tension of molten steel refers to a force between molten steel and its saturated vapor or air interface.

使钢液表面产生缩小倾向的力称为钢液的表面张力,单位为 N/m。实际上,钢液的表面张力就是指钢液和它的饱和蒸汽或空气界面之间的一种力。

The surface tension of molten steel has an effect on the formation of new phase (such as the formation of CO bubble, the formation of crystal core in the solidification process of molten steel, etc.), the interphase reaction (such as the removal of inclusions and gases from molten steel), the separation of slag and steel, and the erosion of molten steel on refractories. There are many factors that affect the surface tension of liquid steel, such as temperature, composition of liquid steel and contact of liquid steel.

钢液的表面张力对新相的生成(如 CO 气泡的产生、钢液凝固过程中结晶核心的形成等)有影响,而且对相间反应(如夹杂物和气体从钢液中排除)、渣钢分离、钢液对耐火材料的侵蚀等也产生影响。影响钢液表面张力的因素很多,主要有温度、钢液成分和钢液的接触物。

The surface tension of liquid steel increases with the increase of temperature. At 1550℃, the surface tension of pure iron liquid is 1.7~1.9N/m. The effect of carbon on the surface tension of liquid steel is complicated. As the structure and density of steel change with the increase of carbon content, its surface tension also changes with the change of carbon content.

钢液的表面张力随着温度的升高而增大,1550℃时,纯铁液的表面张力为 1.7~1.9N/m。碳对钢液表面张力的影响出现复杂的关系,由于钢的结构和密度随着碳含量的增加而发生变化,所以它的表面张力也会随着碳含量的变化而变化。

2.1.1.5　Thermal Conductivity of Steel
2.1.1.5　钢的导热能力

The thermal conductivity of steel can be expressed by thermal conductivity: the heat flowing through unit area in unit time, when the unit temperature gradient maintained in the system. The thermal conductivity of steel is indicated by symbol λ in W/(m·℃).

钢的导热能力可用导热系数来表示:当体系内维持单位温度梯度时,在单位时间内流经单位面积的热量。钢的导热系数用符号 λ 表示,单位为 W/(m·℃)。

The main factors that affect the thermal conductivity of steel are the composition, structure, temperature of molten steel, the content of non-metallic inclusions and the degree of grain refinement.

影响钢导热系数的因素主要有钢液的成分、组织、温度,非金属夹杂物的含量,以及钢中晶粒的细化程度等。

Generally, the more alloy elements in steel, the worse the thermal conductivity of steel. In alloy steel, the order of influences of various alloy elements on the thermal conductivity of steel are: C, Ni, and Cr are the largest; Al, Si, Mn and W is the second; and Zr is the smallest. The thermal conductivity of alloy steel is generally worse than that of carbon steel, and the thermal conductivity of high carbon steel is worse than that of low carbon steel.

通常，钢中合金元素越多，钢的导热能力就越差。在合金钢中，各种合金元素对钢导热能力影响的次序为：C、Ni、Cr 最大，Al、Si、Mn、W 次之，Zr 最小。合金钢的导热能力一般比碳钢差，高碳钢的导热能力比低碳钢差。

The thermal conductivity of various steels varies with temperature. Below 800℃, the thermal conductivity of carbon steel decreases with the increase of temperature; and above 800℃, it slightly increases.

各种钢的导热系数随温度变化的规律不一样。800℃以下，碳钢的导热系数随温度的升高而下降；800℃以上则略有升高。

2.1.2 Physical and Chemical Properties of Slag
2.1.2 炉渣的物理化学性质

In the process of steelmaking, slag plays an important role. In order to obtain the liquid steel meeting the requirements (composition and temperature), improve the internal and surface quality of ingots and billets, it is necessary to have slag meet certain requirements to ensure the smooth progress of the whole steelmaking process. The structure of slag determines the physical and chemical properties of slag, and the physical and chemical properties of slag affect the chemical reaction equilibrium and reaction rate of steel-making. Therefore, in the process of steelmaking, the physical and chemical properties of slag must be controlled and adjusted.

在炼钢过程中，炉渣起着极为重要的作用。为获得符合要求（成分和温度）的钢液，提高钢锭、钢坯的内部质量和表面质量，需要有符合一定要求的炉渣以保证炼钢全过程顺利进行。熔渣的结构决定着炉渣的物理化学性质，而熔渣的物理化学性质又影响着炼钢的化学反应平衡及反应速率。因此在炼钢过程中，必须控制和调整好炉渣的物理化学性质。

2.1.2.1 Functions and Compositions of Slag
2.1.2.1 炉渣的作用与组成

The compositions of steelmaking slag are mainly oxide, and having a small amount of fluoride, phosphide and sulfide. In the process of steelmaking, the roles of slag are mainly reflected in the following aspects:

炼钢炉渣主要是氧化物，还有少量氟化物、磷化物、硫化物。炼钢过程中，炉渣的作用主要体现在以下几方面：

(1) Control the redox reaction of molten steel.
(1) 控制钢液的氧化还原反应。
(2) Remove impurities (S and P) and absorb inclusions.
(2) 脱除杂质（S、P），吸收夹杂物。
(3) Prevent liquid steel from inhaling.
(3) 防止钢液吸气。
(4) Prevent heat dissipation of molten steel to ensure smelting temperature of steel.
(4) 防止钢液散热，以保证钢的冶炼温度。
(5) Stable arc combustion (for EAF and ladle furnace).

（5）稳定电弧燃烧（对电弧炉和钢包炉）。

（6）Slag is a resistance heating body (remelting electroslag).

（6）炉渣是电阻发热体（对电渣重熔）。

（7）Prevent the secondary oxidation of molten steel (ladle/tundish covering agent, mould flux).

（7）防止钢液的二次氧化（钢包/中间包覆盖剂、结晶器保护渣）。

The slag also plays an unfavorable role in the process of steelmaking, mainly in the following aspects: erosion of refractory and reduction of lining life, especially the erosion of low alkalinity slag on the lining is more serious; small particles of metal and unreduced metal oxides are contained in the slag, reducing the metal recovery rate. Therefore, it is an important condition for steelmaking to make good slag. The slag with suitable composition, temperature and refining requirements shall be produced.

炉渣在炼钢过程中也有不利作用，主要表现在：侵蚀耐火材料，降低炉衬寿命，其中低碱度熔渣对炉衬的侵蚀更为严重；熔渣中夹带小颗粒金属及未被还原的金属氧化物，从而降低金属的收得率。因此，造好渣是炼钢的重要条件，应造出成分合适、温度适当并且能够满足某种精炼要求的炉渣。

2.1.2.2 Chemical Properties of Slag
2.1.2.2 炉渣的化学性质

Basicity of Slag
炉渣的酸碱性

The ratio of the total concentration of basic oxide in slag to the total concentration of acid oxide is called basicity of slag, which is usually represented by R. The basicity of slag directly affects the physical and chemical reactions (such as dephosphorization, desulfurization, degassing, etc.) between slag and steel.

炉渣中碱性氧化物浓度的总和与酸性氧化物浓度的总和之比称为炉渣碱度，常用符号 R 表示。炉渣碱度的大小，直接对渣—钢间的物理化学反应（如脱磷、脱硫、去气等）产生影响。

When $R<1.0$, the slag is acid slag. Because of the high content of SiO_2, it can be drawn into fine wire at high temperature, which is called long slag. After cooling, it is in the shape of black and bright glass. When $R>1.0$, it is alkaline slag, which is called short slag. Steel making slag is $R \geqslant 3.0$.

熔渣 $R<1.0$ 时为酸性渣，由于 SiO_2 含量高，高温下可拉成细丝，因此称为长渣，冷却后呈黑亮色玻璃状；$R>1.0$ 时为碱性渣，称为短渣。炼钢熔渣 $R \geqslant 3.0$。

There are different amounts of basic, neutral and acid oxides in the slag of steelmaking. The strength of their acidity and basicity can be arranged as follows:

炼钢熔渣中含有不同数量的碱性、中性和酸性氧化物，它们酸碱性的强弱程度可排列如下：

$$CaO>MnO>FeO>MgO>CaF_2>Fe_2O_3>Al_2O_3>TiO_2>SiO_2>P_2O_5$$

←Basic（碱性）   Neutral（中性）   Acid（酸性）→

Oxidation of Slag
炉渣的氧化性

The oxidizability of slag is also called the oxidizability of slag. It is an important chemical property of slag. The oxidizability of slag refers to the amount of oxygen supplied by slag to molten steel at a certain temperature. Under certain other conditions, the oxidation of slag determines dephosphorization, decarburization and inclusion removal of inclusion. Due to different decomposition pressures of oxides, only (FeO) and (Fe_2O_3) can transfer oxygen to steel, while (Al_2O_3), (SiO_2), (CaO) and (MgO) can't supply oxygen.

炉渣的氧化性也称炉渣的氧化能力,它是炉渣的一个重要的化学性质。炉渣的氧化性是指在一定的温度下,单位时间内炉渣向钢液供氧的数量。在其他条件一定的情况下,炉渣的氧化性决定了脱磷、脱碳以及夹杂物的去除等。由于氧化物分解压不同,只有(FeO)和(Fe_2O_3)才能向钢中供氧,而(Al_2O_3)、(SiO_2)、(CaO)、(MgO)等不能供氧。

2.1.3 Decarburization of Molten Steel
2.1.3 钢液的脱碳

Molten iron is an alloy solution of iron, carbon and some other impurities. Generally, the carbon content of steel-making pig iron is about 4% (by mass), and the carbon content of high phosphorus pig iron is about 3.6% (by mass). Decarburization is a main reaction throughout steelmaking. One of the important tasks of steelmaking is to remove the carbon in the molten pool to the extent required by the steel grade, and the decarburization reaction is also one of the effective means to promote the removal of gas and inclusions. Decarburization is closely related to the oxidation of other elements in steel-making process.

炼钢铁水是铁和碳以及其他一些杂质元素的合金溶液,一般炼钢生铁中的碳含量(质量分数)为4%左右,高磷生铁中的碳含量(质量分数)则为3.6%左右。脱碳反应是贯穿于炼钢始终的一个主要反应。炼钢的重要任务之一就是将熔池中的碳脱除到钢种所要求的程度,同时脱碳反应也是促进气体和夹杂物去除的有效手段之一。脱碳反应与炼钢过程中其他元素的氧化反应有密切的联系。

2.1.3.1 Thermodynamics of Decarbonization
2.1.3.1 脱碳反应的热力学

In modern large-scale steelmaking, oxygen blowing refining is an essential step. Based on the decarburization reaction in the molten pool, there are three basic forms of carbon oxygen reaction as follows:

现代大规模的炼钢生产,吹氧精炼是必不可少的步骤。综合熔池中的脱碳反应,碳氧反应存在如下三种基本形式:

(1) In the process of steelmaking by oxygen blowing, a part of carbon in the liquid metal is oxidized by gas in the reaction zone.

(1) 在吹氧炼钢过程中,金属液中的一部分碳在反应区被气体氧化。

(2) A part of carbon reacts with oxygen dissolved in liquid metal.

(2) 一部分碳与溶解在金属液中的氧进行氧化反应。

(3) A part of carbon reacts with (FeO) in slag to form CO.

(3) 剩下一部分碳与炉渣中的 (FeO) 反应，生成 CO。

At high temperature, [C] is mainly oxidized to CO, and the reaction between [C] and oxygen is as follows.

在高温下，[C] 主要氧化为 CO，[C] 与氧的反应如下：

(1) A part of carbon in the reaction zone at the gas—gold interface is directly oxidized by gas oxygen, expressed by:

(1) 一部分碳在气—金界面上的反应区被气体氧直接氧化，其反应式为：

$$[C]+\frac{1}{2}[O_2] = \{CO\} \qquad (2-1)$$

(2) When the carbon content in the molten pool is high, CO_2 is also the oxidant of carbon, expressed by:

(2) 当熔池中碳含量高时，CO_2 也是碳的氧化剂，其反应式为：

$$[C]+[CO_2] = 2\{CO\} \qquad (2-2)$$

(3) Some of the carbon reacts with the dissolved oxygen in the metal or the oxygen in the slag, mainly on the slag gold interface, expressed by:

(3) 一部分碳与金属中溶解的氧或渣中的氧发生反应，其主要在渣—金界面上发生，反应式为：

$$[C]+(FeO) = \{CO\}+Fe \qquad (2-3)$$

$$[C]+[O] = \{CO\} \qquad (2-4)$$

Among the above reaction formulas, the oxidation capacity of reaction (2-1) is the largest, so it is not suitable in the pool requiring weak oxidation, and it is beneficial to the pool requiring strong oxidation. The oxidation capacity of reaction (2-2) is slightly stronger than that of reactions (2-3) and (2-4). The oxidation capacity of CO_2 is stronger than that of [O]. However, the decarburization reaction (2-2) produces two of CO. If CO_2 is blown into the molten steel, the CO_2 bubble will not only react with [C], but also be thinly released by the product CO, and the concentration will be reduced quickly. CO_2 is a weak oxidant, which has a large stirring capacity for molten steel. When decarburized, it can reduce the loss of Cr and Mn (compared with blowing O_2), and play a role of decarburization and chromium retention.

在上述反应式中，反应式 (2-1) 的氧化能力最大，因此在需要弱氧化的熔池中不宜适用，但适用于需要强氧化的熔池。反应式 (2-2) 的氧化能力比反应式 (2-3)、式 (2-4) 稍强一些，即 CO_2 的氧化能力比 [O] 强。但是脱碳反应式 (2-2) 产生两个 CO，如果向钢液吹入 CO_2，则 CO_2 气泡不仅与 [C] 反应减少，而且还被产物 CO 稀释，浓度降低比较快。CO_2 是一个弱氧化剂，对钢液搅拌能力大，用它脱碳时可减少易氧化元素 Cr、Mn 的损失（与吹入 O_2 相比），起到脱碳保铬的作用。

From the above analysis, It can be seen that the factors affecting the decarbonization reaction are:

从以上分析可知，影响脱碳反应的因素是：

(1) temperature. The direct oxidation of [C] is exothermic, while the indirect oxidation of [C] with (FeO) is endothermic. When ore decarburization is used, due to the heat absorption of ore decomposition and reaction, ore should be added at high temperature, and ore can only be added above 1480℃; and when oxygen blowing decarburization is used, no special requirements are made for temperature.

(1) 温度。[C] 的直接氧化反应是放热反应，[C] 与 (FeO) 的间接氧化反应是吸热反应。当采用矿石脱碳时，由于矿石分解、反应时吸热，因此应在高温下加入矿石，即在 1480℃ 以上时才可加入矿石；而采用吹氧脱碳时，对温度不做特殊要求。

(2) Partial pressure of CO in gas phase. Reducing the partial pressure of CO in the gas phase is beneficial to the further oxidation of [C]. The vacuum decarburization of molten steel is based on this principle.

(2) 气相中 CO 的分压。降低气相中 CO 的分压有利于 [C] 的进一步氧化，钢液的真空脱碳就是依据这一原理进行的。

(3) Oxidation. Strong oxidability can provide oxygen source for decarburization, which is conducive to the process of decarburization.

(3) 氧化性。强的氧化性可以给脱碳反应提供氧源，有利于脱碳反应的进行。

2.1.3.2 Roles of Decarbonization
2.1.3.2 脱碳反应的作用

Carbon oxygen reaction is an extremely important reaction in the process of steelmaking. In modern oxygen converter steelmaking, the main smelting reaction is to remove carbon from molten iron, which runs through the whole steelmaking process. The oxidation reaction of carbon has the following functions in the steelmaking process:

碳氧反应是炼钢过程中极其重要的反应，在现代氧气转炉炼钢中，主要的冶炼反应是除去铁水中的碳，该反应贯穿于整个炼钢过程。碳的氧化反应在炼钢过程中具有如下多方面的作用：

(1) Promote uniformity of both composition and temperature. When CO floats up and is discharged, the pool will generate strong boiling and agitation, which strengthens the heat and mass transfer, and promotes the uniformity of composition and temperature.

(1) 促进熔池成分和温度均匀。CO 上浮排出时，熔池产生强烈沸腾和搅拌，从而强化了热量和质量传递，促进了成分和温度均匀。

(2) The interface between steel and slag is enlarged, and the chemical reaction speed is increased. The strong boiling and stirring of molten pool increase the contact area of slag—gold reaction, which is conducive to the chemical reaction. A large number of CO bubbles pass through the slag layer, which is an important reason for foaming slag and Gas—Slag—Gold three-phase emulsification.

(2) 加大钢—渣界面，提高了化学反应速度。熔池的强烈沸腾和搅拌增加了渣—金反应接触面积，有利于化学反应的进行。大量的 CO 气泡通过渣层是产生泡沫渣和气—渣—

金三相乳化的重要原因。

(3) It is beneficial to the floatation of non-metallic inclusions and the discharge of harmful gases, and reduces the gas content and the number of inclusions in steel. The partial pressure of H_2 and N_2 in CO bubble is very low. For these gases, CO bubble is a small vacuum chamber. Small inclusions will adhere to the surface of CO bubbles and float away, so as to improve the quality of steel.

(3) 有利于非金属夹杂物的上浮和有害气体的排出,降低了钢中气体含量和夹杂物数量。CO 气泡中 H_2 和 N_2 的分压极低,对这些气体来说,CO 气泡是一个小真空室。小颗粒夹杂物会附着在 CO 气泡的表面上浮排除,从而提高钢的质量。

(4) Decarbonization is closely related to other reactions in steelmaking. The oxidation of slag and oxygen content in steel are also affected by decarburization.

(4) 脱碳反应与炼钢中其他反应有着密切的联系。熔渣的氧化性、钢中氧含量等也受脱碳反应的影响。

(5) Cause splashes and spills. The uneven removal of CO bubbles and the rise of molten pool are the main causes of splashing and spilling. Of course, it is closely related to the instability of production operation, which will lead to metal loss and excessive consumption of slagging materials.

(5) 造成喷溅和溢出。CO 气泡排除不均及其造成的熔池上涨,是产生喷溅和溢出的主要原因。当然,这与生产操作不稳定关系很大,会导致金属损失和造渣材料消耗过大。

(6) It is beneficial to the formation of slag.

(6) 有利于熔渣的形成。

(7) Exothermic heating.

(7) 放热升温。

2.1.4 Dephosphorization of Liquid Steel
2.1.4 钢液的脱磷

Dephosphorization is one of the important tasks in steelmaking process. The content of P in molten iron varies with the conditions of iron ore raw materials. The content of P in low phosphorus molten iron is below 0.12% (by mass), while the content of P in high phosphorus molten iron is above 2.0% (by mass). P is a harmful element in most steel grades. Generally, the content of P in molten steel is lower than 0.03% (by mass) or lower, and the content of P in free cutting steel is not more than 0.08%~0.12% (by mass). Phosphorus can be oxidized and reduced in the process of steelmaking, and it will return to phosphorus more or less when tapping. Therefore, it is an important and complex work to control the dephosphorization reaction in the process of steelmaking.

脱磷是炼钢过程的重要任务之一。铁水的 P 含量因铁矿原料条件的不同而不同,低磷铁水的 P 含量(质量分数)在 0.12%以下,高磷铁水的 P 含量(质量分数)则高达 2.0%以上。在绝大多数钢种中 P 都属于有害元素,一般要求钢水 P 含量(质量分数)低于 0.03%或更低,易切削钢中的 P 含量(质量分数)也不得超过 0.08%~0.12%。炼钢过程中，P 既可以被氧化又可以被还原，出钢时或多或少都会发生回磷现象，因此，控制炼钢过程中的脱磷反应是一项重要而又复杂的工作。

2.1.4.1　Effects of Phosphorus on Steel Properties
2.1.4.1　磷对钢材性能的影响

P is a harmful element in steel, which is easy to cause cold embrittlement, especially when carbon content high. The reason is that the effect of solution strengthening formed by the enrichment of phosphorus atoms on the ferrite grain boundary, which results in the increase of the strength between grains and the brittleness.

P 在钢中是有害元素，易使钢发生冷脆现象，碳含量高时影响更明显。其原因是 P 原子富集在铁素体晶界上形成固溶强化作用，造成晶粒间的强度提高，从而产生脆性。

The effects of P on the impact toughness of steel are: that the higher the content of P is, the easier it is to precipitate phosphide at the crystal boundary; reduce the impact value of steel, and make the impact toughness of steel drop sharply at room temperature.

P 对钢的冲击韧性的影响是：P 含量越高，越易在结晶边界析出磷化物；降低钢的冲击值，在室温下使钢的冲击韧性急剧下降。

The beneficial effects of P are that in some steels, P is added in the form of alloy elements, such as shell steel and corrosion-resistant steel. In addition to Cu, the content of P less than 0.1% (by mass) can also be added to increase the atmospheric corrosion resistance of steel.

P 的有益影响是：在某些钢中 P 以合金元素的形式加入，如炮弹钢、耐蚀钢中；除含有 Cu 外，还可加入小于 0.1%（质量分数）的 P，以增加钢的抗大气腐蚀的能力。

2.1.4.2　Oxidative Dephosphorization
2.1.4.2　氧化脱磷

In the process of converter and EAF steelmaking, the main dephosphorization method is oxidative dephosphorization. Oxidative dephosphorization means that dephosphorization is carried out under the condition of oxidizing atmosphere or adding oxidant. P in metal is oxidized to +5 valence and fixed in the slag in the form of phosphate. The stable form of P in steel is Fe_2P, followed by Fe_3P. Fe_3P decomposes at 1166℃ in solid state, so it does not exist in liquid state. P in molten iron exists as Fe_2P, but [P] can also be treated as a component dissolved in molten iron during thermodynamics and kinetics analysis. According to the phase diagram of $CaO-P_2O_5$, there are $4CaO \cdot P_2O_5$ and $3CaO \cdot P_2O_5$ near CaO, but $4CaO \cdot P_2O_5$ will decompose before melting. Therefore, there are mainly $3CaO \cdot P_2O_5$ in the slag, and $4CaO \cdot P_2O_5$ is the second. At present, the product of dephosphorization can be written as $4CaO \cdot P_2O_5$ or $3CaO \cdot P_2O_5$. According to the ion theory of slag, phosphorus exists in the form of PO_4^{3-}.

在转炉和电弧炉炼钢过程中，最主要的脱磷方法是氧化性脱磷。氧化脱磷是指脱磷处理在氧化气氛或添加氧化剂的条件下进行，金属中的磷被氧化为+5 价，以磷酸盐的形式固定在熔渣中。磷在钢中存在的稳定形式是 Fe_2P，其次是 Fe_3P。Fe_3P 在固态下于 1166℃ 时分解，故在液态中不存在。铁液中 P 以 Fe_2P 存在，但在进行热力学和动力学分析时，也可将［P］作为溶于铁液中的一个组分对待。P 在渣中的存在形态由 $CaO-P_2O_5$ 相图可知，靠近 CaO 的一侧主要有 $4CaO \cdot P_2O_5$ 和 $3CaO \cdot P_2O_5$，但 $4CaO \cdot P_2O_5$ 在未熔化前将分解。因此，液渣中主要存在 $3CaO \cdot P_2O_5$，而 $4CaO \cdot P_2O_5$ 次之。目前，脱磷反应产物可写作 $4CaO \cdot P_2O_5$ 或 $3CaO \cdot P_2O_5$。炉渣的离子理论认为，磷以 PO_4^{3-} 形式存在。

In conclusion, the dephosphorization is carried out in the following way:
综上所述可以认为，脱磷反应是按以下方式进行的：

(1) Expression of molecular form, expressed by:
(1) 分子形式的表示法，其化学式为：

$$2[P]+5(FeO)+4(CaO) \Longrightarrow (4CaO \cdot P_2O_5)+5[Fe] \tag{2-5}$$

$$2[P]+5(FeO)+3(CaO) \Longrightarrow (3CaO \cdot P_2O_5)+5[Fe] \tag{2-6}$$

(2) Representation in the form of ions. According to the ion model of slag, it can be written as follows:
(2) 离子形式的表示法。按炉渣的离子模型可写成下式：

$$2[P]+5[O]+3(O^{2-}) \Longrightarrow 2(PO_4^{3-}) \tag{2-7}$$

Lime (CaO dephosphorization) is widely used in steelmaking. In recent years, soda slag and barium oxide slag have been used as dephosphorizing agents for the dephosphorization of ultra-low phosphorus steel and ferroalloy, and the products of dephosphorization are corresponding phosphates.

炼钢中大量使用石灰（CaO 脱磷）。近年来由于冶炼超低磷钢以及铁合金脱磷，苏打渣和氧化钡渣也相继用来作脱磷剂，脱磷产物为相应的磷酸盐。

The main factors affecting slag dephosphorization are as follows:
影响炉渣脱磷的主要因素如下：

(1) Basicity of slag. P_2O_5 is an acid oxide. CaO, MgO and other basic oxides can significantly reduce the activity coefficient of P_2O_5 in slag, so as to improve the dephosphorization capacity of slag and increase the distribution coefficient of phosphorus. The higher alkalinity is, the higher effective concentration of CaO in slag is, and the more complete dephosphorization is. However, the higher the basicity is, the better the basicity is, and the worse the slag is, the slag also becomes sticky, which affects its fluidity and is unfavorable to dephosphorization.

(1) 炉渣的碱度。P_2O_5 是酸性氧化物，CaO、MgO 等碱性氧化物可以显著降低炉渣中 P_2O_5 的活度系数，从而提高炉渣的脱磷能力和增加磷的分配系数。碱度越高，渣中 CaO 的有效浓度越高，脱磷越完全。但是，碱度并非越高越好，碱度高，化渣不好，炉渣变黏，影响其流动性，对脱磷反而不利。

(2) Oxidation of slag. The distribution coefficient of P can be increased by increasing the oxidizability of slag (i.e. the activity of FeO in slag). The content of FeO in slag plays an important role in dephosphorization. FeO in slag is the primary factor for dephosphorization, because P oxidizes first to form P_2O_5, and then reacts with CaO to form $4CaO \cdot P_2O_5$ and $3CaO \cdot P_2O_5$. FeO in slag can also form $3FeO \cdot P_2O_5$ at the early stage of smelting with low basicity, and the reaction is $3(FeO)+(P_2O_5) \Longrightarrow (3FeO \cdot P_2O_5)$. However, FeO has double effects on dephosphorization. As the oxidant of P, the increase of FeO in slag promotes

dephosphorization at a certain basicity. But, as the basic oxide in slag, the dephosphorization ability of FeO is far less than that of CaO. When the content of FeO in slag reaches a certain level, it is equivalent to the dilution of CaO concentration in slag, and the activity of P_2O_5 increases, which makes the dephosphorization effect decrease.

（2）炉渣的氧化性。提高炉渣的氧化性（即提高炉渣中 FeO 的活度）可提高磷的分配系数。熔渣中的 FeO 含量对脱磷反应具有重要作用。渣中 FeO 是脱磷的首要因素，因为 P 首先氧化生成 P_2O_5，然后与 CaO 作用生成 $4CaO \cdot P_2O_5$ 和 $3CaO \cdot P_2O_5$。渣中 FeO 在碱度不太高的熔炼初期也能生成 $3FeO \cdot P_2O_5$，反应为 $3(FeO)+(P_2O_5) = (3FeO \cdot P_2O_5)$。但 FeO 对脱磷有双重影响，其作为磷的氧化剂时，在一定的碱度下，炉渣中 FeO 含量的增加促进了脱磷；但作为炉渣中的碱性氧化物，FeO 的脱磷能力远不及 CaO，当炉渣中的 FeO 含量高到一定程度后，相当于稀释了炉渣中 CaO 的浓度，P_2O_5 的活度有所增加，从而促使脱磷效果有所下降。

（3）Temperature. Dephosphorization is a strong exothermic reaction, so from the thermodynamic point of view, low temperature is conducive to dephosphorization. Because dephosphorization needs the basic slag containing CaO, and adding lime needs a certain temperature to form the liquid slag. Therefore, the effect of temperature on dephosphorization should be understood dialectically. For the other words, the effective dephosphorization can be achieved only when the slag has a certain basicity and fluidity.

（3）温度。脱磷反应是强烈的放热反应，因此从热力学观点来讲，低温有利于脱磷。由于脱磷需要含 CaO 的碱性渣，而加入石灰则需要有一定的温度才能形成液态渣。因此，温度对脱磷的影响应辩证地理解，即在保证炉渣具有一定碱度和流动性的较低温度下才能有效脱磷。

（4）The compositions of the liquid metal. The dephosphorization of molten steel needs a high content of [O]. Therefore, in the actual production process, the high content of [Si], [Mn], [Cr], [C] is not conducive to dephosphorization. Only when the content of elements with high oxygen binding capacity is reduced, dephosphorization can be carried out smoothly.

（4）金属液的成分。钢液脱磷首先要有较高的 [O] 含量，因此实际生产过程中，当 [Si]、[Mn]、[Cr]、[C] 含量高时不利于脱磷，只有当与氧结合能力高的元素含量降低时，脱磷才能顺利进行。

（5）slag volume. With the increase of slag content, the content of P in molten steel can be reduced when L_P is constant, which is beneficial to dephosphorization. Large amount of slag is often used to dephosphorize medium and high phosphorus hot metal. Increasing the slag amount means that the concentration of P_2O_5 is diluted. For the other words, the content of $3CaO \cdot P_2O_5$ is reduced. However, it is difficult to operate if the slag amount is too large at one time. At this time, the double slag method can be used for slag making and dephosphorization. However, repeated slag exchange also increases the loss of molten steel and heat. In consideration of environmental protection and effective utilization of resources, large amount of slag operation should be avoided.

（5）渣量。增加渣量，可以在 L_P 一定时使钢水中的 P 含量降低，从而有利于脱磷。

冶炼中、高磷铁水时，常采用大渣量脱磷。增大渣量意味着稀释了 P_2O_5 的浓度，也就是说 $3CaO \cdot P_2O_5$ 的含量减少。但一次造渣量过大会给操作带来困难，此时可采用双渣法造渣脱磷。但是多次换渣也会增大钢水和热量的损失，从保护环境、有效利用资源方面考虑，应该避免大渣量操作。

In a word, the conditions of dephosphorization are: slag with high basicity, high iron oxide content (oxidizability), good fluidity, sufficient agitation in molten pool, appropriate temperature and large slag volume.

总之，脱磷的条件为：高碱度，高氧化铁含量（氧化性），良好流动性的熔渣，充分的熔池搅动，以及适当的温度和大渣量。

In order to ensure the dephosphorization effect of molten steel, it is necessary to prevent the phenomenon of dephosphorization. Phosphorus recovery refers to the phenomenon that P entering the slag returns to the steel again, which increases the content of P in the steel water. In the process of oxidative dephosphorization, the decrease of oxidizability or basicity of slag, the bad quality of slags and the high temperature may lead to the return of phosphorus. In the process of tapping, factors such as improper addition of deoxidizing alloy, lower slag and higher phosphorus content in the alloy will also lead to higher phosphorus content in the finished steel than that in the end.

要保证钢水脱磷效果，必须防止回磷现象。回磷是指进入炉渣中的 P 又重新回到钢中，使钢水中 P 含量增加的现象。氧化脱磷时，炉渣的氧化性下降或碱度降低、石灰化渣不好、温度过高等因素都可能引起回磷。出钢过程中，脱氧合金加入不当、出钢下渣、合金中磷含量较高等因素也会导致成品钢中磷含量高于终点磷含量。

The measures to avoid phosphorus recovery in molten steel include: preventing slag from coming out of steel, avoiding slag from coming down as far as possible, and slag dropping shall be avoided as far as possible; the basicity of slag before deoxidization increased appropriately; a certain amount of lime added to the slag surface of ladle after tapping to increase the basicity of slag; ladle deoxidization adopted as far as possible instead of in furnace deoxidization; and ladle modifier added.

避免钢水回磷的措施有：挡渣出钢，尽量避免下渣；适当提高脱氧前的炉渣碱度；出钢后向钢包渣面加一定量的石灰，增加炉渣碱度；尽可能采取钢包脱氧而不采取炉内脱氧；加入钢包改质剂。

2.1.4.3　Reduction Dephosphorization
2.1.4.3　还原脱磷

In a reducing atmosphere (low oxidation), P in molten steel reduced to P^{3-}, which can enter the slag or escape from gasification in the form of phosphate, is called reducing dephosphorization.

在还原气氛（低氧化性）下，将钢水中的 P 还原为 P^{3-}，使其以磷酸盐的形式进入炉渣或气化逸出，称为还原脱磷。

In the process of dephosphorization, deoxidizer stronger than Al should be added to make the liquid steel reach deep reduction. In general, strong reducing agents such as Ca, Ba or CaC_2 are added for reducing dephosphorization, and the reaction is as follows:

还原脱磷时,需要加入比 Al 更强的脱氧剂,使钢液达到深度还原。通常加入 Ca、Ba 或 CaC_2 等强还原剂进行还原脱磷,其反应为:

$$3Ca+2[P] = (Ca_3P_2) \quad (2-8)$$
$$3Ba+2[P] = (Ba_3P_2) \quad (2-9)$$
$$3CaC_2+2[P] = (Ca_3P_2)+6[C] \quad (2-10)$$

The commonly used dephosphorizing agents are metal Ca, Mg, Re and alloy containing calcium (such as CaC_2, CaSi).

常用的脱磷剂有金属 Ca、Mg、Re 以及含钙的合金(如 CaC_2、CaSi)等。

In order to increase the stability of the dephosphorization products of Ca_3P_2 and Mg_3P_2 in the slag, it is necessary to add CaF_2, $CaCl_2$, CaO and other fluxes to make slag at the same time of adding strong reducing agent to absorb the reduced dephosphorization products. After the reduction and dephosphorization, the slag should be removed immediately, otherwise P^{3-} in the slag will be oxidized to PO_4^{3-} again, resulting in phosphorus return. The common dephosphorizing agents are Ca-CaF_2, CaC_2-CaF_2 and Mg(Al)-CaF_2.

为增加脱磷产物 Ca_3P_2、Mg_3P_2 在渣中的稳定性,在加入强还原剂的同时还需加入 CaF_2、$CaCl_2$、CaO 等熔剂造渣,以吸收还原的脱磷产物。还原脱磷后的渣应立即去除,否则渣中的 P^{3-} 又会重新氧化成 PO_4^{3-} 而造成回磷。常见的脱磷剂组成有 Ca-CaF_2、CaC_2-CaF_2 和 Mg(Al)-CaF_2 等。

2.1.5 Desulfurization of Molten Steel
2.1.5 钢液的脱硫

The sulfur in steel mainly comes from molten iron, scrap, ferroalloy, slag making agent (such as lime, iron ore, etc.). The harm of sulfur is the hot embrittlement of steel, and the sulfur mainly exists in the solid steel with the melting point of 1190℃. In the process of iron condensation, the sulfide is mainly distributed on the grain boundary, forming a continuous or discontinuous network structure. In the process of rolling or forging, due to the increase of temperature, the structure on the grain boundary will become liquid again, which will cause the sulfur rich liquid phase to slide along the grain boundary under the action of force, causing the fracture of steel.

钢中的硫主要来自铁水、废钢、铁合金、造渣剂(如石灰、铁矿石等)。硫的危害主要表现为产生钢的热脆现象。硫在固体钢中主要以硫化铁存在,其熔点为 1190℃。在铁液冷凝过程中,这种硫化物主要分布在晶界上,形成连续或不连续的网状薄膜结构。在轧制或锻造时,由于温度升高,晶界上的这一结构又会变为液态,在力的作用下会引起富硫液相沿晶界滑动,造成钢材破裂。

The advantages of sulfur in steel are: improving the cutting performance of steel, improving the surface quality of workpiece and saving power.

硫在钢中的有利之处有:改善钢的切削性能,改善工件的表面质量,节省动力。

2.1.5.1 Desulfurization in Molten Metal
2.1.5.1 金属熔体中的脱硫

Hot Metal Desulphurization
铁水脱硫

It can simplify the process of steelmaking and make up for the deficiency of low desulfurization capacity of converter slag. The advantages of hot metal desulfurization are as follows:

采用铁水炉外脱硫可简化炼钢工艺过程，弥补转炉氧化渣脱硫能力低的不足。铁水脱硫的有利因素有如下三方面：

(1) The content of C, Si, P and other elements in molten iron is high, which is helpful to improve the activity coefficient of sulfur in molten iron.

(1) 铁水中 C、Si、P 等元素的含量高，有利于提高硫在铁水中的活度系数。

(2) The high carbon content and low oxygen content in molten iron are conducive to the desulfurization reaction: FeS+MO+C═Fe+CO+MS (where MO represents the metal oxide that plays the role of desulfurization).

(2) 铁水中碳含量高而氧含量低，有利于进行脱硫反应：FeS+MO+C═Fe+CO+MS（式中 MO 代表起脱硫作用的金属氧化物）。

(3) There is no strong oxidizing atmosphere, which is conducive to the direct use of some strong desulfurizers, such as CaC_2, Mg, etc.

(3) 没有强氧化性气氛，有利于直接使用一些强脱硫剂，如 CaC_2、金属 Mg 等。

Desulfurization of Elements in Molten Steel
钢液中元素的脱硫

At present, higher and higher requirements are put forward for the content of S in molten steel, so it is imperative to strengthen desulfurization. On the basis of giving full play to the desulfurization effect of slag, Mn can be added to continue to reduce the harm of S. The content of manganese in general steel is 0.4%~0.8% (by mass). Ca and rare earth elements used in the smelting process can not only reduce the sulfur content in the steel very low, but also change the shape of sulfide inclusions, so as to improve the quality of the steel. Generally, elemental desulfurization should be used after the deoxidization of molten steel is good. Otherwise, the consumption of elements is large, especially the influence of strong deoxidizing elements [Ca], [Ce], [Mg]. The main reason is that the ability of elements and oxygen to generate compounds is higher than that of elements to generate sulfides.

目前对钢液中 S 含量提出了越来越高的要求，强化脱硫势在必行。在充分发挥炉渣脱硫作用的基础上，可加 Mn 继续降低 S 的危害。一般钢中锰含量（质量分数）为 0.4%~0.8%。冶炼过程中使用的 Ca 和稀土元素，除使钢中硫含量降得很低外，还能改变硫化物夹杂的形状，从而提高钢的质量。一般应在钢液脱氧良好之后再用元素脱硫，否则元素的消耗量大，对强脱氧元素 [Ca]、[Ce]、[Mg] 的影响尤为突出，其主要原因是元素与氧生成化合物的能力比元素生成硫化物的能力高。

2.1.5.2 Slag Desulfurization
2.1.5.2 炉渣脱硫

The desulfuration of molten steel is mainly realized by two ways: slag desulfuration and gasification desulfuration. Under the general steel-making operation conditions, slag desulfurization is the main means to reduce the content of S in steel and make it meet the specification requirements. The desulfurization amount accounts for 90% of the total desulfurization amount, while the gasification desulfurization only accounts for about 10%.

钢液的脱硫主要是通过两种途径来实现的,即炉渣脱硫和气化脱硫。在一般炼钢操作条件下,炉渣脱硫占主导,是降低钢中硫含量、使之达到规格要求的主要手段,其脱硫量占总脱硫量的90%,而气化脱硫仅占10%左右。

According to the molecular theory of slag, the desulfurization reactions between slag and steel are as follows:

根据炉渣的分子理论,渣—钢间的脱硫反应如下:

$$[S]+(CaO) = (CaS)+[O] \qquad (2-11)$$

$$[S]+(MnO) = (MnS)+[O] \qquad (2-12)$$

$$[S]+(MgO) = (MgS)+[O] \qquad (2-13)$$

There are three forms of sulfur in liquid metal: [FeS], [S] and $[S^{2-}]$. FeS is soluble in both molten steel and slag. The desulfurization reaction between slag and steel is as follows: firstly, S in molten steel diffuses into slag, i.e., [FeS] → (FeS), and then FeS entering slag combines with free CaO (or MnO) to form stable CaS or MnS.

硫在金属液中存在三种形式,即[FeS]、[S]和$[S^{2-}]$。FeS既溶于钢液也溶于熔渣。渣—钢间的脱硫反应为:首先钢液中的硫扩散至熔渣中,即[FeS] → (FeS),然后进入熔渣中的(FeS)与游离的CaO(或MnO)结合成稳定的CaS或MnS。

According to the ion theory of slag, the desulfurization reaction is as follows:

根据熔渣的离子理论,脱硫反应为:

$$[S]+(O^{2-}) = (S^{2-})+[O] \qquad (2-14)$$

There is almost no free O^{2-} in the acid slag, so the desulfurization effect of acid slag is very small, but the alkaline slag is different, which has strong desulfurization ability.

在酸性渣中几乎没有自由的O^{2-},因此酸性渣的脱硫作用很小;而碱性渣则不同,其具有较强的脱硫能力。

The main factors affecting the desulfurization between steel and slag are as follows:

影响钢—渣间脱硫的因素主要有:

(1) Bath temperature. The desulfurization reaction between steel and slag belongs to endothermic reaction, and the heat absorption is between 108.2~128kJ/mol. The temperature rise is favorable for desulfurization. In addition, increasing the temperature can also accelerate the dissolution of lime and improve the fluidity of slag, so as to improve the desulfurization speed. Therefore, high temperature is conducive to desulfurization reaction.

(1) 熔池温度。钢—渣间的脱硫反应属于吸热反应,吸热量在108.2~128kJ/mol之间，温度升高有利于脱硫。另外，升高温度还可加速石灰的溶解，提高渣的流动性，从而提高脱硫速度。因此，高温有利于脱硫反应进行。

(2) Basicity of slag. Increasing basicity of slag is beneficial to desulfurization, but when basicity is increased, good fluidity of slag should be kept.

(2) 炉渣碱度。炉渣碱度提高有利于脱硫，但提高碱度时，应注意保持炉渣的良好流动性。

(3) Slag oxidability. The slag oxidation decreases. For the other words, the decrease of FeO content in slag, is beneficial to desulfurization.

(3) 炉渣氧化性。炉渣氧化性降低，即渣中的 (FeO) 含量降低，有利于脱硫。

(4) slag volume. Increasing slag content can reduce the content of S in molten steel. In converter steelmaking, due to the high oxidation in the furnace and poor desulfurization effect, the desulfurization rate of single slag operation is only 40%~60%. In the reduction period of electric furnace, the oxidation of slag is low, so the desulfurization effect is good.

(4) 渣量。增大渣量可使钢水中的 P 含量降低。在转炉炼钢中，由于炉内的高氧化性，脱硫效果不好，单渣操作时的脱硫率只有 40%~60%。电炉还原期炉渣的氧化性低，因而脱硫效果好。

2.1.5.3 Gasification Desulfurization
2.1.5.3 气化脱硫

Gasification desulfurization refers to the removal of sulfur in liquid metal in the form of gas $\{SO_2\}$. In the process of steelmaking, liquid metal and slag often contact with the gas phase containing oxygen or oxygen and sulfur. SO_2 is found in the waste gas of steelmaking. The detection of isotope ^{35}S shows that SO_2 also comes from the charge. The research shows that the gasification desulfurization is mainly realized by the gasification of sulfur in slag, expressed by:

气化脱硫是指将金属液中的硫以气态 $\{SO_2\}$ 的形式去除。在炼钢过程中，金属液和熔渣常与含氧的气相或含氧和硫的气相接触，在炼钢废气中发现有 SO_2 存在。同位素 ^{35}S 检测表明，SO_2 也来自炉料。研究表明，气化脱硫主要通过炉渣中硫的气化来实现，即：

$$(S^{2-}) + 3/2\{O_2\} =\!=\!= \{SO_2\} + (O^{2-}) \qquad (2-15)$$

or (或)

$$6(Fe^{3+}) + (S^{2-}) + 2(O^{2-}) =\!=\!= 6(Fe^{2+}) + \{SO_2\} \qquad (2-16)$$

$$6(Fe^{2+}) + 3/2\{O_2\} =\!=\!= 6(Fe^{3+}) + 3(O^{2-})$$

Reaction (2-16) shows that the iron ion in the slag acts as the medium of gasification desulfurization.

反应 (2-16) 表明，渣中的铁离子充当了气化脱硫的媒介。

It should be pointed out that gasification desulfurization is based on slag desulfurization. In terms of gasification desulfurization, slag is required to have high content of Fe_2O_3, which means iron consumption increases. Therefore, for converter steelmaking, high basicity slag desulfurization should be the main operation rather than excessive expectation of gasification desulfurization. In converter steelmaking, about 1/3 of S is removed by gasification desulfurization.

需要指出的是，气化脱硫是以炉渣脱硫为基础的。就气化脱硫而言，要求炉渣有高的 Fe_2O_3 含量，这就意味着铁耗增大。所以对转炉炼钢来说，应以实行高碱度熔渣脱硫操作为主，而不应过分期望气化脱硫。在转炉炼钢中，约有 1/3 的 S 是以气化脱硫的方式去除。

2.1.6 Deoxidation of Liquid Steel
2.1.6 钢液的脱氧

Steelmaking generally goes through the process of oxidation and reduction. To study it is one of the most important topics for metallurgical workers in various countries.

炼钢一般都经历氧化和还原过程。对其进行研究是各国冶金工作者最主要的课题之一。

2.1.6.1 Deoxidation Purpose
2.1.6.1 脱氧目的

Steelmaking is an oxidation refining process. At the end of smelting, the oxygen content of steel is higher, which is 0.02%~0.08% (by mass) and it is also higher than that required by all kinds of steel. When the oxygen content in the steel exceeds the limit, it will affect the quality of the billet (ingot) and the castability of the molten steel, so that the billet (ingot) can not get the correct solidification structure. In addition, it will produce subcutaneous bubbles, looseness and other defects, and aggravate the harm of sulfur, to intensify the hot embrittlement of the steel. At room temperature, the increase of oxygen content in steel will significantly reduce the elongation and reduction of area. When the temperature and oxygen content are very low, the strength and plasticity of steel decrease sharply with the increase of oxygen content. With the increase of oxygen content, the impact resistance of steel will decrease, and the brittle transition temperature will increase rapidly. Therefore, in order to ensure the quality and smooth pouring of the steel, it is necessary to deoxidize the end-point steel so that the oxygen content in the steel is within the normal range required by all kinds of steel.

炼钢是氧化精炼过程，冶炼终点时钢中氧含量较高，为 0.02%~0.08%（质量分数），且氧含量高出各类钢种要求的氧含量。当钢中氧含量超过限度时，会影响铸坯（锭）的质量及钢水的可浇性，使连铸坯（锭）得不到正确的凝固组织结构。此外，还会产生皮下气泡、疏松等缺陷，并加剧硫的危害作用，即加剧钢的热脆。在室温下，钢中氧含量的增加将使钢的延伸率和断面收缩率显著降低。在较低温度和氧含量极低的情况下，钢的强度和塑性随氧含量的增加而急剧降低。随氧含量的增加，钢的抗冲击性能也会下降，脆性转变温度很快升高。因此，为了保证钢的质量和顺利浇注，必须对终点钢进行脱氧，使钢中氧含量在各类钢所要求的正常含量范围内。

2.1.6.2 Methods of Deoxidation
2.1.6.2 脱氧方法

Deoxidization refers to the deoxidization reaction by adding deoxidizer into the molten steel pool or molten steel, and the deoxidized products enter into the slag or discharged as gas phase. According to the different place of deoxidation, deoxidation methods are divided into as follows:

脱氧是指向炼钢熔池或钢水中加入脱氧剂进行脱氧反应,脱氧产物进入渣中或成为气相排出。根据脱氧发生地点的不同,脱氧方法分为:

(1) Precipitation deoxidation. Precipitation deoxidation is also called direct deoxidation. It is a deoxidizing method that the deoxidizing elements react with the oxygen in the steel directly and the deoxidizing products are floated into the slag. Adding ferrosilicon, ferromanganese, aluminum iron or aluminum block to ladle during tapping belongs to precipitation deoxidization. The characteristics of this deoxidizing method are: the deoxidizing reaction is fast (generally exothermic reaction), but the deoxidizing products may be difficult to all float up and remove and become inclusions in steel. So it is necessary to control certain conditions to remove them.

(1) 沉淀脱氧。沉淀脱氧又称为直接脱氧,是将块状脱氧剂加入钢液中,使脱氧元素在钢液内部与钢中的氧直接反应,生成的脱氧产物上浮进入渣中的脱氧方法。出钢时向钢包中加入硅铁、锰铁、铝铁或铝块,属于沉淀脱氧。这种脱氧方法的特点是:脱氧反应速度快(一般为放热反应),但脱氧产物有可能难以全部上浮排除而成为钢中的夹杂物。因此需要控制一定的条件去除。

(2) Diffusion deoxidation. Diffusion deoxidization is also called indirect deoxidization. It is a deoxidization method that powder deoxidizers (such as the powders of C, Fe-Si, Ca-Si and Al) are added to the slag to reduce the oxygen potential in the slag and make the oxygen in the molten steel diffuse to the slag, so as to reduce the oxygen content in the molten steel. The deoxidization by adding powder deoxidizer to slag during reduction period and refining outside the furnace belongs to diffusion deoxidization. Its characteristics are: deoxidation is carried out in slag; oxygen in molten steel is transferred to slag with slow deoxidation speed and long deoxidation time; and non-metallic inclusions will not be formed in steel.

(2) 扩散脱氧。扩散脱氧又称间接脱氧,是将粉状脱氧剂(如 C 粉、Fe-Si 粉、Ca-Si 粉、Al 粉)加到炉渣中,降低炉渣中的氧势,使钢液中的氧向炉渣中扩散,从而降低钢液中氧含量的脱氧方法。在电炉的还原期和炉外精炼中向渣中加入粉状脱氧剂进行的脱氧属于扩散脱氧。其特点是:脱氧反应在渣中进行;钢液中的氧向渣中转移,脱氧速度慢,脱氧时间长;不会在钢中形成非金属夹杂物。

(3) Vacuum deoxidation. Vacuum deoxidation is a method to reduce the oxygen content in molten steel by reducing the pressure of the system. It can only be used for deoxidation with gas as deoxidation product, such as [C]—[O] reaction. RH vacuum treatment, VAD, VD and other refining methods belong to vacuum deoxidation. Vacuum deoxidation will not pollute non-metallic inclusions, but special equipment is needed.

(3) 真空脱氧。真空脱氧是指利用降低系统的压力来降低钢液中氧含量的方法。其只适用于脱氧产物为气体的脱氧,如 [C]—[O] 反应。RH 真空处理、VAD、VD 等精炼方法就属于真空脱氧。真空脱氧不会造成非金属夹杂物的污染,但需要专门的设备。

2.1.6.3 Deoxidizing Characteristics of Deoxidizers and Elements
2.1.6.3 脱氧剂和元素的脱氧特性

The deoxidizing ability of elements in molten steel determines the deoxidizing effect of molten steel. The deoxidizing capacity of elements is expressed by the residual oxygen content in equilibrium with a certain concentration of deoxidizing elements at a certain temperature. Obviously, the lower the oxygen content in equilibrium with a certain concentration of deoxidizing element, the stronger the deoxidizing ability of the element. When the contents of alloy elements in molten steel are fixed, the order of deoxidization ability from strong to weak is as follows:

钢液中元素的脱氧能力决定了钢液脱氧效果的好坏。元素的脱氧能力用一定温度下与一定浓度的脱氧元素相平衡的残余氧量来表示。显然，与一定浓度的脱氧元素平衡存在的氧含量越低，该元素的脱氧能力越强。钢液中合金元素含量一定时，脱氧能力由强到弱的次序为：

$$Ca、Mg、RE（Ce、La）、Al、Ti、Si、V、Mn、Cr$$

The deoxidizing capacity can only represent the deoxidizing state when the content of alloying elements is certain, while the actual oxygen content in molten steel depends on the deoxidizing system, which is determined by the type, quantity, adding time, adding sequence, slag performance, etc.

脱氧能力只能表示合金元素含量一定时的脱氧状态，而钢液中实际的氧含量取决于脱氧制度，即由脱氧剂的种类、数量、加入时间、加入顺序、炉渣性能等来决定。

The commonly used deoxidizing elements are Mn, Si and Al. Mn and Si are often used as deoxidizers in the form of ferroalloys.

常用脱氧元素有 Mn、Si 和 Al。Mn 和 Si 常以铁合金的形式作脱氧剂。

Mn

锰

The deoxidation product of Mn is not pure MnO, but the melt of MnO and FeO. The deoxidation reactions are as follows:

锰的脱氧产物并不是纯 MnO，而是 MnO 与 FeO 的熔体，其脱氧反应为：

$$[Mn]+[O] = (MnO)(l) \tag{2-17}$$

$$[O]+Fe(l) = (FeO) \tag{2-18}$$

$$[Mn]+(FeO) = (MnO)+Fe(l) \tag{2-19}$$

When the content of Mn increases, the amount of $w(MnO)$ in the equilibrium deoxidation product increases. When $w(MnO)$ increases to a certain value, solid FeO·MnO appears in the deoxidation products. Mn is one of the most widely used deoxidizing elements in steelmaking because:

当金属锰含量增加时，与之平衡的脱氧产物中 $w(MnO)$ 也随之增大。当 $w(MnO)$ 增加到一定值时，脱氧产物开始有固态的 FeO·MnO 出现。在炼钢生产中，锰是应用最广泛的一种脱氧元素，这是因为：

(1) Mn can improve the deoxidization ability of aluminum and silicon.

(1) 锰能提高铝和硅的脱氧能力。

(2) Mn is an irreplaceable deoxidizing element in the smelting of rimmed steel.

(2) 锰是冶炼沸腾钢不可替代的脱氧元素。

(3) Mn can reduce the harm of sulfur.

(3) 锰可以减轻硫的危害。

Si

硅

The deoxidation product of silicon is SiO_2 or $FeO \cdot SiO_2$, and the deoxidation reaction is as follows:

硅的脱氧生成物为 SiO_2 或硅酸铁（$FeO \cdot SiO_2$），其脱氧反应为：

$$[Si]+2[O] = SiO_2(s) \tag{2-20}$$

The higher basicity of slag is, the lower residual oxygen is and the better deoxidization effect of silicon is. Various grades of Fe-Si alloys are commonly used deoxidizers.

炉渣碱度越高，残余氧量越低，硅的脱氧效果越好。各种牌号的 Fe-Si 合金是常用的脱氧剂。

Al

铝

Aluminum is a strong deoxidizer, which is commonly used in the final deoxidation of killed steel. The deoxidation reaction is as follows:

铝是强脱氧剂，常用于镇静钢的终脱氧，其脱氧反应为：

$$Al_2O_3(s) = 2[Al]+3[O] \tag{2-21}$$

Note: At 1600℃, the content of deoxygenin in steel is 0.1% (by mass).

注：条件为 1600℃，钢中脱氧元素的含量（质量分数）为 0.1%。

2.1.6.4 Selection Principle of Deoxidizer
2.1.6.4 脱氧剂的选择原则

The selections of deoxidizer shall meet the following principles:

脱氧剂的选择应满足下列原则：

(1) It has a certain deoxidizing ability that the affinity of deoxidizing elements with oxygen is greater than that of iron and carbon.

(1) 具有一定的脱氧能力，即脱氧元素与氧的亲和力比铁和碳大。

(2) The melting point of deoxidizer is lower than that of molten steel, so as to ensure its melting and uniform distribution and even deoxidization.

(2) 脱氧剂的熔点比钢水温度低，从而保证其熔化且均匀分布，进而均匀脱氧。

(3) The deoxidized product is insoluble in steel water and easy to be removed by floating.

(3) 脱氧产物不溶于钢水中，且易于上浮排除。

(4) The deoxidizing elements remained in steel are harmless to the properties of steel.

(4) 残留于钢中的脱氧元素对钢的性能无害。

(5) It is wide source and low price.

(5) 来源广，价格低。

The commonly used deoxidizers in production are Al, Si, Mn, silicon manganese alloy and silicon aluminum alloy composed of them. The order of deoxidizing ability is: Al>Si>Mn.

在生产中常用的脱氧剂为铝、硅、锰及由它们组成的硅锰、硅铝合金等，其脱氧能力次序为：Al>Si>Mn。

2.1.7 Reactions of Hydrogen and Nitrogen
2.1.7 氢、氮的反应

Gas in steel refers to hydrogen and nitrogen dissolved in steel. Gas sources are from:
钢中气体是指溶解在钢中的氢和氮。气体来源有：

(1) Hydrogen and nitrogen in metals, such as scrap and ferroalloys;
(1) 金属料，如废钢及铁合金中的氢和氮；
(2) Water vapor decomposed by wet slagging agent;
(2) 潮湿的造渣剂分解出来的水蒸气；
(3) Hydrogen, with 8%~9% (by mass) in tar, asphalt and resin binder for refractories;
(3) 耐火材料用的焦油、沥青、树脂黏结剂中含有的氢（8%~9%）；
(4) Hydrogen absorbed by liquid steel in contact with air;
(4) 与空气接触的钢液吸收的氢；
(5) Nitrogen contained in impure oxygen for steelmaking.
(5) 炼钢用不纯的氧气中含有的氮气。

The hazards of gas to steel are as follows:
气体对钢的危害包括：

(1) H. H is dissolved in steel in the form of atoms, forming a interstitial solid solution with iron. H causes the steel to produce white spots, also known as cracking, resulting in brittle fracture, which will cause extremely serious accidents in the use process. In the process of condensation, H precipitates due to the decrease of solubility, resulting in point segregation. The quality of steel with point segregation is very poor, which can not be used and discarded. With the increase of the content of H, the tensile strength of steel decreases, while the plasticity and reduction of area decrease sharply.

(1) H。H以原子的形式固溶于钢中，与铁形成间隙式固溶体。H使钢产生白点，又称发裂，导致脆断，在使用过程中将造成极为严重的意外事故。氢在冷凝过程中因溶解度降低而析出，产生点状偏析。具有点状偏析的钢材质量极差，不能使用而报废。随H含量的增加，钢的抗拉强度下降，塑性和断面收缩率急剧降低。

(2) N. N is dissolved in iron to form interstitial solid solution. The solubility of N in α-Fe reached the maximum value at 590℃, about 0.1%, and decreased below 0.001% at room temperature. When the steel with high content of N cools rapidly from high temperature, the ferrite will be saturated by N. At high temperature, N will gradually precipitate in the form of Fe_4N, which will increase the strength and hardness of the steel, and decrease the plasticity and toughness. This phenomenon is called age hardening. N is the main cause of blue brittleness in

steel. N in steel is easy to form bubbles and looseness, and forms brittle inclusions with Ti, Al and other elements in steel.

(2) N。N 固溶于铁中,形成间隙式固溶体。N 在 α-Fe 中的溶解度在 590℃时可达到最大值,约为 0.1%;在室温时则降至 0.001%以下。N 含量高的钢从高温快速冷却时,铁素体会被氮饱和,此种钢在高温下,N 将以 Fe_4N 的形式逐渐析出,使钢的强度和硬度上升,塑性和韧性下降,该现象称为时效硬化。N 是导致钢产生蓝脆现象的主要原因,钢中的 N 易形成气泡和疏松,与钢中的 Ti、Al 等元素形成脆性夹杂物。

N is sometimes used as an alloying element. In common low alloy steel, the formation of vanadium nitride by N and N can refine the grain and strengthen the precipitation. In nitriding steel, N forms nitride with Cr, Al and other alloy elements in the steel surface, which can increase the hardness, strength, wear resistance and corrosion resistance of the steel surface. N can also replace part of Ni in stainless and acid resistant steel.

N 有时也作为合金元素使用。普通低合金钢中,N 和 V 形成氮化钒,可以起到细化晶粒和沉淀强化的作用。渗氮用钢中,N 与钢表层中的 Cr、Al 等合金元素形成氮化物,可增加钢表层的硬度、强度、耐磨性及抗蚀性。N 还可代替部分 Ni 用于不锈耐酸钢中。

Task 2.2 Steel Making Raw Materials
任务 2.2 炼钢原材料

Raw materials are the important material basis of steelmaking, and their quality has a direct impact on the steelmaking process and steel quality. It is one of the important measures to improve the technical and economic indexes of steelmaking and the precondition to realize the automation of smelting process to adopt refined material and ensure its quality stability.

原材料是炼钢的重要物质基础,其质量好坏对炼钢工艺和钢的质量有直接影响。采用精料并保证其质量稳定,是提高炼钢各项技术经济指标的重要措施之一,也是实现冶炼过程自动化的先决条件。

According to the classification of properties, steel-making raw materials can be divided into metal materials, non-metal materials and gas. Metal materials include molten iron (pig iron), scrap steel, ferroalloy, direct reduction iron and iron carbide (for electric furnace). Non-metal materials include lime, dolomite, fluorite, and synthetic slag making agent. And gas includes oxygen, nitrogen, argon, etc. According to the classification of uses, steel-making raw materials can be divided into metal materials, slagging agents, slagging agents, oxidizers, coolants, carburizers, etc.

按性质分类,炼钢原材料可分为金属料、非金属料和气体。金属料包括铁水(生铁)、废钢、铁合金、直接还原铁和碳化铁(电炉使用);非金属料包括石灰、白云石、萤石和合成造渣剂;气体包括氧气、氮气、氩气等。按用途分类,炼钢原材料可分为金属料、造渣剂、化渣剂、氧化剂、冷却剂和增碳剂等。

2.2.1 Metal Materials
2.2.1 金属料

2.2.1.1 Molten Iron
2.2.1.1 铁水

Molten iron is the main raw material of converter steelmaking, generally accounting for 70%~100% of the loading. The physical and chemical heat of molten iron is the main heat source of converter steelmaking. Therefore, there are certain requirements for the temperature and chemical composition of molten iron.

铁水是转炉炼钢的主要原材料,一般占装入量的70%~100%。铁水的物理热和化学热是转炉炼钢的主要热源,因此,对入炉铁水的温度和化学成分有一定的要求。

Temperature of Hot Metal

铁水温度

The temperature of molten iron is a sign of its physical heat content, which accounts for about 50% of the heat revenue of converter. Too low temperature of molten iron will lead to insufficient heat in the furnace, which will affect the temperature rise of molten pool and the process of element oxidation. Chinese enterprises generally stipulate that the temperature of molten iron entering the furnace should be higher than 1250℃, and keep stable.

铁水温度是铁水含物理热多少的标志,铁水物理热占转炉热量收入的50%左右。铁水温度过低会导致炉内热量不足,影响熔池升温和元素氧化进程。中国企业一般规定铁水入炉温度应高于1250℃,并且保持稳定。

Chemical Composition of Molten Iron

铁水化学成分

The chemical composition of molten iron is as follows:

铁水化学成分主要包括:

(1) Si. Si is one of the important heating elements. With the increase of the content of Si in molten iron, the chemical heat in the furnace increases. When the content of Si in molten iron is too low, less than 0.3% (by mass), the chemical heat of molten iron is too small, the scrap ratio is low, and the lime is difficult to dissolve. When the content of Si in molten iron is too high, higher than 1.0% (by mass), the slag quantity is large, the lime consumption is increased, which is easy to cause splashing and the metal recovery rate is reduced. The increase of SiO_2 in slag will intensify the erosion of furnace lining. And the content of Si can reduce the slag forming speed, which will affect the removal of P and S.

(1) Si。Si是重要的发热元素之一。铁水Si含量升高,炉内的化学热增加。铁水中Si含量(质量分数)过低(低于0.3%)时,铁水所含化学热过少,废钢比低,石灰溶解困难。铁水中硅含量(质量分数)过高(高于1.0%)时,渣量大,石灰消耗增加,易引起喷溅,金属收得率降低;渣中(SiO_2)增多,会加剧对炉衬的侵蚀。Si含量还能降低成渣速度,影响去P和S。

(2) Mn. Mn is a weak heating element, which is generally considered to be a beneficial element in molten iron. MnO formed by the oxidation of Mn in molten iron can promote the dissolution of lime, accelerate the slag formation, and reduce the amount of flux and lining erosion. At the same time, if the content of Mn in molten iron is high, the amount of residual manganese in the end-point molten steel will be increased, which can reduce the content of ferromanganese during alloying and improve the cleanliness of molten steel.

(2) Mn。Mn 是弱发热元素,一般认为 Mn 在铁水中是有益元素。铁水中的 Mn 氧化后生成的 MnO 能促进石灰溶解,加速成渣,减少助熔剂的用量和炉衬侵蚀。同时,铁水 Mn 含量高,终点钢水中的余锰量提高,可以减少合金化时的锰铁含量,有利于提高钢水的洁净度。

(3) P. P is a strong heating element and a harmful element for general steel grades. The lower the content of P in molten iron, the better, because P can not be removed in blast furnace smelting. Only the content of P in molten iron entering converter is required to be as stable as possible. The dephosphorization rate of oxygen top blown converter is between 84% and 94%. For low phosphorus hot metal, single slag operation can be used; for medium phosphorus hot metal, double slag or residual slag operation can be used; for high phosphorus hot metal, multiple slag replacement operation or lime powder spraying process must be used, but the technical and economic indexes of converter will be deteriorated. With the development of hot metal pretreatment technology, the hot metal entering the converter is generally treated by 'Three Dephosphorization' (desilication, dephosphorization and desulfurization) to reduce the content of P in the hot metal entering the converter and simplify the converter operation.

(3) P。P 是强发热元素,对一般钢种来说是有害元素,因此铁水 P 含量越低越好。由于 P 在高炉冶炼中不能去除,因此只能要求进入转炉的铁水 P 含量尽可能稳定。氧气顶吹转炉的脱磷率为 84%~94%。对于低磷铁水,可以采用单渣操作;对于中磷铁水,可以采用双渣或留渣操作;而对于高磷铁水,则必须采用多次换渣操作或喷石灰粉工艺,但会恶化转炉的技术经济指标。随着铁水预处理技术的发展,目前进入转炉内的铁水一般经过"三脱"(脱硅、脱磷、脱硫)处理,以降低进入转炉铁水的 P 含量,从而简化转炉操作。

(4) S. Except for S free cutting steel, P is a harmful element in most steel. In the oxidizing atmosphere of converter, the desulfurization rate is only 30%~60%, so it is difficult to desulfurize. Therefore, the content of S in molten iron into converter is required. Generally, the content of S into converter is required to be less than 0.05% (by mass).

(4) S。除了含 S 易切削钢外,S 在大多数钢中都是有害元素。在转炉的氧化性气氛下,脱硫率只有 30%~60%,脱硫比较困难,因此对转炉入炉铁水的 S 含量有要求。通常要求进入转炉的硫含量(质量分数)低于 0.05%。

Amount of Slag in Hot Metal

铁水带渣量

The content of S, SiO_2 and Al_2O_3 in the blast furnace slag is high. Too much blast furnace slag entering the converter will lead to a large amount of converter slag, increasing in lime consumption, and easy to cause splashing. Therefore, the amount of slag in molten iron mixed into converter should not exceed 0.5% (by mass).

高炉渣中 S、SiO_2、Al_2O_3 含量较高，过多的高炉渣进入转炉内会导致转炉渣量大，石灰消耗增加，且容易造成喷溅。因此，兑入转炉的铁水要求带渣量（质量分数）不得超过 0.5%。

2.2.1.2　Scrap
2.2.1.2　废钢

Scrap is one of the main metal materials for converter steelmaking and the basic raw material for EAF steelmaking. Due to the surplus of heat, scrap of 10%~30% can be added to the oxygen converter, which is the coolant with stable cooling effect. Increasing the amount of waste steel can reduce the cost, energy consumption and auxiliary material consumption.

废钢是转炉炼钢的主要金属料之一，也是电炉炼钢的基本原料。由于氧气转炉热量有富余，可加入 10%~30% 的废钢，废钢是冷却效果比较稳定的冷却剂。增加转炉废钢用量可以降低转炉炼钢的成本、能耗和炼钢辅助材料的消耗。

From the perspective of rational use and smelting process, the requirements for scrap are as follows:

从合理使用和冶炼工艺的角度出发，对废钢有如下要求：

（1）The surface of scrap steel shall be clean, dry and less rusty. The impurities such as soil, sand, refractory and slag also shall be avoided as far as possible.

（1）废钢表面应清洁、干燥、少锈，尽量避免带入泥土、沙石、耐火材料和炉渣等杂质。

（2）The scrap steel shall be carefully inspected before being put into the furnace to prevent the mixture of explosives, inflammables, closed containers, drugs, and the mixture of Cu, Pb, Sn, Sb, As and other non-ferrous metal elements.

（2）废钢在入炉前应仔细检查，严防混入爆炸物、易燃物、密闭容器和毒品，严防混入 Cu、Pb、Sn、Sb、As 等有色金属元素。

（3）Scrap of different properties shall be stacked in different categories to avoid loss of precious elements and smelting of scrap.

（3）不同性质的废钢应分类堆放，以避免贵重元素损失和熔炼出废品。

（4）Scrap steel shall have proper block size and overall dimension. The shape and block of scrap steel should ensure that it can be smoothly added into converter from furnace mouth. The length of scrap steel should be less than 1/2 of converter mouth diameter, and the weight of single piece should not exceed 300kg generally. According to the national standard, the length of scrap is not more than 1000mm, and the maximum weight of single piece is not more than 800kg.

（4）废钢要有合适的块度和外形尺寸。废钢的外形和块度应能保证其从炉口顺利加入转炉。废钢的长度应小于转炉炉口直径的 1/2，单件质量一般不应超过 300kg。国标要求废钢的长度不大于 1000mm，最大单件质量不大于 800kg。

2.2.1.3 Pig Iron
2.2.1.3 生铁

Compared with molten iron, pig iron has no sensible heat and its composition is similar to molten iron. In general, a large amount of pig iron is rarely used as charge in converter, and pig iron can be used as auxiliary raw material when hot metal is insufficient. High quality pig iron can also be used for carburization and predeoxidation before the end of converter smelting.

与铁水相比,生铁没有显热,成分与铁水相似。一般情况下,转炉很少用大量生铁作炉料,在铁水不足时可用生铁作为辅助原料。优质生铁还可以在转炉冶炼终点前用于增碳和预脱氧。

Pig iron is used in electric furnace, its main purpose is to improve the content of C in furnace charge or steel, and to solve the problem of shortage of scrap sources. The requirements of electric furnace steel for pig iron are higher, generally the content of S and P is lower, the content of Mn cannot be higher than 2.5%, and the content of Si cannot be higher than 1.2%.

生铁在电炉中使用,其主要目的在于提高炉料或钢中的 C 含量,并解决废钢来源不足的困难。电炉钢对生铁的要求较高,一般要求 S、P 含量低,Mn 含量(质量分数)不能高于 2.5%,Si 含量(质量分数)不能高于 1.2%。

2.2.1.4 Direct Reduction Iron
2.2.1.4 直接还原铁

DRI (Direct Reduction Iron) is a kind of metal iron product, which uses iron ore or concentrate pellet as raw material and uses gas (CO and H_2) or solid carbon as reducing agent under the temperature lower than the melting point of furnace charge to directly reduce iron oxide.

直接还原铁是以铁矿石或精矿粉球团为原料,在低于炉料熔点的温度下,以气体(CO 和 H_2)或固体碳作还原剂,直接还原铁的氧化物而得到的金属铁产品。

There are three forms of direct reduction iron products as follows:

直接还原的铁产品有以下三种形式:

(1) Sponge iron: Sponge iron refers to the sponge like metallic iron obtained by direct reduction of lump ore in shaft furnace or rotary kiln.

(1) 海绵铁:海绵铁是指块矿在竖炉或回转窑内直接还原得到的海绵状金属铁,称为海绵铁。

(2) Metallized pellets: Metallized pellets are the direct reduced iron, which can keep the shape of pellet, obtained by pelletizing iron concentrate first, drying and then reducing it directly in shaft furnace or rotary kiln.

(2) 金属化球团:金属化球团是使用铁精矿粉先造球,干燥后在竖炉或回转窑内直接还原得到的保持球团外形的直接还原铁。

(3) Hot pressed block iron (HBI). HBI is a kind of block iron with a certain size, which is usually $100mm \times 50mm \times 30mm$, by forming the sponge iron or metallized pellet just reduced under hot pressure.

（3）热压块铁（HBI）。热压块铁是将刚刚还原出来的海绵铁或金属化球团趁热加压成形，使其成为具有一定尺寸的块状铁，一般尺寸多为100mm×50mm×30mm。

2.2.1.5　Ferroalloy
2.2.1.5　铁合金

Ferroalloy is mainly used to adjust the composition of molten steel and remove impurities from steel. And it is mainly used as deoxidizer and alloy element additive in steelmaking. The types of ferroalloys can be divided into iron-based alloy, pure metal alloy, composite alloy, rare earth alloy and oxide alloy.

铁合金主要用于调整钢液成分和脱除钢中杂质，主要作为炼钢的脱氧剂和合金元素添加剂。铁合金的种类可分为铁基合金、纯金属合金、复合合金、稀土合金和氧化物合金。

The requirements for ferroalloys are as follows：

对铁合金有如下要求：

(1) When using block ferroalloy, the block size should be appropriate, and it is better to control it at 10~40mm, which is beneficial to reduce the burning loss and ensure the uniform composition of molten steel.

（1）使用块状铁合金时，块度要合适，以控制在10~40mm为宜，这有利于减少烧损和保证钢水成分均匀。

(2) The composition of ferroalloy shall be in accordance with the technical standards to avoid steel-making operation errors. For example, the content of Al and Ca in ferrosilicon and the content of Si in ferromanganese for deoxidization of rimmed steel directly affect the deoxidization degree of molten steel.

（2）铁合金成分应符合技术标准规定，以避免炼钢操作失误。例如，硅铁中的Al、Ca含量以及沸腾钢脱氧用锰铁的Si含量，都直接影响钢水的脱氧程度。

(3) Ferroalloys shall be strictly classified and stored according to their composition to avoid mixing.

（3）铁合金应按其成分严格分类保管，避免混杂。

(4) The content of non-metallic inclusions, gases and harmful impurities of P and S in ferroalloys is less.

（4）铁合金中非金属夹杂物、气体以及有害杂质P、S的含量要少。

2.2.2　Slagging Materials
2.2.2　造渣材料

2.2.2.1　Lime
2.2.2.1　石灰

Lime is the slag making material with the largest consumption and low price. It has a strong ability of dephosphorization and desulfuration without damaging the furnace lining. There are the following basic requirements for lime for steelmaking：

石灰是炼钢用量最大且价格便宜的造渣材料。它具有很强的脱磷、脱硫能力，不损坏炉衬。对炼钢用石灰有下列基本要求：

(1) The content of CaO in lime is higher, and the content of SiO_2 and S is lower. The high content of SiO_2 and S in lime will reduce the effective the content of CaO in lime. In order to ensure a certain basicity of slag, the consumption of lime needs to be increased, but the increase of slag quantity will worsen the technical and economic indexes of converter.

(1) 石灰中 CaO 含量要高，SiO_2 和 S 含量要低。石灰中 SiO_2 和 S 的含量高会降低石灰中的有效 CaO 含量。为保证一定的炉渣碱度，需增加石灰消耗，但渣量增加，将恶化转炉技术经济指标。

(2) Lime shall be clean, dry and fresh. It is easy for lime to absorb water and turn into $Ca(OH)_2$. The newly burnt lime should be used as much as possible, and it should be stored and transported in a closed container, which is particularly important for the electric furnace steelmaking plant. The lime used in the oxidation and reduction periods of the electric furnace should be baked at 700℃. When the super high power electric furnace is used to smelt foamed slag, some small pieces of limestone can be used for slag making.

(2) 石灰应保证清洁、干燥、新鲜。石灰容易吸水粉化变成 $Ca(OH)_2$，应尽量使用新烧的石灰，并采用密闭的容器储存和输送，这对于电炉炼钢厂尤其重要，电炉氧化期和还原期用的石灰要在 700℃ 高温下烘烤使用。超高功率电炉采用泡沫渣冶炼时，可用部分小块石灰石造渣。

(3) The burning rate of lime should be controlled at about 3%. The high burning rate indicates that the high burning rate of lime will significantly reduce the thermal efficiency, and make slagging, temperature control and end-point control difficult.

(3) 石灰的灼减率应控制在 3% 左右。灼减率高表明石灰的生烧率高，会使热效率显著降低，且使造渣、温度控制和终点控制遇到困难。

(4) The lime shall have suitable lumpiness. If the block is too large, the dissolution will be slow, and even it will not be dissolved until the end of blowing, which will affect the slag forming speed and fail to play a role. If the block is too small, the lime will be easily taken away by the furnace gas, resulting in waste.

(4) 石灰应具有合适的块度。块度过大，溶解缓慢，甚至到吹炼终点还来不及溶解，影响成渣速度且不能发挥作用；过小的石灰则容易被炉气带走，造成浪费。

(5) The activity of lime should be high. The activity of lime refers to the ability of lime to react with other substances, expressed by the dissolution rate of lime. The solubility of lime in high temperature slag is called thermal activity, which has not been determined in the laboratory. Therefore, the reaction between lime and water, (the water activity of lime), is generally used to approximately reflect the dissolution rate of lime in slag. The higher the activity is, the faster the lime dissolves, the faster the slag forms and the stronger the reaction ability.

(5) 石灰活性度要高。石灰的活性是指石灰与其他物质发生反应的能力，用石灰的溶解速度来表示。石灰在高温炉渣中的溶解能力称为热活性，目前在实验室还没有条件测定。因此，一般用石灰与水的反应（即石灰的水活性）来近似地反映石灰在炉渣中的溶解速度。活性度越大，石灰溶解越快，成渣越迅速，反应能力越强。

2.2.2.2　Dolomite
2.2.2.2　白云石

The main component of raw dolomite is $CaCO_3 \cdot MgCO_3$. For many years, raw dolomite or light burned dolomite has been widely used in oxygen converter to replace part of lime for slagging. The practice shows that dolomite slagging can increase the content of MgO in slags, reduce the erosion of slags on the furnace lining, accelerate the dissolution of lime, and keep the content of MgO in slags to reach saturation or supersaturation, so that the final slags can meet the requirements of slag splashing operation.

生白云石的主要成分为 $CaCO_3 \cdot MgCO_3$。多年来，氧气转炉采用生白云石或轻烧白云石代替部分石灰造渣得到了广泛的应用。实践证明，采用白云石造渣可以提高渣中 MgO 含量，减轻炉渣对炉衬的侵蚀，还可以加速石灰的溶解，同时也可保持渣中 MgO 含量达到饱和或过饱和，使终渣达到溅渣操作的要求。

2.2.2.3　Fluorite
2.2.2.3　萤石

The main component of fluorite is CaF_2, with a melting point of 930℃. It is used as a flux in steelmaking.

萤石的主要成分是 CaF_2，熔点约为 930℃，在炼钢中作助熔剂使用。

CaF_2 in fluorite can form a eutectic with CaO, and its melting point is 1362℃. It can reduce the melting point of $2CaO \cdot SiO_2$ shell, accelerate the dissolution of lime and improve the fluidity of slag. Fluorite is characterized by fast action and short time, but it will cause serious splashing and intensify the erosion of furnace lining if it is widely used.

萤石中的 CaF_2 能与 CaO 组成共晶体，其熔点为 1362℃。它能使阻碍石灰溶解的 $2CaO \cdot SiO_2$ 外壳的熔点降低，加速石灰溶解，迅速改善炉渣的流动性。萤石助熔的特点是作用快、时间短，但大量使用会造成严重喷溅，加剧对炉衬的侵蚀。

Fluorite for steelmaking requires high content of CaF_2, low content of SiO_2, S and other impurities, with suitable block size.

炼钢时萤石需要满足 CaF_2 含量高，SiO_2、S 等杂质含量低的要求，且具有合适的块度。

2.2.2.4　Synthetic Slagging Agent
2.2.2.4　合成造渣剂

The synthetic slagging agent is to make the low melting point slagging material out of the furnace with lime and flux in advance. And then it is used for slagging in the furnace to move part or even all of the slagging process of the lime block in the furnace out of the furnace. Obviously, it is an effective measure to improve slag forming speed and metallurgical effect.

合成造渣剂是将石灰和熔剂预先在炉外制成低熔点的造渣材料，然后用于炉内造渣，即把炉内石灰块造渣过程的一部分甚至全部移到炉外进行。显然，这是一种提高成渣速度、改善冶金效果的有效措施。

The materials used as fluxes in the synthetic slagging agent include iron oxide, manganese oxide and other oxides, fluorite, etc. One or several of them can be prefabricated together with lime powder at low temperature. This kind of preformed material has low melting point, high alkalinity, small particles and uniform composition, and is easy to break under high temperature. So it is a good slag forming material. High basicity sinter or pellet can also be used as synthetic slag making agent. Its chemical and physical components are stable and the slag making effect is good. In recent years, some steel plants in China have made composite slag making agent with converter sludge as the base material, which has also achieved good use effect and economic benefit.

在合成造渣剂中作为熔剂的物质有氧化铁、氧化锰及其他氧化物、萤石等，可用其中的一种或几种与石灰粉一起在低温下预制成形。这种预制料一般熔点较低、碱度高、颗粒小、成分均匀，在高温下容易碎裂，是效果较好的成渣料。高碱度烧结矿或球团矿也可作合成造渣剂使用，它的化学成分和物理成分稳定，造渣效果良好。近年来，国内一些钢厂以转炉污泥为基料制备复合造渣剂，也取得了较好的使用效果和经济效益。

2.2.2.5 Magnesite
2.2.2.5 菱镁矿

Magnesite is also a natural mineral. Its main component is $MgCO_3$. It is used as refractory after roasting. It is also a slag adjusting agent for slag splashing and furnace protection of converter at present.

菱镁矿也是天然矿物，主要成分是 $MgCO_3$，焙烧后用作耐火材料，它也是目前转炉溅渣护炉的调渣剂。

2.2.3 Oxidants, Coolants and Carburizers
2.2.3 氧化剂、冷却剂和增碳剂

2.2.3.1 Oxidant
2.2.3.1 氧化剂

The main oxidants in converter steelmaking are as follows:
在转炉炼钢时使用的主要氧化剂包括：

(1) Oxygen. Oxygen is the main oxygen source of converter steelmaking. Its purity should be more than 99.5%, pressure should be stable, and water and soap should be removed.

(1) 氧气。氧气是转炉炼钢的主要氧源，其纯度应大于99.5%，压力要稳定，还应脱除水分和皂液。

(2) Iron ore. The existing forms of iron oxide in iron ore are Fe_2O_3, Fe_3O_4 and FeO, the oxygen content of which were 30.06%, 27.64% and 22.28% respectively. The iron content of iron ore for electric furnace is higher, because the higher the iron content and the higher the

density, the easier it will pass through the slag layer and contact with the molten steel directly after entering the furnace, so as to accelerate the oxidation reaction. The requirements for ore composition are: $w(TFe) \geqslant 55\%$, $w(SiO_2) < 8\%$, $w(S) < 0.1\%$, $w(P) < 0.10\%$, $w(Cu) < 0.2\%$, $w(H_2O) < 0.5\%$, and the block size is 30~100mm.

（2）铁矿石。铁矿石中铁氧化物的存在形式为 Fe_2O_3、Fe_3O_4 和 FeO，其氧含量分别为 30.06%、27.64% 和 22.28%。电炉用铁矿石的铁含量要高，因为铁含量越高，密度越大，入炉后越容易穿过渣层直接与钢液接触，从而加速氧化反应的进行。对矿石成分的要求为：$w(TFe) \geqslant 55\%$，$w(SiO_2) < 8\%$，$w(S) < 0.1\%$，$w(P) < 0.10\%$，$w(Cu) < 0.2\%$，$w(H_2O) < 0.5\%$，块度为 30~100mm。

（3）Oxide scale. Oxide scale is also called scale, which is a by-product of steel rolling workshop, with the content of Fe of 70%~75% (by mass), and it is helpful for converter slagging and cooling. Slag making with iron oxide scale can improve the fluidity of slag and improve the dephosphorization ability of slag.

（3）氧化铁皮。氧化铁皮也称铁鳞，是轧钢车间的副产品，Fe 含量（质量分数）为 70%~75%，有帮助转炉化渣和冷却的作用。电炉用氧化铁皮造渣，可以改善炉渣的流动性，提高炉渣的去磷能力。

2.2.3.2 Coolant
2.2.3.2 冷却剂

Generally, the heat of oxygen converter is surplus. According to the heat balance calculation, a certain amount of coolant can be added. The coolant used in oxygen converter includes scrap, pig iron, iron ore, scale, sinter, pellet, limestone and raw dolomite. Among them, scrap, iron ore and scale are the main ones to be used.

通常氧气转炉的热量有富余，根据热平衡计算，可加入一定数量的冷却剂。氧气转炉用的冷却剂有废钢、生铁块、铁矿石、氧化铁皮、烧结矿、球团矿、石灰石和生白云石等，其中主要使用废钢、铁矿石和氧化铁皮。

Scrap is the main coolant. Its advantages are: stable cooling effect, high utilization rate, less slag, and not easy to cause splashing. Its disadvantages are: the smelting time occupied when adding, and not convenient to adjust the process temperature.

废钢是最主要的一种冷却剂。其优点是：冷却效果稳定，利用率高，渣量少，不易造成喷溅；其缺点是：加入时占用冶炼时间，调节过程温度时不方便。

Iron ore and scale are not only coolant, but also slagging agent and oxidizer. Iron ore is often used as the coolant for natural rich ore and pellets, and its main components are Fe_2O_3 and Fe_3O_4. When iron ore and scale are melted, the iron in them is reduced and absorbs heat, which can adjust the temperature of molten pool. Compared with scrap, this kind of coolant does not take up smelting time and has high cooling effect. It is convenient to adjust process temperature and reduce consumption of iron and steel materials. However, high gangue content in ore will increase consumption of lime and slag, which can not be added too much at the same time at one time, otherwise it is easy to cause splashing. In addition, the fine body of iron oxide scale is light and easy to float in the slag. Increasing the content of iron oxide in the slag is conducive to slag melting. It not only plays the role of coolant, but also plays the role of flux.

铁矿石和氧化铁皮既是冷却剂,又是化渣剂和氧化剂。铁矿石作为冷却剂时常采用天然富矿和球团矿,其主要成分为 Fe_2O_3 和 Fe_3O_4。铁矿石和氧化铁皮熔化后,其中的铁被还原,吸收热量,能起到调节熔池温度的作用。与废钢相比,这类冷却剂加入时不占用冶炼时间,冷却效应高,用于调节过程温度方便,还可以降低钢铁料消耗;但矿石中脉石含量高,会增加石灰消耗和渣量,一次同时加入量不能过多,否则容易引起喷溅。此外,氧化铁皮细小体轻,容易浮在渣中,增加渣中氧化铁的含量,有利于化渣。它不仅起到冷却剂的作用,还起到助熔剂的作用。

2.2.3.3 Carburizer
2.2.3.3 增碳剂

In the process of EAF smelting, the content of C can not meet the expected requirements due to the improper batching or charging and excessive decarburization. So it is necessary to increase the content of C in molten steel. When using carburizing method to smelt medium and high carbon steel in oxygen converter, carburizing agent should also be used.

由于电炉冶炼时配料或装料不当,以及脱碳过量等原因,冶炼过程中碳含量达不到预期要求,因此必须对钢液增碳。氧气转炉用增碳法冶炼中、高碳钢时,也要使用增碳剂。

The main carburizers include pitch coke powder, electrode powder, coke powder, pig iron, etc.

常用的增碳剂有沥青焦粉、电极粉、焦炭粉、生铁等。

The carburizer used in converter requires high and stable fixed content of C $[w(C) \geqslant 96\%]$, low sulfur content $[w(S) \leqslant 0.5\%]$ and moderate particle size (1~5mm).

转炉所用的增碳剂要求固定 C 含量高且稳定 $[w(C) \geqslant 96\%]$,S 含量应尽可能低 $[w(S) \leqslant 0.5\%]$,粒度应适中(1~5mm)。

Task 2.3 Converter Steelmaking Process
任务 2.3 转炉炼钢工艺

2.3.1 Change Rules of Metal Composition in Converter Converting Process
2.3.1 转炉吹炼过程中金属成分的变化规律

2.3.1.1 Oxidation Law of Si
2.3.1.1 Si 的氧化规律

In the early stage of blowing, the affinity between [Si] and oxygen in molten iron is large, and the oxidation reaction of [Si] is exothermic, which is favorable for the reaction at low temperature. Therefore, [Si] is oxidized in a large amount in the early stage of blowing. Generally, it is oxidized to a very low level within 5min, and silicon reduction will not occur until the end of blowing. The reaction formula can be expressed as follows:

在吹炼初期,铁水中[Si]与氧的亲和力大,而且[Si]氧化反应为放热反应,低温下有利于反应的进行。因此,[Si]在吹炼初期就被大量氧化,一般在5min内即被氧化到很低的程度,一直到吹炼终点也不会发生硅的还原。其反应式表示如下:

$$2(CaO)+(2FeO \cdot SiO_2) = (2CaO \cdot SiO_2)+2(FeO) \quad (2-22)$$

When the product is unstable and changes with the basicity of slag, the reaction is as follows:

当产物不稳定,随炉渣碱度的提高而转变时,其反应式为:

$$2(FeO)+(SiO_2) = (2FeO \cdot SiO_2) \quad (2-23)$$

The interface reaction is expressed by:
界面反应为:

$$[Si]+2(FeO) = (SiO_2)+2[Fe] \quad (2-24)$$

The reaction in molten pool is expressed by:
溶池内发生的反应为:

$$[Si]+2[O] = (SiO_2) \quad (2-25)$$

The reaction of direct oxidation of oxygen is expressed by:
氧气直接氧化的反应式为:

$$[Si]+\{O_2\} = (SiO_2) \quad (2-26)$$

The oxidation of Si in molten steel affects the temperature of molten pool, basicity of slag and the oxidation of other elements. [Si] oxidation can raise the temperature of the molten pool, which is one of the main heat sources. After [Si] oxidation, it will generate (SiO_2), which will reduce the basicity of the molten slag, and is not conducive to dephosphorization and desulfurization. It also will erode the furnace lining, reduce the oxidation of the slag, and increase the consumption of slag. The oxidation reaction of C in the molten pool can be intense only when $w[Si]<0.15\%$.

钢液中Si的氧化对熔池温度、熔渣碱度和其他元素的氧化产生影响。[Si]氧化可使熔池温度升高,因此是主要热源之一。[Si]氧化后生成(SiO_2),会降低熔渣碱度,不利于脱磷、脱硫,同时还会侵蚀炉衬,降低炉渣的氧化性,增加渣料消耗。熔池中C的氧化反应只有在$w[Si]<0.15\%$时才能激烈进行。

2.3.1.2　Oxidation Law of Mn
2.3.1.2　Mn 的氧化规律

Mn is oxidized rapidly in the early stage of blowing, but it is not as fast as Si. The reaction equations are as follows:

锰在吹炼初期被迅速氧化,但不如硅氧化得快,其反应方程式为:

(1) Early stage of blowing (吹炼前期):

$$(MnO)+(SiO_2) = (MnO \cdot SiO_2) \quad (2-27)$$

(2) Interface reaction (界面反应):

$$[Mn]+(FeO) = (MnO)+[Fe] \quad (2-28)$$

(3) Reaction in molten pool (熔池内反应):
$$[Mn]+[O] = (MnO) \qquad (2\text{-}29)$$
(4) Direct oxidation reaction (直接氧化反应):
$$[Mn]+1/2\{O_2\} = (MnO) \qquad (2\text{-}30)$$

With the process of blowing, the content of CaO and basicity of slag increase, and $2(CaO)+(MnO \cdot SiO_2) = (MnO)+(2CaO \cdot SiO_2)$ will take place. Most of (MnO) is in free state. At the later stage of converting, when the furnace temperature increases, (MnO) is reduced, and $[C]+(MnO)=[Mn]+\{CO\}$ will take place.

随着吹炼的进行,渣中 CaO 含量增加,炉渣碱度升高,会发生反应 $2(CaO)+(MnO \cdot SiO_2) = (MnO)+(2CaO \cdot SiO_2)$,大部分(MnO)呈自由状态。吹炼后期炉温升高后,(MnO)被还原,会发生反应 $[C]+(MnO)=[Mn]+\{CO\}$。

At the end of blowing, the content of Mn in steel is called residual manganese content or residual manganese content. The high content of residual manganese can reduce the harm of S in steel and the amount of alloy. However, when smelting industrial pure iron, the lower the residual manganese content, the better.

吹炼结束时,钢中的 Mn 含量称为余锰量或残锰量。残锰量高,可以降低钢中 S 的危害,减少合金用量。但冶炼工业纯铁时,则要求残锰量越低越好。

The oxidation of Mn is also one of the heat sources, but not the main one. In the early stage of blowing, MnO is formed by manganese oxidation, which can help to melt slag and reduce the erosion of acid slag on furnace lining. In the process of steelmaking, the oxidation of Mn should be controlled as much as possible to increase the amount of residual manganese in molten steel.

锰的氧化也是吹氧炼钢的热源之一,但不是主要的。在吹炼初期,锰氧化生成 MnO,可帮助化渣,减轻初期酸性渣对炉衬的侵蚀。在炼钢过程中,应尽量控制锰的氧化,从而提高钢水残锰量。

2.3.1.3 Oxidation of C
2.3.1.3 C 的氧化规律

The oxidation law of C is mainly shown as the oxidation rate of C in the blowing process. The oxidation reactions formula of C are as follows:

碳的氧化规律主要表现为吹炼过程中碳的氧化速度。碳的氧化反应式如下:

(1) Reaction in emulsion (乳浊液内反应):
$$[C]+(FeO) = [Fe]+\{CO\} \qquad (2\text{-}31)$$
(2) Interface reaction (界面反应):
$$[C]+(FeO) = [Fe]+\{CO\} \qquad (2\text{-}32)$$
(3) Only when $w[C]<0.05\%$ can the reaction take place on the rough surface of molten pool. The reaction is expressed by:

(3) 只有当 $w[C]<0.05\%$ 时,熔池粗糙表面上才发生以下反应:
$$[C]+2[O] = \{CO_2\} \qquad (2\text{-}33)$$
(4) Direct oxidation in jet impingement zone (射流冲击区,直接氧化反应):

$$[C]+1/2\{O_2\} = \{CO\} \tag{2-34}$$

The C-O reaction mainly occurs at the interface between bubble and metal. The main factors affecting the rate of carbon oxidation are bath temperature, bath metal composition, $\sum w(FeO)$ in the slag and stirring intensity in the furnace. These factors change with the time of the blowing process during the whole blowing process, so as to reflect the different carbon oxidation rate in each stage of the blowing process.

C-O 反应主要发生在气泡与金属的界面上。影响碳氧化速度的主要因素有熔池温度、熔池金属成分、熔渣中 $\sum w(FeO)$ 和炉内搅拌强度。在吹炼的前、中、后期，这些因素随吹炼过程的进行时刻在发生变化，因此能够体现出吹炼各期不同的碳氧化速度。

2.3.1.4 Law of Oxidation of P
2.3.1.4 P 的氧化规律

The oxidation law of P mainly shows the dephosphorization speed in the blowing process, and the dephosphorization reactions formula are as follows:

磷的氧化规律主要表现为吹炼过程中的脱磷速度，脱磷反应式如下：

（1）Reaction of middle and later stage of blowing（吹炼中、后期）：

$$n(CaO)+(3FeO \cdot P_2O_5) = (nCaO \cdot P_2O_5)+3(FeO) \tag{2-35}$$

Note: $n=3$ or 4.

注：$n=3$ 或 4。

（2）Reaction of early stage of blowing（吹炼前期）：

$$(3FeO)+(P_2O_5) = 3FeO \cdot P_2O_5 \tag{2-36}$$

（3）Interface reaction（界面反应）：

$$2[P]+5(FeO) = (P_2O_5)+5[Fe] \tag{2-37}$$

（4）
$$2[P]+5/2[O_2] = (P_2O_5) \tag{2-38}$$

The main factors affecting dephosphorization rate are bath temperature, metal content of P in the bath, $\sum w(FeO)$ in the slag, basicity of the slag, stirring strength of the bath and decarbonization rate. In the early, middle and later stages of blowing, these factors are different, and change with the process of blowing. So the dephosphorization rate of each stage of blowing will change.

影响脱磷速度的主要因素有熔池温度、熔池金属 P 含量、熔渣中 $\sum w(FeO)$、熔渣碱度、熔池的搅拌强度和脱碳速率。在吹炼的前、中、后期，这些影响因素是不同的，而且随吹炼过程的进行又时刻发生变化。因此，吹炼各期的脱磷速度会发生变化。

It is hoped that dephosphorization will be completed in the oxygen top blown converter. The factors that are not conducive to dephosphorization are as follows: the basicity of the slag in the early stage is low, and the slag with $R>2$ should be formed as soon as possible; The content of $\sum w(FeO)$ in the slag in the middle stage is low, and the content of $\sum w(FeO)$ in the slag should be controlled to be 10%~12%, so as to prevent the slag from drying back; The temperature of the molten pool in the later stage is high, and the terminal temperature should be prevented from being too high.

在氧气顶吹转炉中,希望全程脱磷。吹炼各期不利于脱磷的因素是:前期炉渣碱度较低,应尽快形成 $R>2$ 的炉渣;中期渣中 $\sum w(\text{FeO})$ 较低,应控制渣中 $\sum w(\text{FeO})=10\%\sim12\%$,避免炉渣返干;后期熔池温度高,应防止终点温度过高。

2.3.1.5　Law of Changes of S
2.3.1.5　S 的变化规律

The variation of S is mainly shown as the desulfurization speed in the blowing process. According to the slag ion theory, the desulfurization reaction can be expressed as follows:

S 的变化规律主要表现为吹炼过程中的脱硫速度。按熔渣离子理论,脱硫反应可表示为:

$$[S]+(O^{2-}) \Longleftrightarrow (S^{2-})+[O] \tag{2-39}$$

The main factors affecting the desulfurization speed are bath temperature, the content of S in the bath, $\sum w(\text{FeO})$ in the slag, basicity of the slag, stirring strength of the bath and decarburization speed.

影响脱硫速度的主要因素有熔池温度、熔池硫含量、熔渣中 $\sum w(\text{FeO})$、熔渣碱度、熔池的搅拌强度和脱碳速度。

2.3.2　Variation of Slag Composition and Bath Temperature during Converter Blowing
2.3.2　转炉吹炼过程中熔渣成分和熔池温度的变化规律

2.3.2.1　Variation of Slag Composition
2.3.2.1　熔渣成分的变化规律

In the process of converter blowing, the slag composition and temperature in the molten pool affect the oxidation and removal of elements, while the oxidation and removal of elements also affect the change of slag composition.

转炉吹炼过程中,熔池内的炉渣成分和温度影响着元素的氧化和脱除规律,而元素的氧化和脱除又影响着熔渣成分的变化。

After the start of blowing, the content of SiO_2 in the slag increased rapidly to more than 30% (by mass), due to the rapid oxidation of Si and the fact that lime had not yet entered the slag. After that, the content of CaO in the slag increasing as the lime gradually enters into the slag, and the absolute content of SiO_2 in the slag no longer increases because Si in the metal has been oxidized and there are only traces left. So the relative concentration decreases and the basicity of the slag increases gradually. The slag with high basicity and good fluidity can be obtained in the middle and later stage of blowing.

吹炼开始后,由于 Si 的迅速氧化和石灰尚未入渣,渣中的 SiO_2 含量(质量分数)迅速升高到 30% 以上。其后由于石灰逐渐入渣,渣中 CaO 含量不断升高。由于金属中的 Si 已经全部氧化,仅余痕迹,渣中 SiO_2 的绝对含量不再增加,因而相对浓度降低,熔渣碱度逐渐升高。到吹炼中、后期,可得到高碱度、流动性良好的炉渣。

In the early stage of blowing, the slag with high gun position is generally used, so the content of FeO in the slag can rapidly increase to 20% (by mass) or even higher soon after blowing. With the increase of decarburization rate, the content of FeO in slag decreases gradually, and it can be reduced to about 10% at the peak of decarburization. In the later stage of blowing, especially in the process of low carbon steel blowing and slagging before the end point, the content of FeO in the slag obviously rises again.

吹炼初期一般采用高枪位化渣，所以开吹后不久，渣中 FeO 含量（质量分数）可迅速升高到 20%，甚至更高。随着脱碳速度的增加，渣中 FeO 含量逐渐下降，到脱碳高峰期可降到 10%左右。到吹炼后期，特别是在吹炼低碳钢和终点前提枪化渣时，渣中 FeO 含量又明显回升。

2.3.2.2　Law of Temperature Changes in Molten Pool
2.3.2.2　熔池温度的变化规律

The changes of bath temperature are related to the heat source and heat consumption.
熔池温度的变化与熔池的热量来源和热量消耗有关。

In the early stage of blowing, the temperature of molten iron mixed into the furnace is generally about 1300℃. The higher the temperature of molten iron is, the higher the heat will be brought into the furnace. And the elements such as [Si], [Mn], [C], [P] will oxidize and release heat. However, adding scrap steel can reduce the temperature of molten iron mixed into the furnace, and the added slag will absorb a lot of heat in the early stage of blowing. As a result, the temperature of molten pool can be raised to about 1500℃ at the end of blowing period.

吹炼初期，兑入炉内的铁水温度一般为 1300℃左右，铁水温度越高，带入炉内的热量就越高，[Si]、[Mn]、[C]、[P]等元素氧化放热。但加入废钢可使兑入的铁水温度降低，加入的渣料在吹炼初期大量吸热。综合作用的结果是，吹炼前期终了时，熔池温度可升高至 1500℃左右。

In the middle stage of blowing, a large amount of oxidation and exothermic heat of [C] and [P] in the molten pool continue to increase the temperature of the molten pool. But at this time, a large amount of waste steel and the added two batches of feed liquid melts and absorbs heat. The results show that the temperature of molten pool can reach 1500~1550℃.

吹炼中期，熔池中的[C]和[P]继续大量氧化放热，使熔池温度提高。但此时废钢大量熔化吸热，加入的二批料液也熔化吸热。综合作用的结果是，熔池温度可达 1500~1550℃。

In the later stage of blowing, the temperature of molten pool is close to the tapping temperature, up to 1650~1680℃, which depends on steel type and furnace size.

吹炼后期，熔池温度接近出钢温度，可达 1650~1680℃，具体因钢种、炉子大小而异。

In the whole process of steelmaking, the temperature of molten pool is increased by 350℃.
在整个一炉钢的吹炼过程中，熔池温度约提高 350℃。

In conclusion, after the top blown oxygen converter is started, the bath temperature, slag composition and metal composition change one after another. And their respective changes interact with each other, forming the complex physical and chemical changes of multi-phase and multi-component at high temperature. Figure 2-1 shows the change of metal and slag composition during the actual blowing of a furnace of steel by top blown converter.

综上所述，顶吹氧气转炉开吹以后，熔池温度、炉渣成分和金属成分相继发生变化，它们各自的变化又彼此相互影响，形成高温下多相、多组元极其复杂的物理化学变化。如图 2-1 所示为顶吹转炉实际吹炼一炉钢的过程中，金属和炉渣成分的变化。

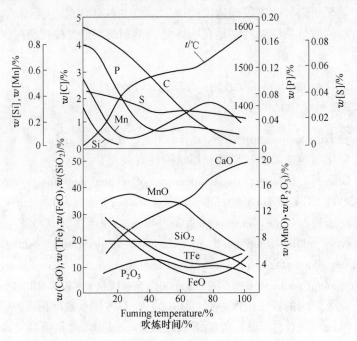

Figure 2-1 Composition change in top blown converter
图 2-1 顶吹转炉炉内成分变化

2.3.3 Oxygen Top Blown Converter Steelmaking Operation System
2.3.3 氧气顶吹转炉炼钢操作制度

2.3.3.1 Loading System
2.3.3.1 装入制度

The charging system is to determine the proper charging amount of converter and the scrap ratio of molten iron.

装入制度就是要确定转炉合适的装入量以及铁水废钢比。

Determination of Loading Capacity
装入量的确定

The loading capacity refers to the total weight of the metal material loaded into the converter for each heat, mainly including the loading quantity of molten iron and scrap steel.

装入量是指转炉每炉次装入金属料的总质量，主要包括铁水和废钢的装入数量。

Production practice shows that each converter has its proper loading capacity. Too much loading will lead to poor mixing of the molten pool and difficulty in slag melting, which may lead to splashing and metal loss, shorten the service life of the furnace cap. Too little loading will reduce the output, and the furnace bottom will be easily damaged by the impact of oxygen jet. Therefore, the following factors should be taken into account when determining the converter loading capacity:

生产实践证明,每座转炉都有其合适的装入量。装入量过多时,会使熔池搅拌不良,化渣困难,有可能导致喷溅和金属损失,缩短炉帽部分的使用寿命;装入量过少时,则产量降低,炉底易受到氧气射流的冲击而损坏。因此,在确定转炉装入量时要考虑以下因素:

(1) Appropriate furnace capacity ratio. Furnace volume ratio refers to the ratio of volume V of free space in the converter to metal loading volume T, expressed in V/T in m^3/t, which usually fluctuates between $0.75 \sim 1.0 m^3/t$. The appropriate furnace volume ratio is summarized from the production practice, which is related to the hot metal composition, nozzle structure, oxygen supply intensity and other factors. When the content of Si and P in molten iron is high, the oxygen supply intensity is high, and the number of blowholes is small. When iron ore or iron oxide scale is used as coolant, the converter capacity ratio should be larger when the converter capacity is small, otherwise it should be smaller.

(1) 合适的炉容比。炉容比是指转炉新砌砖后,炉内自由空间的容积 V 与金属装入量 T 之比,以 V/T 表示,单位为 m^3/t,通常在 $0.75 \sim 1.0 m^3/t$ 之间波动。合适的炉容比是从生产实践中总结出来的,它与铁水成分、喷头结构、供氧强度等因素有关。当铁水中的Si、P含量较高时,供氧强度大,喷孔数少。用铁矿石或氧化铁皮作冷却剂,当转炉容量小时,炉容比应取大一些,反之则取小一些。

(2) A certain depth of molten pool. In order to ensure the safety of production and prolong the life of furnace bottom, it is necessary to ensure that the molten pool has a certain depth. The depth of the molten pool must be greater than the maximum penetration depth h of the oxygen jet to the molten pool. For single hole spray gun, it is generally considered that $h/H \leq 0.7$ is reasonable. For porous spray gun, it is generally considered that $h/H = 0.25 \sim 0.4$ is reasonable.

(2) 一定的熔池深度。为了保证生产安全和延长炉底寿命,要保证熔池具有一定的深度。熔池深度必须大于氧气射流对熔池的最大穿透深度 h。对于单孔喷枪,一般认为 $h/H \leq 0.7$ 是合理的;而对于多孔喷枪,一般认为 $h/H = 0.25 \sim 0.4$ 是合理的。

(3) The loading capacity should be compatible with the ladle capacity, the lifting capacity of the crane and the tilting moment of the converter.

(3) 装入量应与钢包容量、行车的起重能力和转炉的倾动力矩相适应。

Types of Loading System
装入制度的类型

The charging system refers to the arrangement of charging quantity in a furnace service life. The charging system of oxygen top blown converter includes quantitative charging, fixed depth charging and staged quantitative charging, which are discussed as follows:

装入制度是指一个炉役期中装入量的安排方式。氧气顶吹转炉的装入制度现分别讨论如下:

(1) Quantitative loading. Quantitative loading is to keep the loading amount of each furnace unchanged during the service life of the furnace. It has the advantages of simple production organization, stable supply of raw materials and automatic control of the process. The disadvantages of the converter are: more loading in the early stage, deeper molten pool, less loading in the later stage, shallower molten pool, and the production capacity of the converter not brought into full play. This loading system is only suitable for large converter.

(1) 定量装入。定量装入就是在整个炉役期内，保持每炉的装入量不变。其优点是：生产组织简便，原材料供给稳定，有利于实现过程的自动控制。其缺点是：炉役前期装入量偏多、熔池偏深，后期装入量偏少、熔池较浅，转炉的生产能力得不到较好的发挥。该装入制度只适合大型转炉。

(2) Deep mount. Fixed depth loading is to keep the bath depth unchanged during the whole furnace service life. For the other words, with the continuous expansion of the furnace, the loading amount increases gradually. It has the advantages of stable operation of oxygen lance, improving oxygen supply intensity, reducing splashing, protecting furnace bottom and giving full play to production capacity of converter. This loading system has advantages for the workshop with full continuous casting. But when using the mold casting production, the ingot mold is difficult to cooperate, which brings difficulties to the production organization.

(2) 定深装入。定深装入就是在整个炉役期内，保持熔池深度不变，即随着炉膛的不断扩大，装入量逐渐增加。其优点是：氧枪操作稳定，有利于提高供氧强度和减少喷溅，可保护炉底，充分发挥转炉的生产能力。这种装入制度对于采用全连铸的车间具有优越性，但当采用模铸生产时，锭型难以配合，给生产组织带来困难。

(3) Quantitative loading in stages. According to the expansion degree of furnace, the whole service life of furnace can be divided into several stages, in which molten iron and scrap steel can be loaded quantitatively. In this way, it can not only keep the proper depth of molten pool in the whole furnace service period, but also keep the relative stability of the loading amount in each stage. It can also increase the loading amount and facilitate the organization of production, which is a kind of loading system with strong adaptability. This loading system is widely used in medium and small converter steelworks in China.

(3) 分阶段定量装入。分阶段定量装入就是根据炉膛的扩大程度，将整个炉役期划分为几个阶段，每个阶段定量装入铁水和废钢。这样既大体上保持了整个炉役期中具有比较合适的熔池深度，又保持了各个阶段中装入量的相对稳定；既能增加装入量，又便于组织生产，是一种适应性较强的装入制度。中国各中、小转炉炼钢厂普遍采用这种装入制度。

2.3.3.2 Oxygen Supply System
2.3.3.2 供氧制度

The content of oxygen supply system includes choosing reasonable nozzle structure, oxygen supply intensity, oxygen pressure and gun position. Oxygen supply is the key operation to ensure the removal speed of impurities, the heating speed of molten pool, the rapid slag formation, the reduction of splashing and the removal of gas and inclusions in steel, which is related to the end point control and lining life. It also has an important impact on the technical and economic indicators of smelting a furnace of steel.

供氧制度的内容包括选择合理的喷头结构、供氧强度、氧压和枪位。供氧是保证杂质去除速度、熔池升温速度、快速成渣、减少喷溅、去除钢中气体与夹杂物的关键操作，关系到终点控制和炉衬寿命，对冶炼一炉钢的技术经济指标产生重要影响。

Oxygen Lance

氧枪

Oxygen lance is the main equipment for oxygen supply of converter, which is composed of nozzle, gun body and tail structure. The nozzle is usually made of copper with good thermal conductivity by forging and cutting, and some are also made of pressure casting. The gun body is made of three concentric casings, and the central pipe is aerated. The middle pipe is the inlet channel of cooling water, and the outer pipe is the outlet channel. The nozzle and the central tube are welded together to form an oxygen gun.

氧枪是转炉供氧的主要设备，它是由喷头、枪身和尾部结构组成的。喷头通常用导热性能良好的紫铜经锻造和切割加工制成，有的也用压力浇注制成。枪身由三层同心套管套配而成，中心管道通氧。中间管是冷却水的进水通道，外层管是出水通道。喷头与中心管焊接在一起成为氧枪。

The nozzle is the core of the oxygen lance. The requirements for the selection of the nozzle in the oxygen converter are as follows: The supersonic air flow should be obtained, which is conducive to the improvement of oxygen utilization rate; The reasonable impact area should be provided, so that the slag on the liquid surface of the molten pool is fast, and the erosion on the furnace lining is small; It is conducive to the improvement of the thermal efficiency in the furnace; It is convenient for processing and manufacturing, and has a certain service life.

喷头是氧枪的核心，氧气转炉对喷头的选择要求有：应获得超声速气流，有利于氧气利用率的提高；具有合理的冲击面积，使熔池液面化渣快，对炉衬冲刷小；有利于提高炉内的热效率；便于加工制造，有一定的使用寿命。

The types of sprayers are Laval type, straight cylinder type and spiral type. At present, the most widely used is the Laval nozzle. Using Laval nozzle can get supersonic jet flow, which is helpful to improve the working conditions of oxygen lance and the technical and economic indexes of steelmaking.

喷头的类型有拉瓦尔型、直筒型和螺旋型等。目前应用最多的是多孔的拉瓦尔型喷头。使用拉瓦尔型喷头可得到超声速射流，有利于改善氧枪的工作条件和炼钢技术经济指标。

Oxygen Supply Parameters

供氧参数

The main oxygen supply parameters are as follows:

供氧参数主要包括：

(1) Oxygen pressure. The working oxygen pressure stipulated in the oxygen supply system

refers to the oxygen pressure at the measuring point, expressed as p_{using}, which is the pressure in the pipeline before the oxygen enters the spray gun. At present, the working oxygen pressure of some small converter in China is $(4\sim8)\times10^5$Pa, while that of some large converter is $(8.4\sim11)\times10^5$Pa.

(1) 氧气压力。供氧制度中规定的工作氧压是指测定点的氧压,以 $p_{用}$ 表示,是氧气进入喷枪前管道中的压力。目前国内一些小型转炉的工作氧压为 $(4\sim8)\times10^5$Pa,一些大型转炉则为 $(8.4\sim11)\times10^5$Pa。

(2) Oxygen flow. Oxygen flow rate refers to the quantity of oxygen supplied to the molten pool in unit time (volume measurement under common standard state) in m³/min or m³/h. The oxygen flow rate is determined according to the amount of oxygen required for converting each ton of metal, the amount of metal loaded, the oxygen supply time and other factors.

(2) 氧气流量。氧气流量是指单位时间内向熔池供氧的数量(常用标准状态下的体积量度),单位为 m³/min 或 m³/h。氧气流量是根据吹炼每吨金属所需要的氧气量、金属装入量、供氧时间等因素来确定。

(3) Oxygen supply intensity. Oxygen supply intensity refers to the oxygen consumption per ton of metal in unit time in m³/(min·t). The oxygen supply intensity is determined according to the nominal tonnage and furnace capacity ratio of converter.

(3) 供氧强度。供氧强度是指单位时间内每吨金属的氧耗量,单位为 m³/(min·t)。供氧强度的大小根据转炉的公称吨位和炉容比来确定。

The strength of oxygen supply mainly depends on the splashing in the furnace, which should be controlled at a higher limit without splashing. At present, the oxygen supply intensity of domestic small converter is $2.5\sim4.5$m³/(min·t), and that of converter above 120t is $2.8\sim3.6$m³/(min·t). The oxygen supply intensity of converter abroad fluctuates in the range of $2.5\sim4.0$m³/(min·t).

供氧强度的大小主要取决于炉内喷溅情况,通常在不产生喷溅的情况下应控制在高限。目前国内小型转炉的供氧强度为 $2.5\sim4.5$m³/(min·t),120t 以上转炉的供氧强度为 $2.8\sim3.6$m³/(min·t)。国外转炉的供氧强度波动在 $2.5\sim4.0$m³/(min·t) 范围内。

(4) Gun position. The height of the oxygen lance which is the lance position, refers to the distance between the lance nozzle and the surface of the static molten pool. The gun position have important influences on the oxidation rate of elements, slagging rate, heating rate and oxidation of slag. The principles to determine the position of the blowing gun are: incinerating the slag, turning the slag well and removing more phosphorus. Generally, high gun position operation is adopted. The control principles of process gun position are: good slag, no splashing, rapid decarburization, and uniform temperature rise of molten pool. Generally, lower gun position operation is adopted. After adding the second batch of slag, the shotgun slag shall be lifted. When the amount of iron oxide scale, ore and fluorite are added in the slag proportion (or the lime activity is high), the slag is easy to be melted well, and the lower gun position operation can also be adopted. In the period of intense oxidation of carbon, decarburization at a lower lance position is generally adopted. If slag is found to be 'Back Dry', the lance slag or flux slag shall be added to prevent metal splashing. In the later stage of blowing, the gun position operation shall ensure

that the tapping temperature and the final carbon content are accurate.

(4) 枪位。氧枪高度即为枪位，是指氧枪喷头与静止熔池表面之间的距离。枪位对元素的氧化速度、化渣速度、升温速度和炉渣的氧化性有重要的影响。开吹枪位确定的原则是：早化渣，化好渣，多去磷。一般采用较高枪位操作。过程枪位的控制原则是：化好渣，不喷溅，快速脱碳，熔池均匀升温。一般采用较低枪位操作。在加入二批渣料后应提枪化渣，当渣料配比中氧化铁皮、矿石、萤石加入量较多（或石灰活性较高）时，炉渣易于化好，也可采用较低枪位操作。在碳的激烈氧化期一般采用较低枪位脱碳，若发现炉渣"返干"，应提枪化渣或加入助熔剂化渣，以防止金属喷溅。吹炼后期的枪位操作要保证达到出钢温度、拉准碳。

Oxygen Supply Operation
供氧操作

Oxygen supply operation refers to the operation of regulating oxygen pressure or gun position to achieve the purpose of regulating oxygen flow, jet outlet air pressure and the interaction between jet and molten pool, so as to control the chemical reaction process. The operation of oxygen supply can be divided into constant pressure variable gun, constant pressure variable gun and staged constant pressure variable gun. In China, most of them adopt the operation method of changing gun by stages and constant pressure.

供氧操作是指调节氧压或者枪位，达到调节氧气流量、喷头出口气流压力及射流与熔池相互作用程度的目的，以控制化学反应进程的操作。供氧操作分为恒压变枪、恒枪变压和分阶段恒压变枪等方法。在国内，大多采用分阶段恒压变枪操作法。

2.3.3.3 Slagging System
2.3.3.3 造渣制度

Slagging is to add slagging materials—lime and fluxes (fluorite, bauxite, dolomite and iron oxide scale) to the converter smelting process, so as to combine them with the oxides in the blowing process to form a kind of slag with good physical properties, which is good to make a proper basicity, viscosity and oxidizing slag, to satisfy the dephosphorization and desulfurization, reduce the furnace lining erosion, reduce and prevent metal evaporation, splash, slag discharge and reduce the requirement of the finish oxidizing slag. So in a sense, steelmaking is slag making.

造渣就是在转炉冶炼过程中加入造渣材料——石灰和助熔剂（萤石、铁矾土、白云石、氧化铁皮），使之与吹炼过程中的氧化物相结合而形成一种具有良好物理性质的炉渣，也就是要造好具有适当碱度、黏度和氧化性的炉渣，以满足脱磷、脱硫、减少炉衬侵蚀、防止金属蒸发、喷溅、溢渣及降低炉渣终点氧化性的要求。所以从某种意义上来说，炼钢就是炼渣。

The slag making system is to determine the appropriate slag making methods, the quantity and time of slag materials, and how to accelerate slag forming.

造渣制度就是要确定合适的造渣方法、渣料的加入数量和时间以及如何加速成渣。

Dissolving Mechanism of Lime and Factors Affecting Dissolving Speed of Lime
石灰的溶解机理及影响石灰溶解速度的因素

The dissolution of lime in slag is a complex multi-phase reaction, and its dissolution processes are divided into the following three steps:

石灰在炉渣中的溶解是复杂的多相反应,其溶解过程分为以下三个步骤:

(1) FeO, MnO and other oxides or other fluxes in the liquid slag diffuse to the surface of the lime block through the diffusion boundary layer (external mass transfer). The liquid slag migrates to the inside of the lime block along the pores and cracks in the lime block, and its oxide ions further diffuse to the lime lattice (internal mass transfer).

(1) 液态炉渣中 FeO、MnO 等氧化物或其他熔剂通过扩散边界层向石灰块表面扩散(外部传质),并且液态炉渣沿石灰块中的孔隙、裂缝向石灰块内部迁移,同时其氧化物离子进一步向石灰晶格中扩散(内部传质)。

(2) CaO reacts with slag to form a new phase. The reaction is not only carried out on the outer surface of the lime block, but also on the surface of the internal pores of the lime block. The reaction products are generally solid solutions and compounds with lower melting point than CaO.

(2) CaO 与炉渣进行化学反应,形成新相。反应不仅在石灰块的外表面进行,而且也在石灰块内部孔隙的表面上进行。其反应生成物一般都是熔点比 CaO 低的固溶体及化合物。

(3) The reaction products leave the reaction zone and transfers to the slag melt through the diffusion boundary layer.

(3) 反应产物离开反应区,通过扩散边界层向炉渣熔体中传递。

The main factors affecting the dissolution rate of lime are the quality of lime, slag composition, bath temperature, mixing strength of the bath, etc., which are discussed as follows:

影响石灰溶解速度的主要因素有石灰质量、炉渣成分、熔池温度、熔池搅拌强度等,现分别讨论如下:

(1) Lime quality. Lime quality mainly refers to the reaction ability of lime, for the ability of lime to absorb, absorbing slag and reacting with it. The practice shows that the reactive capacities of the active lime with small particle size, high porosity and large specific surface area are stronger than that of the hard burned lime, the slag forming speed is faster in blowing, and the effect of P and S removal is better.

(1) 石灰质量。石灰质量主要是指石灰的反应能力,即石灰吸附、吸收炉渣及与之反应的能力。实践证明,粒度细小、孔隙率高、比表面积大的活性石灰的反应能力比硬烧石灰强,吹炼中成渣速度快,去 P、S 效果好。

(2) Slag composition. The influences of slag composition on lime dissolution rate can be expressed as follows:

(2) 炉渣成分。炉渣成分对石灰溶解速度的影响可用下式表述:

$$J_{CaO} \approx k[w(CaO) + 1.35w(MgO) - 1.09w(SiO_2) + 2.75w(FeO) + 1.9w(MnO) - 39.1]$$

(2-40)

Where J_{CaO}——the dissolution rate of lime in slag in $kg/(m^2 \cdot s)$;

$w(CaO)$, $w(MgO)$, $w(SiO_2)$, $w(FeO)$, $w(MnO)$ ——the content of MgO, SiO$_2$, FeO and MnO (by mass) respectively in slag in %;

k——the proportion coefficient.

式中 J_{CaO}——石灰在渣中溶解速度，kg/(m²·s)；

$w(CaO)$，$w(MgO)$，$w(SiO_2)$，$w(FeO)$，$w(MnO)$ ——渣中 CaO、MgO、SiO$_2$、FeO 和 MnO 含量（质量百分数），%；

k——比例系数。

It can be seen from equation (2-40) that for the common slag system in production, the increase content of FeO, MnO, MgO and CaO (within their general variation range) has a decisive influence on the slagging of lime. The main flux of lime is FeO under the condition of oxygen converter.

从式（2-40）可以看出，对生产中常见的炉渣体系而言，FeO、MnO、MgO、CaO 含量的提高（在它们一般的变化范围内）对石灰渣化具有决定性的影响。在通常的氧气转炉炼钢条件下，石灰的主要熔剂是 FeO。

(3) Bath temperature. If the temperature of the molten pool is higher than the melting point of the slag, the viscosity of the slag will be reduced, the penetration of the slag into the lime block will be accelerated, and the resulting lime block shell compound will melt rapidly and fall off into slag.

（3）熔池温度。当熔池温度高于熔渣熔点以上时，可以使熔渣黏度降低，加速熔渣向石灰块的渗透，使生成的石灰块外壳化合物迅速熔融而脱落成渣。

(4) Specific slag quantity. Specific slag quantity refers to the ratio of the slag of the smelting furnace to the amount of the unmelted lime. The production practice shows that it is beneficial to promote the dissolution of lime by using the method of slag retention and adding the second batch of lime.

（4）比渣量。比渣量是指已熔炉渣和未熔石灰量之比。生产实践表明，采用留渣法、少量多批加入第二批石灰的方法有利于促进石灰溶解。

(5) Pool stirring. It is an important dynamic condition for lime dissolution that the molten pool is stirred strongly and uniformly. The mass transfer process of lime dissolution, and the reaction interface and the speed of lime dissolution can be improved by strengthening the mixing in the molten pool.

（5）熔池搅拌。熔池搅拌强烈而均匀是石灰溶解的重要动力学条件。加强熔池搅拌，可以改善石灰溶解的传质过程，增加反应界面，提高石灰溶解速度。

Measures for Rapid Slagging
快速成渣的措施

The basic characteristics of oxygen top blown converter are fast speed and short cycle. At present, the blowing time of large converter has reached 15~18min. In order to ensure the normal operation of smelting in this short period of more than ten min, slagging must be accelerated. Therefore, the problem of slag forming speed is the central link of controlling slag forming in oxygen top blown converter. The specific measures to improve slag forming speed are as follows：

氧气顶吹转炉的基本特点是速度快、周期短，目前大转炉的吹炼时间已达 15~18min，要在这短短的十几分钟内保证冶炼正常进行，必须加速化渣。因此，成渣速度问题是氧气顶吹转炉控制造渣的中心环节。提高成渣速度的具体措施主要有以下几个方面：

(1) Active lime is used for slagging. Compared with ordinary lime, active lime has higher reaction capacity. C₂S shell deposited on the surface is not dense and easy to peel off, which can accelerate the dissolution of lime.

(1) 采用活性石灰造渣。活性石灰与普通石灰相比,具有更高的反应能力,表面沉积的 C_2S 外壳不致密、易剥落,可加速石灰的溶解。

(2) Avoid depositing C₂S on the surface of the lime block. It can be seen from the phase diagram of CaO-SiO₂-FeO ternary system that the basicity of slag can be increased along the route of $w(FeO)/w(SiO_2)>2$, the deposition area of C₂S can be avoided, and the melting of lime can be accelerated.

(2) 避免在石灰块表面沉积 C_2S。从 CaO-SiO₂-FeO 三元系相图上可以看出,沿着 $w(FeO)/w(SiO_2)>2$ 的路线提高炉渣碱度,可避开 C_2S 的沉积区,加快石灰的熔化。

(3) Synthetic slag is used. For example, It will achieve good results to use the synthetic slag of CaO + Al₂O₃ + Fe₂O₃, converter dust mixed with lime powder, raw dolomite powder, cold solidified pellet made of rolled iron oxide scale, lime infiltrated with FeO and dolomite infiltrated with FeO.

(3) 采用合成渣料。例如,采用 CaO+Al₂O₃+Fe₂O₃ 合成渣料,转炉烟尘拌加石灰粉、生白云石粉、轧钢氧化铁皮制成的冷固球团,以及渗 FeO 的石灰和渗 FeO 的白云石,都取得了很好的效果。

(4) Slag retention method is adopted, to shorten the lag time of lime dissolution. The main measures are as follows: firstly, add 1/3~2/5 lime and all scrap steel at the completion of steel tapping in the upper furnace, with preheating them, so that when the molten iron is blown and the lime enters the slag, there will be no condensation shell around the slag. Then, reduce the lime block as much as possible, and add the lime with the particle size of 10~30mm into the reaction zone continuously, which can shorten the delay of lime dissolution Period.

(4) 采用留渣法操作,缩短石灰溶解的滞止期。其主要的措施有:首先,在上炉出钢完毕时立即加入 1/3~2/5 的石灰和全部废钢,并预热它们,这样在兑铁水开吹后、石灰进入炉渣时,其周围就不会形成炉渣的冷凝外壳;其次,尽量减小石灰块度,采用粒度为 10~30mm 的石灰连续加入一次反应区,缩短石灰溶解的滞止期。

(5) Prevent lime from agglomerating during blowing. Once a large lime lump is formed, its dissolution in the slag will be very difficult. The reason of lime agglomerating is that the amount of liquid slag is small, the viscosity is large, and the mixing of molten pool is insufficient.

(5) 防止开吹期石灰成团。一旦形成很大的石灰团,它在炉渣中的溶解就会很困难。石灰块成团的原因包括液渣数量少、黏度大和熔池搅拌不足。

(6) Increase bath temperature. Any measures to increase the temperature of the molten pool can promote slag formation.

(6) 提高熔池温度。任何提高熔池温度的措施都能促进化渣。

(7) The stirring movement of the molten pool in the early stage of strengthening. The slagging can be accelerated by adopting the technology of double flow combined oxygen lance and top bottom combined blowing.

(7) 强化前期的熔池搅拌运动。采用双流复合氧枪及顶底复吹技术可加速化渣。

Slagging Methods

造渣方法

According to the different composition of molten iron and the requirements of the steel, the slag making methods can be divided into single slag method, double slag method and double slag retention method, which are discussed as follows:

根据铁水成分不同和对所炼钢种的要求,造渣方法可分为单渣法、双渣法和双渣留渣法。其介绍如下:

(1) Single slag method. The single slag method refers to the process of making slag only once in the smelting process, without slag pouring or raking, until the end of tapping. This slag making method is suitable for the situation that the content of Si, P and S in molten iron is low, the requirements of steel grades on the content of P and S are not strict, and low-carbon steel grades are smelted. The single slag process is simple in operation, short in blowing time, good in working conditions and easy to realize automatic control. Its dephosphorization rate is about 90% and desulfurization rate is about 35%.

(1) 单渣法。单渣法是指在冶炼过程中只造一次渣,中途不倒渣、不扒渣,直到终点出钢的方法。这种造渣方法适用于铁水 Si、P、S 含量较低,钢种对 P、S 含量要求不严格,以及冶炼低碳钢种的情况。单渣法操作工艺简单,吹炼时间短,劳动条件好,易于实现自动控制,其脱磷率在90%左右,脱硫率在35%左右。

(2) Double slag method. Double slag method which is the operation of changing slag, refers to the process of pouring out or raking out $1/2 \sim 2/3$ slag in one (or several) time(s) during the blowing process, and then adding slag material to make slag again. The slag making method is suitable for the production of high-quality steel with low content of P when the content of Si in molten iron is more than 1.0%, the content of P is more than 0.5% (or the content of P in raw material is less than 0.5%), for the blowing of medium and high carbon steel and alloy steel with a large amount of oxidizable elements. The advantages of the method are: the effect of P and S removal is better, the dephosphorization rate can reach 92%~95%, and the desulfurization rate is about 50%; the splashing caused by large amount of slag can be eliminated; and some acid slag can be poured out, which can reduce the erosion of furnace lining and the consumption of lime.

(2) 双渣法。双渣法就是换渣操作,即在吹炼过程中分一次(或几次)倒出或扒出 $1/2 \sim 2/3$ 的炉渣,然后加渣料重新造渣。这种造渣方法适合在铁水 Si 含量(质量分数)大于1.0%(或 P 含量(质量分数)大于0.5%),或原料 P 含量(质量分数)小于0.5%,但要求生产低磷的优质钢;吹炼中、高碳钢以及需在炉内加入大量易氧化元素的合金钢时采用。此法的优点是:去除 P、S 的效果较好,其脱磷率可达92%~95%,脱硫率约为50%;可消除大渣量引起的喷溅;倒出部分酸性渣,可减轻对炉衬的侵蚀,减少石灰消耗。

(3) Double slag method. The double slag method is to leave a part of the end slag of the previous smelting furnace in the furnace after tapping for the use of the next smelting furnace as part of the initial slag. And then pour it out at the end of the early stage of blowing and make slag

again. This method is suitable for blowing medium and high phosphorus hot metal ($w[P] > 1.5\%$). Due to the high basicity, high slag temperature, high content of (FeO) and good fluidity of the final slag, it is helpful for the melting of lime in the early stage of lower furnace blowing. It can accelerate the formation of the initial slag, and improve the early dephosphorization, desulfurization rate and furnace thermal efficiency. At the same time, it can also reduce the consumption of lime, reduce iron loss and oxygen consumption.

(3) 双渣留渣法。双渣留渣法是出钢后将上一炉冶炼的终点炉渣留一部分在炉内,供下一炉冶炼时做部分初期渣使用,然后在吹炼前期结束时倒出,重新造渣。这种方法适用于吹炼中、高磷铁水($w[P]>1.5\%$)。由于终渣碱度高、渣温高、(FeO)含量较高、流动性好,有助于下炉吹炼前期石灰的熔化,可加速初期渣的形成,提高前期脱磷、脱硫率和炉子热效率。同时,还可以减少石灰的消耗,降低铁损和氧耗。

Slag Addition Time
渣料加入时间

The slag of top blown converter is usually added in two (or three) batches. The first batch of slag is added before mixing with molten iron (or when blowing), the amount of which is $1/2 \sim 2/3$ of the total slag amount, and dolomite is all added into the furnace. The addition time of the second batch of slag is $1/3 \sim 1/2$ of the total slag amount after the first batch of slag is well formed and the oxidation of Si and Mn in molten iron is basically completed. In case of double slag operation, the second batch of slag is added after slag pouring, and it is also added in small batches for many times. The addition of several times is beneficial to lime dissolution, and small batch of slag can be used to control the overflow of slag in the furnace. Whether to add the third batch of slag depends on the removal of P and S in the furnace. The quantity and time of adding should be determined according to the actual situation of blowing. No matter how many batches of slag are added, the last small batch of slag must be added 3min before carbon drawing, otherwise slag will not be melted.

通常情况下,顶吹转炉渣料分两批(或三批)加入。第一批渣料在兑铁水前(或开吹时)加入,加入量为总渣量的$1/2 \sim 2/3$,并将白云石全部加入炉内。第二批渣料加入时间是在第一批渣料化好且铁水中Si、Mn氧化基本结束后,其加入量为总渣量的$1/3 \sim 1/2$。若是双渣操作,则在倒渣后加入第二批渣料。第二批渣料分小批多次加入,多次加入对石灰溶解有利,也可用小批渣料来控制炉内泡沫渣的溢出。第三批渣料视炉内P、S的去除情况来决定是否加入,其加入数量和时间均应根据吹炼实际情况而定。无论加几批渣,最后一小批渣料都必须在拉碳前3min加完,否则来不及化渣。

Foamed Slag
泡沫渣

In the process of converter blowing, many metal droplets are produced due to the impact of oxygen jet and the stirring of molten pool. These metal droplets fall into the slag and react with (FeO) to produce a large number of CO bubbles and disperse them into molten slag to form a Steel-Slag-Air mixed emulsion and produce foamed slag. In the oxygen top blown converter steelmaking process, due to the full development of foamed slag, the contact area of Steel-Slag-

Air has been greatly increased, and the reactions such as decarburization and dephosphorization have been accelerated. Therefore, a certain degree of foaming slag in blowing process is an important technological measure to shorten the smelting time and improve the quality of products.

转炉吹炼过程中,由于氧气射流的冲击和熔池搅拌,产生了许多金属液滴。这些金属液滴落入炉渣后,与(FeO)作用生成大量的 CO 气泡并分散于熔渣之中,形成了钢—渣—气密切混合的乳浊液,并产生泡沫渣。在氧气顶吹转炉炼钢中,由于泡沫渣较为充分地发展,大大增加了钢—渣—气之间的接触面积,加速了脱碳、脱磷等反应的进行。因此,在吹炼过程中造成一定程度的泡沫渣是缩短冶炼时间、提高产品质量的一个重要工艺措施。

It has been proved that foam slag always occurs in the process of oxygen converter steelmaking. After forming, it should be controlled within the proper range so as to make the blowing smooth and achieve the requirements of tapping and carbon pulling.

实践证明,氧气转炉炼钢过程中泡沫渣总是要发生的,形成后应将其控制在合适的范围内,以使吹炼平稳,并达到出钢拉碳的要求。

For steelmaking operations, it is necessary to build normal foam residue of 'Unsaturated Type'. The key is: the slag should be incipient in the initial stage, and the content of $\sum w(\text{FeO})$ in the slag should be kept at 10%~20% in the middle stage; at the same time, the lance position should be ensured to work under the appropriate submerged blowing conditions, and the second batch of materials should be added according to the system of small amount and many times.

对炼钢操作来说,要造的是"非饱和型"的正常泡沫渣。其关键是:初期要早化渣,中期要保持渣中 $\sum w(\text{FeO}) = 10\% \sim 20\%$;同时要保证枪位在合适的"淹没"吹炼条件下工作,二批料应按少量多次的制度加料。

2.3.3.4 Temperature Regime
2.3.3.4 温度制度

The temperature system includes process temperature control and terminal temperature control. For converter converting process, temperature is not only an important thermodynamic parameter, but also an important dynamic parameter. It has an important influence on the reaction in the furnace, slag melting, lining life and molten steel quality. The purpose of process temperature control is to make the blowing process evenly warm up to ensure that the blowing is stable and accurate to the end temperature. The purpose of terminal temperature control is to ensure the proper tapping temperature. There are certain requirements for the temperature range of the end point in converting any steel.

温度制度包括过程温度控制和终点温度控制。对于转炉吹炼过程,温度既是重要的热力学参数,又是重要的动力学参数,它对炉内反应、渣料熔化、炉衬寿命、钢水质量都有重要影响。过程温度控制的目的是使吹炼过程均衡升温,保证吹炼平稳及准确达到终点温度。终点温度控制的目的是保证合适的出钢温度。吹炼任何钢种都对终点温度范围有一定的要求。

Heat Source and Heat Expenditure
热量来源与热量支出

The heat source of oxygen converter steelmaking is the physical heat and chemical heat of molten iron. Physical heat refers to the heat brought in by molten iron, which is directly related to the temperature of molten iron. Chemical heat refers to the heat released after the oxidation of various elements in molten iron, which is directly related to the chemical composition of molten iron, in which C and Si are the main heating elements of converter steelmaking.

氧气转炉炼钢的热量来源是铁水的物理热和化学热。物理热是指铁水带入的热量,与铁水温度有直接关系;化学热是指铁水中各元素氧化后放出的热量,与铁水化学成分直接相关,其中C、Si两大元素为转炉炼钢的主要发热元素。

The heat expenditure of converter includes two parts. One is the heat directly used for steelmaking, i.e. the heat used for heating molten steel and slag. The other is the heat not directly used for steelmaking, including the heat taken away by exhaust gas and smoke, the heat taken away by cooling water, the heat dissipation loss of furnace mouth and shell, the heat absorption of coolant, etc.

转炉的热量支出包括两部分:一部分是直接用于炼钢的热量,即用于加热钢水和熔渣的热量;另一部分是未直接用于炼钢的热量,包括废气、烟尘带走的热量,冷却水带走的热量,炉口炉壳的散热损失和冷却剂的吸热等。

Determination of Tapping Temperature
出钢温度的确定

The tapping temperature is affected by the steel type, ingot shape and pouring method. The determination principle is as follows:

出钢温度的高低受钢种、锭型和浇注方法的影响,其确定原则是:

(1) It shall be ensured that the pouring temperature is 60~100℃ higher than the solidification temperature of the steel type (upper limit of small furnace and lower limit of large furnace).

(1) 应保证浇注温度高于所炼钢种凝固温度60~100℃(小炉子偏上限,大炉子偏下限)。

(2) The process of tapping, transportation and sedation time of molten steel should be considered. The temperature drop of molten steel during argon blowing is generally 40~80℃.

(2) 应考虑出钢过程及钢水运输和镇静时间,钢液吹氩时的温降一般为40~80℃。

(3) The casting method and the size of the casting ingot shall be considered. When pouring small ingots, the tapping temperature should be higher. If continuous casting is adopted, the tapping temperature should also be higher (20~50℃ higher than that of die casting).

(3) 应考虑浇注方法和浇注锭型的大小。浇注小钢锭时,出钢温度要偏高些。若采用连铸,其出钢温度也要高些(比模铸高20~50℃)。

Temperature Control in Blowing Process
吹炼过程的温度控制

The main method of temperature control is to timely add the required amount of coolant to control the process temperature and provide guarantee for the direct hit to the terminal temperature. The addition time of the coolant varies according to the conditions. It is not convenient to add scrap steel in blowing before blowing. When using ore or iron sheet as coolant, their adding time is often considered at the same time as slagging, because they are slagging agents at the same time, and most of them are added in batches.

温度控制的办法主要是适时加入需要数量的冷却剂,以控制好过程温度,并为直接命中终点温度提供保证。冷却剂的加入时间因条件而异。废钢在吹炼时加入不方便,通常在开吹前加入。利用矿石或铁皮作冷却剂时,由于它们同时又是化渣剂,其加入时间往往与造渣同时考虑,大多采用分批加入方式。

The change of the content of Si in molten iron, steel grades, furnace lining and empty furnace time should be taken into account when adding coolant.

冷却剂的加入量需考虑铁水的 Si 含量、所炼钢种、炉衬和空炉时间的变化。

2.3.3.5 End Point Control and Tapping
2.3.3.5 终点控制和出钢

Terminal point control mainly refers to the control of terminal point temperature and composition.

终点控制主要是指终点温度和成分的控制。

Sign of End Point

终点的标志

After the converter is mixed with molten iron, through a series of operations such as oxygen supply, slag making and a series of physical and chemical reactions, the time when the molten steel reaching the composition and temperature of the steel is called the end point. The specific signs to reach the destination are:

转炉兑入铁水后,通过供氧、造渣等一系列操作,经过一系列物理化学反应,使钢水达到所炼钢种的成分和温度的时刻,称为终点。到达终点的具体标志是:

(1) The content of C in the steel reaches the control range of the steel.

(1) 钢中 C 含量达到所炼钢种的控制范围。

(2) The content of P and S in steel is lower than the lower limit of specification.

(2) 钢中 P、S 含量低于规格下限的一定范围。

(3) The tapping temperature can ensure the smooth refining and pouring.

(3) 出钢温度能保证顺利进行精炼和浇注。

(4) For the boiling steel, the molten steel has certain oxidability.

(4) 对于沸腾钢,钢水有一定的氧化性。

End point control is an important operation in the later stage of converter converting. As the removal of P and S is usually more complex than decarbonization, S and P should be removed as early as possible to the range required by the end point. The end point control is simplified as the control of the content of C and molten steel temperature. The end point control is also commonly known as carbon pulling. Improper terminal control will cause a series of hazards.

终点控制是转炉吹炼后期的重要操作。由于 P、S 的去除通常比脱碳复杂,应尽可能地使 S、P 提早脱除到终点要求的范围内，这样终点控制就简化为 C 含量和钢水温度的控制，所以终点控制也俗称拉碳。终点控制不当，会造成一系列的危害。

End Point Control Methods
终点控制方法

The end point control methods are divided into experience control method and automatic control method. For medium and small converter, experience control method is mainly used at present. There are three methods of end-point carbon empirical control including one-time carbon pulling method, carburizing method and high pulling supplementary blowing method, which are discussed as follows:

终点控制方法分为经验控制方法和自动控制方法。对于中小转炉，目前采用的主要是经验控制方法。终点碳经验控制的方法有三种，即一次拉碳法、增碳法和高拉补吹法。其介绍如下：

(1) One pull carbon method. Blowing shall be carried out according to the end point carbon and end point temperature required by tapping. When the requirements are met, the gun shall be lifted to stop blowing oxygen. In this method, the carbon and temperature of the end point hit the target at the same time at the end of the blowing process. The operation technology level is high, and other methods are generally difficult to achieve. The method also has the following advantages:

(1) 一次拉碳法。按出钢要求的终点碳和终点温度进行吹炼，当达到要求时提枪停止吹氧。这种方法在吹炼终点时终点碳和终点温度同时命中目标，操作技术水平高，其他方法一般很难达到。该方法还具有如下优点：

1) The content of TFe in final slag is low, the recovery rate of molten steel is high, and the erosion of lining is small.

1) 终渣 TFe 含量低，钢水收得率高，对炉衬侵蚀量少。

2) There are few harmful gases in the molten steel, no carburizer is added, and the molten steel is clean.

2) 钢水中有害气体少，不加增碳剂，钢水洁净。

3) There are high residual manganese and low alloy consumption.

3) 余锰量高，合金消耗少。

4) There are high less oxygen consumption, saving carburizer.

4) 氧耗量少，节约增碳剂。

(2) Carbon addition method. When the content of C in the steel is more than 0.08% (by mass), the gun is lifted when the content of C is 0.05% ~ 0.06% (by mass), and then the carbon is added in the ladle according to the specification of the steel. The quality of carburizer should be strictly guaranteed when adopting carburizing method. The carbon powder used in the carburizer requires high purity, low content of sulfur and ash. And sometimes the content of N is required, otherwise the molten steel will be polluted.

(2) 增碳法。当吹炼碳含量（质量分数）大于 0.08% 的钢种时，均在吹炼到 $w[C]$ = 0.05% ~ 0.06% 时提枪，然后按照所炼钢种的规格要求在钢包内增碳。采用增碳法时应严格保证增碳剂的质量。增碳剂所用炭粉要求纯度高，硫和灰分含量要很低，有时对其氮含量也有要求，否则会污染钢水。

(3) High pull make-up blow method. When smelting medium and high carbon steel, the end point is slightly higher according to the specification for carbon drawing. After temperature measurement and sampling, the time for supplementary blowing is determined according to the difference between the analysis result and the specification. In the range of the content of C in medium and high carbon steel, the decarburization speed is fast, and the flame has no obvious change. It is not easy to judge from the spark, and it is difficult to judge accurately by manual one-time carbon pulling at the end point, so the method of high pulling and supplementary blowing is adopted. The high drawing supplementary blowing method is only applicable to the blowing of medium and high carbon steel.

(3) 高拉补吹法。当冶炼中、高碳钢时，终点按规格稍高些进行拉碳，待测温、取样后，按分析结果与规格的差值决定补吹时间。由于在中、高碳钢的碳含量范围内，脱碳速度较快，火焰没有明显的变化，从火花上也不易判断，终点人工一次拉碳很难判断准确，所以采用高拉补吹的方法。高拉补吹法只适用于中、高碳钢的吹炼。

Tapping
出钢

In the tapping process of converter, in order to reduce the gas absorption of molten steel and to make the alloy mix evenly after being added into the ladle, a proper tapping duration is needed. The tapping duration is 1~4min for the converter of less than 50t, 3~6min for the converter of 50~100t and 4~8min for the converter of more than 100t. Since the invention of slag retaining tapping method in Japan in 1970, there have been many kinds of tapping methods, the purpose of which is to control the composition of molten steel accurately, reduce the return phosphorus of molten steel and improve the refining effect of ladle. At present, slag blocking and tapping methods include slag blocking cap method, slag blocking ball method, slag blocking plug method, pneumatic slag stopper method, pneumatic slag blowing method and electromagnetic slag blocking method.

在转炉出钢过程中，为了减少钢水吸气和有利于合金加入钢包后搅拌均匀，需要有适当的出钢持续时间。小于50t 转炉的出钢持续时间为1~4min，50~100t 的转炉为3~6min，大于100t 的转炉为4~8min。自 1970 年日本发明挡渣出钢法后，先后又出现多种出钢方式，其目的是：利于准确控制钢水成分，减少钢水回磷，提高钢包精炼效果。目前采用的挡渣出钢法有挡渣帽法、挡渣球法、挡渣塞法、气动挡渣器法、气动吹渣法和电磁挡渣法等。

2.3.3.6　Slag Splashing for Furnace Protection
2.3.3.6　溅渣护炉

Slag splashing is a new technology developed in recent years to improve the life of the furnace. It is a protection technology which is based on the widely used slag hanging technology (adding MgO slag making agent to slag to make slime slag) in the 1970s, using oxygen gun to spray high-pressure nitrogen, splashing the residual slag left in the furnace after steel tapping on the whole surface of the converter lining in 2~4min to form the slag protection layer surgery. The technology was first developed by Praxair Gas Co., Ltd. in the Great Lakes branch of the United States Republic steel company. It has not been promoted since the implementation of the technology in the Great Lakes branch and the Granite City branch.

溅渣护炉是近年来开发的一项提高炉龄的新技术，是在20世纪70年代广泛应用过的挂渣补炉技术（向炉渣中加入含MgO的造渣剂造黏渣）的基础上，采用氧枪喷吹高压氮气，在2~4min内将出钢后留在炉内的残余炉渣喷溅涂敷在转炉内衬整个表面上，生成炉渣保护层的护炉技术。该技术最先是在美国共和钢公司的大湖分厂，由普莱克斯气体有限公司开发的，在大湖分厂和格棱那也特市分厂实施后并没有得到推广。

The basic principle of slag splashing for furnace protection is as follows: a layer of slag splashing layer with high melting point is formed on the surface of furnace lining through the splashing of high-pressure nitrogen with the end slag saturated or supersaturated by MgO content, and it is well sintered and adhered with furnace lining. The slag splashing layer has good corrosion resistance, thus protecting the lining brick, slowing down its damage degree and improving the lining life. The main process is as follows: after the end-point molten steel is cleaned, part of the end-point slag with the content MgO reaching saturation or supersaturation will be left, and high pressure nitrogen will be blown into the place 0.8~2m above the theoretical level of the molten pool by the spray gun to make the slag splash and stick on the surface of the furnace lining, at the same time, the slag protective layer formed. The slag splashing position can be adjusted by moving the spray gun up and down, and the slag splashing time is generally about 3min.

溅渣护炉的基本原理是：利用MgO含量达到饱和或过饱和的炼钢终点渣，通过高压氮气的吹溅，在炉衬表面形成一层高熔点的溅渣层，并与炉衬很好地烧结附着。这个溅渣层耐蚀性较好，从而保护了炉衬砖，减缓其损坏程度，炉衬寿命得到提高。其工艺过程主要是：在吹炼终点钢水出净后，留部分MgO含量达到饱和或过饱和的终点炉渣，通过喷枪在熔池理论液面以上0.8~2m处吹入高压氮气，使炉渣飞溅黏挂在炉衬表面，与此同时形成炉渣保护层。通过喷枪上下移动可以调整溅渣的部位，溅渣时间一般为3min左右。

The features of slag splashing technology are as follows:

溅渣护炉技术的特点包括：

(1) Easy to operate. After the composition is adjusted according to the slag viscosity, the oxygen supply is changed to nitrogen supply by using the oxygen gun and automatic control system, and the lance can be lowered for slag splashing operation.

(1) 操作简便。根据炉渣黏稠程度调整成分后，利用氧枪和自动控制系统，将供氧气改为供氮气，即可降枪进行溅渣操作。

(2) Low cost. The technology makes full use of the converter high basicity final slag and the by-product nitrogen of the oxygen plant. The slag splashing can be realized by adding a small amount of slag adjusting agent (such as magnesite ball, light burning dolomite, etc.) and the consumption of per ton of steel lime can also be reduced.

(2) 成本低。该技术充分利用了转炉高碱度终渣和制氧厂副产品氮气，加少量调渣剂（如菱镁球、轻烧白云石等）就可实现溅渣，还可以降低吨钢石灰消耗。

(3) Short time. Generally, it only takes 3~4min to complete slag splashing operation without affecting normal production.

(3) 时间短。一般只需 3~4min 即可完成溅渣护炉操作,不影响正常生产。

(4) The splashed slag is evenly covered on the inner wall of the whole furnace, and the shape of the furnace is basically unchanged.

(4) 溅渣均匀覆盖在整个炉膛内壁上,基本上不改变炉膛形状。

(5) Workers have low labor intensity and no environmental pollution.

(5) 工人劳动强度低,无环境污染。

(6) The furnace temperature is relatively stable, and the lining brick has no thermal shock change.

(6) 炉膛温度较稳定,炉衬砖无热震变化。

(7) Due to the increase of furnace life, the time of building furnace is saved, which is beneficial to the increase of steel output, balance and coordination of production organization.

(7) 由于炉龄提高,节省了修砌炉时间,有利于提高钢产量和平衡、协调生产组织。

(8) Due to the increase of converter operation rate and single furnace output, conditions are created for converter to realize the production mode of 'Two Blowing and Two Blowing' or 'One Blowing and One Blowing'.

(8) 由于转炉作业率和单炉产量提高,为转炉实现"二吹二"或"一吹一"的生产模式创造条件。

2.3.3.7 Spatter
2.3.3.7 喷溅

Splashing is a common phenomenon in the process of top blown converter converting. It is usually called splashing that slag and metal are carried away by furnace gas, and overflowed from furnace mouth or ejected. The splashing will cause a lot of metal and heat loss, intensify the erosion of furnace lining, and even cause the slag hanging on the stick gun, burning gun, furnace mouth and smoke hood, so as to increase the labor intensity of slag cleaning. As a result of ejecting a large amount of slag, the stability of dephosphorization, desulfurization and operation will also be affected, which limits the improvement of oxygen supply intensity. Therefore, it is very important to prevent splashing during converter operation. During the blowing period of converter, there are mainly the following types of splashes as follows:

喷溅是顶吹转炉吹炼过程中经常发生的一种现象,通常将被炉气携走、从炉口溢出或喷出炉渣和金属的现象称为喷溅。喷溅的产生会造成大量的金属和热量损失,对炉衬的冲刷加剧,甚至造成黏枪、烧枪、炉口和烟罩挂渣,增大清渣处理的劳动强度。由于喷出大量的熔渣,还会影响脱磷、脱硫及操作的稳定性,限制了供氧强度的提高。因此,在转炉操作过程中防止喷溅是十分重要的。在转炉的吹炼时期,喷溅主要有以下几种类型:

(1) Metal splashing: When the slag has not been formed in the early stage of blowing or the slag is dried back in the middle stage of blowing, the solid or high viscosity slag is pushed to the furnace wall by the top blowing oxygen jet and the CO gas discharged from the reaction zone. In

this case, the metal surface is exposed. Due to the impact of the oxygen jet, the metal droplets are ejected from the furnace mouth. This phenomenon is called metal splashing.

(1) 金属喷溅：吹炼初期炉渣尚未形成或吹炼中期炉渣返干时，固态或高黏度炉渣被顶吹氧射流和从反应区排出的 CO 气体推向炉壁。在这种情况下，金属液面裸露，由于氧气射流冲击力的作用，使金属液滴从炉口喷出，这种现象称为金属喷溅。

(2) Foam slag splashing: During the blowing process, the slag is foaming seriously due to the more active surfactant in the slag. When a large amount of CO gas is discharged from the furnace, a large amount of foam slag spilled from the furnace mouth is called foam slag splashing.

(2) 泡沫渣喷溅：吹炼过程中，由于炉渣中表面活性物质较多，使炉渣泡沫化严重。在炉内 CO 气体大量排出时从炉口溢出大量泡沫渣的现象，称为泡沫渣喷溅。

(3) Explosive splash: In the process of blowing, when there is more FeO accumulation in the slag, the temperature of the molten pool will be reduced due to the addition of too much slag material or coolant; Or when the viscosity of the slag is too large to prevent the CO gas from being discharged due to improper operation, once the temperature rising, the carbon and oxygen in the molten pool will react violently, generate a large amount of CO gas and be discharged rapidly. And at the same time, a large amount of metal and slag will be ejected from the furnace mouth, which is a sudden phenomenon. The phenomenon is called explosive splashing.

(3) 爆发性喷溅：吹炼过程中，当炉渣中（FeO）积累较多，由于加入渣料或冷却剂过多而造成熔池温度降低；或是由于操作不当，使炉渣黏度过大而阻碍 CO 气体排出时，一旦温度升高，熔池内碳与氧则剧烈反应，产生大量 CO 气体并急速排出，同时也使大量金属和炉渣喷出炉口，这种突发的现象称为爆发性喷溅。

(4) Other splashes: In some special cases, splashing may occur due to improper handling. For example, in the operation of slag retention, the oxidation of slag is strong. If the speed is too fast when mixing molten iron, the carbon in molten iron may react with the oxygen in slag, resulting in splashing of molten iron. For example, in the later stage of blowing, splashing may also be caused when the hot metal is added.

(4) 其他喷溅：在某些特殊情况下，由于处理不当也会产生喷溅。例如，在采用留渣操作时，渣的氧化性强，兑铁水时如果速度过快，可能使铁水中的碳与炉渣中的氧发生反应，引起铁水喷溅。又比如在吹炼后期，采用补兑铁水时也可能造成喷溅。

Task 2.4　EAF Steelmaking Process
任务 2.4　电弧炉炼钢工艺

The traditional EAF smelting process can be divided into three types: oxidation process, back blowing process and non oxidation process. The characteristics of the oxidation process are that the smelting process has a complete oxidation period and a complete reduction period, which can decarbonize, dephosphorize, desulfurize, degass and remove inclusions, and there is no special requirement for furnace burden, which is conducive to the improvement of steel quality. So far, the domestic oxidation smelting process is still the main method of EAF steelmaking. This section focuses on the oxidation smelting process and introduces the basic process of EAF smelting.

传统电弧炉冶炼工艺可分为氧化法、返回吹氧法和不氧化法三种类型。氧化法的特点是：冶炼过程有完整的氧化期和完整的还原期，能脱碳、脱磷、脱硫、去气、去夹杂，对炉料无特别要求，有利于钢质量的提高。到目前为止，国内氧化法冶炼工艺仍是电弧炉炼钢的主要方法。本节以氧化法冶炼工艺为主，介绍电弧炉冶炼的基本工艺。

2.4.1　Size and Classification of Electric Arc Furnace
2.4.1　电弧炉的大小与分类

Generally speaking, three parameters are used to express the size of EAF: tapping amount, rated power of transformer and furnace shell diameter. In recent years, with the development of EAF to ultra-high power and large scale, the distinction between large and small is also changing. Generally, the EAF below 40t/4.6m is regarded as small EAF, and the EAF above 50t/5.2m is regarded as large EAF. In terms of large-scale EAF, the United States leads the world trend. There are many 200st EAFs (1st = 0.907t), and 6 EAFs above 350st. In 1971, 400st/9.8m/162MV·A EAF was put into operation to produce ingots. In 2000, northwest steel wire rod company of the United States put into production the world's largest 415t electric arc furnace. The largest electric arc furnace in China is 150t. The ultra-high power and large scale of EAF improve the productivity and reduce the cost of steelmaking.

通常采用出钢量、变压器额定功率与电炉炉壳直径三个参数来表示电弧炉的大小。近年来，随着电弧炉向超高功率化、大型化发展，其大小的区分界限也在改变，通常把40t/4.6m以下的电弧炉看作小电弧炉，把50t/5.2m以上的电弧炉看作大电弧炉。就电弧炉大型化而言，美国领导世界潮流，200st级的电弧炉很多（1st=0.907t），350st以上的电弧炉有6座，并于1971年投产了400st/9.8m/162MV·A电弧炉以生产钢锭。2000年，美国西北钢线材公司投产世界最大的415t电弧炉。中国最大电弧炉为150t。电弧炉的超高功率化、大型化提高了生产率，降低了炼钢成本。

In the development process of EAF, ultra-high power and large scale have played a positive role. At present, the capacity of more EAFs is between 60~120t, and the corresponding capacity is between 300,000~800,000t/y. It is because that not only the single technology of EAF in this tonnage range is relatively perfect and mature, but also the matching and connection between EAF and refining, continuous casting, rolling, etc. are easier to optimize and more reasonable economically.

在电弧炉发展过程中，超高功率化、大型化起到了积极促进作用。目前来看，较多的电弧炉容量在60~120t之间，相应能力在30万~80万吨/年之间。这不仅是由于该吨位范围内的电弧炉本身单体技术比较完善和成熟，更重要的是由于其与精炼、连铸、轧制等在工程上的匹配与衔接更容易优化，经济上也更合理。

The classification methods of EAF are as follows:
电弧炉的分类方法具体如下：

（1）According to the properties of refractory lining, it can be divided into acid and alkaline arc furnace.

(1) 按炉衬耐火材料的性质,可分为酸性和碱性电弧炉。

(2) According to the current characteristics, it can be divided into AC and DC arc furnaces.

(2) 按电流特性,可分为交流和直流电弧炉。

(3) According to the power level, it can be divided into ordinary power, high power and ultra-high power arc furnaces.

(3) 按功率水平,可分为普通功率、高功率和超高功率电弧炉。

(4) According to scrap preheating, it can be divided into shaft furnace, double shell furnace and continuous preheating electric arc furnace.

(4) 按废钢预热,可分为竖炉、双壳炉和炉料连续预热电弧炉等。

(5) According to the way of tapping, it can be divided into slot tapping, eccentric bottom tapping (EBT), center bottom tapping (CBT) and horizontal tapping (HOT) electric arc furnace, etc.

(5) 按出钢方式,可分为槽式出钢、偏心底出钢、中心底出钢和水平出钢电弧炉等。

(6) According to the form of bottom electrode, it can be divided into contact type, conductive furnace bottom type and metal rod type DC furnace.

(6) 按底电极形式,可分为触针式、导电炉底式和金属棒式直流炉。

2.4.2 Traditional EAF Steelmaking Process
2.4.2 传统电炉炼钢工艺

The traditional oxidation smelting process consists of six stages: furnace filling, charging, melting, oxidation, reduction and tapping. It is mainly divided into three stages: melting stage, oxidation stage and reduction stage, commonly known as the 'Old Three Stages'. The 'Old Three Stage' process can not meet the development of modern metallurgical industry because of its low equipment utilization, low productivity and high energy consumption. It must be reformed, but it is the basis of EAF steelmaking.

传统的氧化法冶炼工艺操作过程由补炉、装料、熔化、氧化、还原与出钢六个阶段组成,其主要分为熔化期、氧化期和还原期,俗称"老三期"。传统电炉"老三期"工艺因其设备利用率低、生产率低、能耗高等缺点,满足不了现代冶金工业的发展。因此必须进行改革,但它是电炉炼钢的基础。

2.4.2.1 Furnace Repair
2.4.2.1 补炉

Furnace Repair Position
补炉部位

The working conditions of each part of the furnace lining are different, and the damage conditions are different too. The main part of lining damage is slag line on the furnace wall, which is seriously damaged by the radiation of high temperature electric furnace, chemical erosion and mechanical erosion of slag steel, smelting operation, etc. Because of the erosion of slag steel near the tap hole, it is easy to thin. The damage is also serious due to the action of frequent thermal shock on both sides of the furnace door, the scouring of slag and the collision between operation and tools. Therefore, after tapping, the slag line, tap hole, furnace door and other parts of the electric furnace should be repaired. No matter for gunning or patching, these parts should be repaired.

炉衬各部位的工作条件不同,损坏情况也不一样。炉衬损坏的主要部位是炉壁渣线,渣线受到高温电炉的辐射、渣钢的化学侵蚀与机械冲刷,以及冶炼操作等的作用损坏严重。出钢口附近因受渣钢的冲刷也极易减薄。炉门两侧由于常受热震的作用、流渣的冲刷及操作与工具的碰撞等,损坏也比较严重。因此,一般电炉在出钢后要对渣线、出钢口及炉门附近等部位进行修补,无论进行喷补或投补,均应重点补好这些部位。

Principles of Furnace Repair
补炉原则

The principles of furnace mending are high temperature, fast mending and thin mending. Remelting is to spray remelting materials to the lining damage, with the help of the residual heat in the furnace to make the new refractory and the original lining sintering into a whole under high temperature. And this sintering needs a high temperature to complete. After the electric furnace tapped, the temperature of the lining surface drops rapidly. So we should seize the time to make up quickly while it is hot. The purpose of thin patching is to ensure good sintering of refractories. Experience shows that the thickness of the new patching should not be more than 30mm at a time, and it should be layered for many times when it needs to be thicker.

补炉的原则是高温、快补、薄补。补炉是将补炉材料喷投到炉衬损坏处,并借助炉内的余热在高温下使新补的耐火材料和原有的炉衬烧结成为一个整体,而这种烧结需要很高的温度才能完成。电炉出钢后,炉衬表面温度下降很快,因此应该抓紧时间趁热快补。薄补的目的是为了保证耐火材料良好的烧结。经验表明,新补的厚度一次不应大于30mm,需要补得更厚时应分层多次进行。

2.4.2.2 Loading
2.4.2.2 装料

At present, the top charging basket is widely used in electric furnaces, and the charging of each furnace is divided into 1~3 times. The quality of charging affects lining life, smelting time, power consumption, electrode consumption and alloy element burning loss, etc. Therefore, it is required to charge reasonably, and the quality of charging depends on whether the charging is reasonable or not in the furnace basket.

目前电炉广泛采用炉顶料筐装料,每炉钢的炉料分1~3次加入。装料的好坏影响着炉衬寿命、冶炼时间、电耗、电极消耗以及合金元素的烧损等。因此要求装料合理,而装料的好坏取决于炉料在炉筐中布料的合理与否。

The orders of reasonable cloth are as follows:
合理布料的顺序如下:

(1) The large, medium and small materials must be reasonably distributed when

loading. Generally, a layer of lime (except for steel operation, conductive furnace bottom, etc.) shall be evenly paved on the furnace bottom, which is 2%~3% of the charging amount, so as to protect the furnace bottom from the slag in advance.

（1）装料时必须将大、中、小块料合理布料。一般先在炉底上均匀地铺一层石灰（留钢操作、导电炉底等除外），为装料量的2%~3%，以保护炉底，同时可提前造渣。

（2）If the furnace bottom is normal, small pieces of material are paved on the lime, which is about 1/2 of the total amount of small pieces of material, so as to avoid the large pieces of material directly impacting the furnace bottom.

（2）如果炉底正常，在石灰上面铺小块料，约为小块料总量的1/2，以免大块料直接冲击炉底。

（3）Large block and refractory materials are loaded on the small block and arranged in the high temperature area of the arc to accelerate the melting. Fill medium and small materials between large materials to improve the loading density. The medium block material is generally installed on and around the large block material, which can not only fill the gap around the large block material, but also accelerate the melting of the furnace material near the furnace wall.

（3）小块料上再装大块料和难熔料，并布置在电弧高温区，以加速熔化。在大块料之间填充中、小块料，以提高装料密度。中块料一般装在大块料的上面及四周，不仅可填充大块料周围的空隙，也可加速靠炉壁处的炉料熔化。

In a word, during the distribution, it should be compact at the bottom, loose at the top, high in the middle and low around, and no big materials at the furnace door, so that the well can be penetrated quickly after power transmission without bridging, which is conducive to the smooth melting.

总之，布料时应做到下致密、上疏松、中间高、四周低、炉门口无大料，使得送电后穿井快，不搭桥，有利于熔化的顺利进行。

2.4.2.3　Melting Period
2.4.2.3　熔化期

The melting period of traditional process accounts for 50%~70% of the whole smelting time, and the power consumption accounts for 60%~80%. Therefore, the length of melting period affects the productivity and power consumption, and the operation of melting period affects the oxidation period and reduction period.

传统工艺的熔化期占整个冶炼时间的50%~70%，电耗占60%~80%。因此，熔化期的长短影响生产率和电耗的高低，熔化期的操作影响氧化期、还原期的顺利与否。

Main Tasks of Melting Period
熔化期的主要任务

The main tasks of melting period are as follows:
熔化期的主要任务是:

（1）The massive solid charge is melted rapidly and heated to the oxidation temperature.

（1）将块状的固体炉料快速熔化，并加热到氧化温度。

(2) Early slagging and early dephosphorization.

(2) 提前造渣,早期去磷。

(3) Reduce liquid steel suction and volatilization.

(3) 减少钢液吸气与挥发。

Operation during Melting Period
熔化期的操作

The operation contents of melting period are reasonable power supply, oxygen blowing in time and slagging in advance. Among them, a reasonable power supply system is an important guarantee for the smooth progress of the melting period.

熔化期的操作内容主要是合理供电、及时吹氧和提前造渣。其中,合理供电制度是使熔化期顺利进行的重要保证。

After charging, it can be electrified and melted. However, the electrode shall be adjusted before power supply to ensure that the electrode will not be switched during the whole smelting process. And the furnace cooling system and insulation shall be checked as necessary. The melting process of furnace charge can be roughly divided into four stages as follows:

装料完毕即可通电熔化。但在供电前应调整好电极,保证整个冶炼过程中不切换电极,并对炉子冷却系统及绝缘情况进行必要的检查。炉内炉料的熔化过程大致可分为以下四个阶段:

(1) Arcing period. At the beginning of power on, under the action of electric arc, a small number of elements volatilize are oxidized by furnace gas to generate red brown smoke, which escapes from the furnace. The arc initiation period (2~3min) is the depth of $1.5d_{electrode}$ drop from the power supply initiation to the electrical extreme. The current is unstable and the arc burns radiation near the top of the furnace. During this period, the electric current is unstable, and the arc burns and radiates near the furnace top. In order to protect the top of the furnace, some light and thin materials are placed on the top of the furnace, so that the electrode can be inserted into the material quickly to reduce the radiation of the arc on the top of the furnace. Low voltage and current are used in power supply.

(1) 起弧期。通电开始,在电弧的作用下,一少部分元素挥发并被炉气氧化,生成红棕色的烟雾,从炉中逸出。从送电起弧至电极端部下降 $1.5d_{电极}$ 深度,为起弧期(2~3min)。此期电流不稳定,电弧在炉顶附近燃烧辐射。为了保护炉顶,在炉上部布一些轻薄小料,以便使电极快速插入料中,以减少电弧对炉顶的辐射。供电方面采用较低的电压和电流。

(2) Well penetration period. From the end of the arc to the end of the electrode to the bottom of the furnace, it is the well penetrating period. In this period, although the arc is covered by the furnace charge, the arc combustion is unstable due to the continuous occurrence of material collapse. In the aspect of power supply, large secondary voltage, large current or high voltage live reactance operation shall be adopted to increase the diameter and speed of well penetration. However, it should be paid attention to the protection of the furnace bottom by taking lime as the bottom before charging, spreading large and heavy scrap steel in the middle of the

furnace and adopting a reasonable furnace type.

（2）穿井期。从起弧完毕至电极端部下降到炉底，为穿井期。此期虽然电弧被炉料遮蔽，但因不断出现塌料现象，电弧燃烧不稳定。供电方面采取较大的二次电压、大电流或采用高电压带电抗操作，以增加穿井的直径与穿井的速度。但应注意保护炉底，其办法是：加料前采取石灰垫底，炉中部布大、重废钢，以及采用合理的炉型。

（3）Main melting period. The main melting period begins when the electrode starts to pick up after falling to the furnace bottom. With the continuous melting of the charge, the electrode gradually rises. When the charge is basically melted (more than 80%), only a small amount of charge exists near the furnace slope and slag line. When the arc begins to be exposed to the furnace wall, the main melting period ends. In the main melting period, because the arc is buried in the furnace charge, the arc is stable, the thermal efficiency is high, and the heat transfer condition is good. So the maximum power supply should be adopted, and the maximum voltage and current supply should be adopted. The main melting period accounts for 70% of the whole melting period.

（3）主熔化期。电极下降至炉底后开始回升时，主熔化期开始。随着炉料不断地熔化，电极逐渐上升，至炉料基本熔化（大于80%）时，仅炉坡、渣线附近存有少量炉料。电弧开始暴露给炉壁时，主熔化期结束。在主熔化期内，由于电弧埋入炉料中，电弧稳定，热效率高，传热条件好，故应以最大功率供电，即应采用最高电压和最大电流供电。主熔化期时间占整个熔化期的70%。

（4）Temperature rise period at the end of melting. From the beginning of the arc exposure to the furnace wall to the melting of the furnace charge, it is the temperature rise period of the end of fusion. In this stage, the furnace wall is exposed, especially the hot spot area of the furnace wall, which is strongly radiated by the arc. So it should be paid attention to the protection. At this point, low voltage and high current can be used in power supply, otherwise foam slag submerged arc process should be adopted.

（4）熔末升温期。从电弧开始暴露给炉壁至炉料全部熔化，为熔末升温期。此阶段因炉壁暴露，尤其是炉壁热点区的暴露，受到电弧的强烈辐射，故应注意保护。此时供电方面可采取低电压和大电流，否则应采取泡沫渣埋弧工艺。

2.4.2.4　Oxidation Period
2.4.2.4　氧化期

In order to remove phosphorus, gas and inclusions from steel, it is necessary to use oxidation smelting. Oxidation period is the main process of oxidation smelting. In the traditional smelting process, when the slag such as scrap is completely melted and reaches the oxidation temperature with phosphorus removed by more than 70%, it will enter the oxidation period, which will end when the oxidation slag is removed. In order to ensure the metallurgical reaction, the starting temperature of oxidation should be 50~80℃ higher than the melting point of molten steel.

要去除钢中的磷、气体和夹杂物，必须采用氧化法冶炼。氧化期是氧化法冶炼的主要过程。传统冶炼工艺中，当废钢等炉料完全熔化并达到氧化温度、磷脱70%以上时便进入氧化期，这一阶段到扒完氧化渣时结束。为保证冶金反应的进行，氧化开始温度应高于钢液熔点 50~80℃。

Main Tasks of Oxidation Period

氧化期的主要任务

The main tasks of oxidation period are as follows：

氧化期的主要任务是：

(1) Further reduce the content of P in the molten steel to make it lower than half of the finished product specification. Considering the reduction period and the possibility of phosphorus recovery in the ladle, the general steel type requires $w[P]=0.015\% \sim 0.01\%$.

(1) 进一步降低钢液中的磷含量，使其低于成品规格的一半。考虑到还原期及钢包中可能回磷，一般钢种要求 $w[P]=0.015\% \sim 0.01\%$。

(2) Remove gas and non-metallic inclusions from molten steel. Degassing and inclusion removal of molten steel are carried out in the oxidation period. With the help of C-O reaction and CO bubble floatation, the molten pool will boil violently to remove gas and inclusions, and even the composition and temperature. Therefore, it is necessary to control the decarburization reaction speed to ensure that the pool has a certain time of intense boiling.

(2) 去除钢液中气体和非金属夹杂物。电炉炼钢钢液去气、去夹杂物是在氧化期内进行的。它是借助 C-O 反应和 CO 气泡的上浮使熔池产生激烈沸腾，促使气体和夹杂物去除，并均匀成分与温度。为此，一定要控制好脱碳反应速度，保证熔池有一定的激烈沸腾时间。

(3) Heating and uniform molten steel temperature. The temperature at the end of oxidation should be 20~30℃ higher than the tapping temperature, which is mainly considered as follows：

(3) 加热和均匀钢水温度。应使氧化末期温度高于出钢温度 20~30℃，这主要考虑以下两点：

1) Slagging, making new slags and adding alloy will make the molten steel cool down.

1) 扒渣、造新渣以及加合金将使钢液降温。

2) It is not allowed to raise the temperature of molten steel in the reduction period, otherwise the molten steel under the arc will overheat, the furnace lining will be damaged, and the molten steel will be inhaled due to the reflection of high current arc light.

2) 不允许钢液在还原期升温，否则电弧下的钢液过热，大电流弧光反射会损坏炉衬以及使钢液吸气。

(4) Oxidation and decarbonization. According to the different sources of oxygen in the molten pool, the operation methods in the oxidation period are divided into three kinds：ore oxidation method, blowing oxygen oxidation method and ore oxygen comprehensive oxidation method. In recent years, the practice of strengthening oxygen use shows that oxygen blowing oxidation is used, except for the comprehensive oxidation method of mineral oxygen when the content of P in steel is particularly high. Especially when the dephosphorization task is not heavy, the carbon content in steel should be reduced by strengthening oxygen blowing oxidation liquid steel.

(4) 氧化与脱碳。按照熔池中氧来源的不同，氧化期操作方法分为矿石氧化法、吹氧氧化法和矿氧综合氧化法三种。近年来通过强化用氧的实践表明，除了钢中 P 含量特别高时采用矿氧综合氧化法外，均采用吹氧氧化。尤其是当脱磷任务不重时，应通过强化吹氧氧化钢液来降低钢中碳含量。

Process Operation in Oxidation Period
氧化期的工艺操作

- Slagging System
- 造渣制度

The requirements for the slag in the oxidation period are sufficient oxidation performance, appropriate basicity, slag quantity and good physical properties, so as to ensure the successful completion of the oxidation period. The characteristics of dephosphorization and decarbonization should be taken into account in the slagging of oxidation process. The common requirement of the two is that the slag has good fluidity and high oxidation ability. The differences between the two are that dephosphorization requires a large amount of slag, continuous slag flow and new slag formation, and the basicity is 2.5~3; decarbonization requires a thin slag layer, which is convenient for CO bubbles to escape through the slag layer, and the basicity of slag is about 2.

对氧化期炉渣的要求是具有足够的氧化性能、合适的碱度与渣量以及良好的物理性能，从而保证能够顺利完成氧化期的任务。氧化过程的造渣应兼顾脱磷和脱碳的特点。两者共同的要求是：炉渣的流动性良好，且有较高的氧化能力。两者的不同是：脱磷要求渣量大，不断流渣和造新渣，碱度以 2.5~3 为宜；脱碳要求渣层薄，便于 CO 气泡穿过渣层逸出，炉渣碱度约为 2。

The amount of slag in oxidation period is determined according to dephosphorization task. When the dephosphorization task is completed, the amount of slag is suitable to stabilize the arc combustion. In general, the amount of slag in oxidation period should be controlled in the range of 3%~5%.

氧化期的渣量是根据脱磷任务而确定的。在完成脱磷任务时，渣量以能稳定电弧燃烧为宜。一般氧化期的渣量应控制在 3%~5% 范围内。

- Temperature Regime
- 温度制度

Temperature control is very important for thermodynamics and kinetics of metallurgical reaction. From the later stage of melting, temperature conditions should be created for the oxidation period to ensure high temperature oxidation and lay a good foundation for the reduction period.

温度控制对于冶金反应的热力学和动力学都是十分重要的。从熔化后期就应该为氧化期创造温度条件，以保证高温氧化并为还原期打好基础。

Since the decarbonization reaction can only be carried out smoothly under certain temperature conditions, the starting oxidation temperature is specified in the field, whether ore oxidation method, ore oxygen comprehensive oxidation method or blowing oxygen oxidation method. The temperature at the end of oxidation (slag raking temperature) is generally 40~60℃ higher than that at the beginning of oxidation, because many elements in steel have been oxidized, which makes the melting point of steel increase. In addition, there is a great heat loss in slag raking, and heat is also required for melting reduction slag and alloy, so the temperature at the end of oxidation is generally controlled at 110~130℃ above the melting point of steel (1470~1520℃). The tapping temperature of the electric furnace should be 90~110℃ higher than the melting point of the steel grade that the temperature of slagging at the end of oxidation should generally be 10~20℃ higher than the tapping temperature of the steel grade.

由于脱碳反应必须在一定的温度条件下才能顺利进行,在现场中无论是采用矿石氧化法、矿氧综合氧化法还是吹氧氧化法,都规定了开始氧化的温度。氧化终了的温度(扒渣温度)一般应比开始氧化的温度高出40~60℃,其原因是钢中许多元素已经氧化,使钢的熔点有所升高;另外,扒除氧化渣有很大的热量损失,而熔化还原渣料和合金料也需要热量,所以氧化结束时的温度一般控制在高出钢熔点(1470~1520℃)110~130℃以上。电炉出钢温度应高出钢种熔点90~110℃,即氧化末期扒渣温度一般应高于该钢种的出钢温度10~20℃。

- Oxidation Operation
- 氧化操作

The process operation methods in the oxidation stage are divided into ore oxidation method, blowing oxygen oxidation method and ore oxygen comprehensive oxidation method, which are discussed as follows:

氧化期的工艺操作方法分为矿石氧化法、吹氧氧化法和矿氧综合氧化法。其介绍如下:

(1) Ore oxidation: Ore oxidation is a kind of indirect oxidation method, which is to add the high valent iron oxide (Fe_2O_3 or Fe_3O_4) in iron ore into the molten pool to make it change into low valent iron oxide (FeO). A small part of FeO remains in the slag, and most of it is used for the oxidation of carbon and phosphorus in molten steel. This method can be used in small factories where there is a lack of oxygen. However, dephosphorization and decarbonization are easy to cooperate with each other.

(1) 矿石氧化法:矿石氧化法是一种间接氧化法,它是将铁矿石中的高价氧化铁(Fe_2O_3或Fe_3O_4)加入到熔池中,使其转变成低价氧化铁(FeO)。FeO小部分留在渣中,大部分用于钢液中碳和磷的氧化。此法可应用于缺乏氧气的地方小厂。矿石氧化法炉内冶炼温度较低,致使氧化时间延长,但脱磷和脱碳反应容易相互配合。

(2) Oxygen blowing oxidation method: Oxygen blowing oxidation is a direct oxidation method, that oxygen is directly blown into the molten pool, and carbon and other elements in the steel are oxidized. Under the same carbon content, the content of FeO in slag is much lower than that in ore oxidation when oxygen is used alone. Therefore, it is easier to stabilize the molten pool after stopping oxygen blowing than ore used for oxidation. The temperature of the molten pool is relatively high, and the oxidation loss of W, Cr, Mn and other elements in the steel is also less, but it is not conducive to dephosphorization. So it is not suitable to use when the content of P is high after melting.

(2) 吹氧氧化法:吹氧氧化法是一种直接氧化法,即直接向熔池吹入氧气,氧化钢中碳等元素。单独采用氧气进行氧化操作时,在碳含量相同的情况下,渣中FeO含量远远低于矿石氧化时的含量。因此,停止吹氧后熔池比用矿石氧化时更容易趋于稳定,熔池温度比较高,钢中 W、Cr、Mn 等元素的氧化损失也较少,但不利于脱磷。所以在熔清后 P 含量高时不宜采用。

(3) Comprehensive oxidation of mineral oxygen: Ore oxygen comprehensive oxidation method refers to adding ore in the early stage of oxidation and blowing oxygen in the later stage. Both of them complete the task of oxidation stage together, which is a commonly used method in production.

(3) 矿氧综合氧化法:矿氧综合氧化法是指氧化前期加矿石,后期吹氧,两者共同完成氧化期的任务。该方法是生产中常用的一种方法。

When dealing with the relationship between dephosphorization and decarburization, the following process operation rules should be followed: in the oxidation sequence, first phosphorus and then carbon; in the temperature control, first low temperature and then high temperature; in the slag making, first large amount of slag dephosphorization and then decarburization of thin slag layer; in the oxygen supply, first ore and then oxygen.

在处理脱磷和脱碳的关系时,应遵守以下工艺操作制度:在氧化顺序上,先磷后碳;在温度控制上,先低温后高温;在造渣上,先大渣量去磷后薄渣层脱碳;在供氧上,先矿后氧。

2.4.2.5 Reduction Period
2.4.2.5 还原期

The period from the end of oxidation to tapping is called reduction period. Reduction period is one of the important characteristics of EAF steelmaking process.

从氧化末期扒渣完毕到出钢这段时间称为还原期。电炉有还原期是电炉炼钢法的重要特点之一。

Restore Tasks
还原期的任务

The tasks of the restore period are as follows:

还原期的任务是:

(1) Deoxidize the molten steel and remove the dissolved oxygen (no more than 0.003%) and oxide inclusions in the molten steel as much as possible.

(1) 使钢液脱氧,尽可能地去除钢液中溶解的氧量(不大于 0.003%)和氧化物夹杂。

(2) Remove the sulfur from the steel to the steel specification.

(2) 将钢中的硫去除至钢种规格要求。

(3) Adjust the composition of liquid steel alloy to ensure that the content of all elements in the finished steel meets the standard requirements.

(3) 调整钢液合金成分,保证成品钢中所有元素的含量都符合标准要求。

(4) The composition of slag is adjusted so that the basicity of slag is suitable and the fluidity is good.

(4) 调整炉渣成分,使炉渣碱度合适、流动性良好,有利于脱氧和去硫。

(5) Adjust the temperature of molten steel to ensure normal smelting and good pouring temperature.

These tasks are closely related to each other. Generally speaking, deoxidation is the core, temperature is the condition and slagging is the guarantee.

(5) 调整钢液温度,确保冶炼正常进行并有良好的浇注温度。

这些任务互相之间有着密切的联系,一般认为:脱氧是核心,温度是条件,造渣是保证。

Temperature Control

温度控制

The temperature control during reduction period is particularly important. Considering the temperature loss from tapping to pouring, the tapping temperature should be 100~140℃ higher than the melting point of steel.

还原期的温度控制尤为重要,考虑到出钢到浇铸过程中的温度损失,出钢温度应比钢的熔点高出 100~140℃。

Since the temperature of molten steel at the end of oxidation is 20~30℃ higher than that of tapping, the temperature control of reduction period after slagging is actually a heat preservation process. If the temperature rises greatly in the reduction period, the reasons are that: the liquid steel sucks heavily; the high temperature arc aggravates the erosion of the furnace lining; and the local liquid steel is overheated. Therefore, it is necessary to avoid heating up operation during reduction period.

由于氧化末期控制钢液温度高于出钢温度 20~30℃,扒渣后还原期的温度控制实际上是保温过程。如果还原期大幅度升温,其原因包括以下三种:钢液吸气严重;高温电弧加重对炉衬的侵蚀;局部钢水过热。因此,应避免在还原期时进行升温操作。

Deoxygenation Operation

脱氧操作

The comprehensive deoxidization method of ore oxygen is commonly used in electric furnace, in which deoxidization is the core of reduction operation. The composite deoxidizers commonly used in steelmaking include silicon manganese, silicon calcium, silicon manganese aluminum and other alloys, as well as carbon powder and calcium carbide (CaC_2). The brief description is as follows:

电炉常用矿氧综合脱氧法,其中还原操作以脱氧为核心,炼钢中常用的复合脱氧剂有硅锰、硅钙、硅锰铝等合金以及炭粉和电石(CaC_2),简述如下:

(1) When the temperature, the content of P and C in the molten steel meet the requirements, the amount of slagging is more than 95%.

(1) 当钢液的温度、P 含量和 C 含量符合要求时,扒渣量大于95%。

(2) Pre deoxidization (precipitation deoxidization) with Fe-Mn and Fe-Si blocks was added.

(2) 加 Fe-Mn、Fe-Si 块等预脱氧(沉淀脱氧)。

(3) Add lime, fluorite or brick to make thin slag.

(3) 加石灰、萤石或砖块，造稀薄渣。

(4) After the formation of the thin slag, deoxidization (diffusion deoxidization) is carried out by adding carbon powder and Fe-Si powder, which is divided into 3~5 batches with a time of 7~10min/batch (This is the reason for the long reduction period of "Old Third Stage" steelmaking).

(4) 稀薄渣形成后还原，加炭粉、Fe-Si 粉等脱氧（扩散脱氧），分 3~5 批，时间为 7~10min/批（这就是"老三期"炼钢还原期时间长的原因）。

(5) Mixing, sampling and temperature measurement.

(5) 搅拌，取样，测温。

(6) Adjust composition (alloying).

(6) 调整成分（即合金化）。

(7) Add final deoxidation (precipitation deoxidation) such as Al or Ca-Si block.

(7) 加 Al 或 Ca-Si 块等终脱氧（沉淀脱氧）。

Alloying of Liquid Steel
钢液的合金化

The process of adjusting the composition of molten steel in steelmaking is called alloying. The alloying of traditional EAF steelmaking can be carried out in the process of charging, oxidation and reduction. And the alloy can also be added to the ladle during tapping. Generally, pre alloying is carried out at the end of oxidation and reduction. The fine adjustment of alloy composition is carried out at the end of reduction before tapping or during tapping. Alloying operation mainly refers to the determination of alloy addition time and quantity.

炼钢过程中调整钢液合金成分的过程称为合金化。传统电炉炼钢的合金化可以在装料、氧化、还原过程中进行，也可在出钢时将合金加到钢包里。一般是在氧化末期、还原初期进行预合金化，在还原末期、出钢前或出钢过程中进行合金成分微调。合金化操作主要是指确定合金加入时间与加入数量。

The adding principles of alloy elements are as follows. According to the binding capacity of alloy elements and oxygen, the adding time in furnace is determined. For alloy elements that are not easy to oxidize (such as Co, Ni, Cu, Mo, W, etc.), most of them are loaded with the furnace charge, and a few are added in the oxidation or reduction period. When W is added in the oxidation process, it is usually added with the thin slag. For elements that are easy to oxidize, such as Mn and Cr (less than 2%), they are usually added in the early stage of reduction. FEV (less than 0.3%) is added 5~8min before tapping. For alloy elements that are easy to oxidize, such as Al, Ti, B and rare earth, they shall be added before tapping or in the molten steel tank. Generally speaking, alloy elements with large amount should be added early, and those with small amount should be added late.

合金元素的加入原则为：根据合金元素与氧的结合能力大小，决定其在炉内的加入时间。对不易氧化的合金元素（如 Co、Ni、Cu、Mo、W 等），多数随炉料装入，少量在氧化期或还原期加入。氧化法加 W 元素时，一般随稀薄渣料加入。对较易氧化的元素，如

Mn、Cr（小于2%），一般在还原初期加入。钒铁（小于0.3%）在出钢前5~8min加入。对极易氧化的合金元素，如Al、Ti、B、稀土，在出钢前或在钢水罐中加入。一般来说，合金元素加入量大的应早加，加入量小的宜晚加。

Tapping Operation

出钢操作

In the traditional EAF steelmaking process, when the chemical composition of liquid steel is qualified, the temperature meets the requirements, the deoxidization is good, the basicity and fluidity of slag are suitable, and the steel can be produced. In the process of tapping, the contact between steel and slag can further deoxidize and desulfurize, so the tapping method of 'Big Mouth, Deep Drawing and Steel—Slag Mixing' is required.

传统电炉炼钢工艺中，钢液经氧化、还原后，当其化学成分合格、温度合乎要求、脱氧良好、炉渣碱度与流动性合适时即可出钢。因出钢过程中钢与渣接触可进一步脱氧与脱硫，故要求采取"大口、深冲、钢—渣混合"的出钢方式。

The traditional 'Old Three Stage' smelting process of electric furnace integrates melting, refining and alloying into one furnace, including melting, oxidation and reduction stages. In the furnace, the melting of scrap steel, dephosphorization, desulfurization, degassing, inclusion removal and temperature rise of molten steel, deoxidization, desulfurization, alloying and temperature, and composition adjustment of molten steel are required, so the smelting cycle is very long. It is not only difficult to meet the more and more strict quality requirements for steel, but also limits the improvement of furnace productivity.

传统电炉"老三期"冶炼工艺操作集熔化、精炼和合金化于一炉，包括熔化期、氧化期和还原期。该工艺在炉内既要完成废钢的熔化，钢液的脱磷、脱硫、去气、去夹杂以及升温，又要进行钢液的脱氧、脱硫、合金化以及温度和成分的调整，因而冶炼周期很长。因此，该工艺既难以满足对钢材越来越严格的质量要求，又限制了电炉生产率的提高。

Task 2.5　Refining Outside the Furnace
任务2.5　炉外精炼

2.5.1　Basic Means of Refining Outside the Furnace
2.5.1　炉外精炼的基本手段

So far, in order to create the best metallurgical reaction conditions, the basic means are slag washing, stirring, vacuum, heating, spraying and feeding. At present, a variety of off-furnace refining methods are also different combinations of these basic means.

到目前为止，为了创造最佳的冶金反应条件，所采用的基本手段不外乎渣洗、搅拌、真空、加热、喷吹及喂丝等几种。目前，大多数炉外精炼方法也都是这些基本手段的不同组合。

2.5.1.1　Slag Washing
2.5.1.1　渣洗

Slag washing method is to use the impact of steel flow to mix the synthetic slag in ladle with steel liquid when steel is produced. The traditional arc furnace has an early reduction period, and the white slag is used to reduce and refine the molten steel, which can be considered as a typical synthetic slag washing method.

渣洗法是在出钢时利用钢流的冲击作用使钢包中的合成渣与钢液混合，以精炼钢液的方法，也是最早出现的炉外精炼方法。传统电弧炉还原期早，白渣对钢液进行还原精炼，因此可认为是一种典型的合成渣洗方法。

According to the different synthetic slag refining methods, slag washing can be divided into the Co-slag washing and Heterogeneous slag washing. Co-slag washing is a process in which the liquid slag used for slag washing and molten steel are first refined in the same container, and the molten steel is made to have the composition and properties of synthetic slag. Heterogeneous slag washing is the process of refining the slag with a certain ratio into liquid slag with certain composition and metallurgical properties.

根据合成渣炼制方式的不同，渣洗可分为同炉渣洗和异炉渣洗。同炉渣洗是先将用于渣洗的液渣和钢液在同一容器内炼制，并使钢液具有合成渣的成分和性质，然后通过出钢最终完成渣洗钢液的过程。异炉渣洗是将配比一定的渣料炼制成具有一定成分和冶金性质的液渣，出钢时钢液冲入事先盛有渣的钢包内实现渣洗。

The metallurgical purpose and metallurgical effect of slag washing are different due to different refining methods. The following metallurgical effects can be achieved comprehensively: strengthening deoxidation and desulfurization; removing the inclusion in the steel and partially changing the inclusion shape; preventing molten steel from being aspirated; reducing the temperature loss of molten steel; foam slag formed to achieve the purpose of submerged arc heating.

由于炉外精炼方法不同，渣洗的冶金目的和冶金效果也不同，综合起来可达到以下冶金效果：强化脱氧，强化脱硫；去除钢中的夹杂物，部分改变夹杂物的形态；防止钢液吸气；减少钢水温度散失；形成泡沫渣，以达到埋弧加热的目的。

2.5.1.2　Stir
2.5.1.2　搅拌

Most of the reactions in the metallurgical process are controlled by mass transfer. So in order to speed up the metallurgical reaction, it is necessary to strengthen the molten steel agitation. Stirring of molten steel is the most basic and important means of refining outside the furnace, which can improve the metallurgical reaction kinetics conditions, strengthen the mass transfer and heat transfer of the reaction system. and accelerate the metallurgical reaction, the composition and temperature of molten steel. The steel is also conducive to the polymerization and removal of inclusions.

冶金过程中的绝大多数反应都是由传质控制的，因此为了加快冶金反应的进行，首先要强化钢液搅拌。对钢液进行搅拌是炉外精炼最基本、最重要的手段，其可改善冶金反应动力学条件，强化反应体系的传质和传热，加速冶金反应，均匀钢液的成分和温度，有利于夹杂物的聚合长大和上浮排除。

The stirring methods in refining outside the furnace mainly include gas stirring, electromagnetic stirring, gravity (or negative pressure) driven stirring and mechanical stirring. Although mechanical stirring, electromagnetic stirring, gravity (or negative pressure) driven stirring have been successfully applied in various stirring methods of off-furnace refining, they are only used in a small number of off-furnace refining. The most widely used stirring methods are various gas stirring methods.

炉外精炼中的搅拌方式主要有气体搅拌、电磁搅拌、重力（或负压）驱动搅拌和机械搅拌四类。在炉外精炼的各种搅拌方法中，虽然机械搅拌、电磁搅拌、重力（或负压）驱动搅拌都有十分成功的应用实例，但却只在少数炉外精炼中使用。应用最广泛的搅拌方法是各种形式的气体搅拌方法。

Gas stirring, also known as bubble stirring, usually comes in two forms:

气体搅拌也称为气泡搅拌，通常有如下两种形式：

(1) Bottom blowing argon. Bottom blowing argon is mostly through the installation of the bottom of the ladle at a certain position of the plug blowing argon. The advantages of this method are: uniform temperature and composition of molten steel, and good effect of removing inclusions; simple equipment, easy to operate, and without occupying a fixed operation site; and argon blown in the process of steel production or in transit. The form is the most common way.

(1) 底吹氩。底吹氩大多数是通过安装在钢包底部一定位置的透气砖吹入氩气。这种方法的优点是：均匀钢水温度和成分，去除夹杂物的效果好；设备简单，操作灵便，不需占用固定操作场地；可在出钢过程或运输途中吹氩。此种方式最为常用。

(2) Top blow argon. Top blowing argon is to blow argon gun from the top of the ladle with molten steel to blow argon stirring, it is required to set up a fixed blowing argon station. The method is stable in operation and can also be sprayed with powder. However, top blowing argon stirring effect is not as good as bottom blowing argon.

(2) 顶吹氩。顶吹氩是将吹氩枪从钢包上部浸入钢水来进行吹氩搅拌，要求设立固定吹氩站。该方法操作稳定，也可喷吹粉剂。但是，顶吹氩的搅拌效果不如底吹氩好。

Electromagnetic stirring is to use the principle of electromagnetic induction, the electromagnetic induction agitator installed outside the ladle generates a directional electromagnetic stirring force in the molten steel, so as to achieve the purpose of liquid steel circulation stirring. For electromagnetic stirring, parts of the ladle housing close to the induction stirring coil shall be made of austenitic stainless steel.

电磁搅拌是利用电磁感应原理，用装置在钢包外的电磁感应搅拌器在钢液中产生一个定向的电磁搅拌力，以达到钢液循环搅拌的目的。为进行电磁搅拌，靠近电磁感应搅拌线圈的部分钢包壳应由奥氏体不锈钢制造。

Gravity (or negative pressure) driving stirring is to use the drop to make the molten steel under the action of gravity or the use of negative pressure in the action of driving gas, with a certain impact kinetic energy into the ladle or container to achieve the purpose of stirring or mixing. Typical gravity (or negative pressure) driven stirring method is vacuum pouring method (VC method), using the combined effect of gravity and negative pressure to produce the mixing of the furnace refining method, such as RH method and DH method, also known as the circular stirring method.

重力或负压驱动搅拌是利用落差使钢水在重力作用下或利用负压在驱动气体作用下，以一定的冲击动能冲入钢包或容器中，以达到搅拌或混合的目的。典型的重力（或负压）驱动搅拌法有真空浇注法（VC 法），利用重力和负压综合作用而产生搅拌的炉外精炼方法有 RH 法和 DH 法，也有人称其为循环搅拌法。

Mechanical stirring is to achieve the purpose of stirring and mixing materials by rotating blades or propellers or rotating, vibrating, rotating containers and other mechanical methods. In the metallurgical high-temperature system, only a few examples of mechanical stirring method for mixing and blending.

机械搅拌是通过叶片或螺旋桨等部件的旋转或旋转、振动、转动容器等机械方法，达到搅拌、混匀物料的目的。在冶金高温体系中，只有很少量的例子采用机械搅拌方式进行搅拌、混匀。

2.5.1.3 Vacuum
2.5.1.3 真空

Vacuum is an important treatment method widely used in molten steel refining. Nearly 2/3 of the more than 40 refining methods are equipped with vacuum equipment. Vacuum has a significant impact on the reactions in which gases participate, mainly including the carbon dissolved in molten steel and the reaction in which CO is formed, and the dissolution and removal of gases (H_2, N_2) in molten steel. Blowing oxygen refining under vacuum can improve the deoxidization capacity of carbon, so as to enhance the process of decarbonization and deoxidization of carbon.

真空是钢水炉外精炼中广泛应用的一种重要的处理手段。目前采用的 40 余种炉外精炼方法中，将近2/3配置了真空设备。真空对有气体参加的有关反应产生重大影响，其中主要包括溶解于钢液中的碳参与并生成 CO 的反应、气体（H_2，N_2）在钢液内的溶解与脱除反应。在真空下吹氧精炼可提高碳的脱氧能力，从而强化脱碳与碳脱氧反应的进行。可用于冶炼低碳及超洁净钢、真空去气、合金元素的挥发、夹杂物的去除等。

Blowing argon into the liquid steel, each small bubble floating from the liquid steel is equivalent to a small vacuum chamber. The partial pressure of H_2, N_2 and CO in the bubble is close to zero, and the [H], [N] and the production of CO in reaction of carbon and oxygen in the steel will diffuse into the small bubble and then float up. Therefore, blowing argon has the effect of 'Air Washing' on molten steel. For example, the return oxygen blowing method of smelting stainless steel by electric arc furnace can hardly reduce $w[C]$ to a very low value at 1873K. In the AOD method, continuously changing the ratio of Ar to O_2 gas is blown into the molten steel to reduce the partial pressure of CO generated in the carbon and oxygen reaction, so that the content C in of liquid steel can easily reach the ultra-low carbon level.

向钢液中吹入氩气,从钢液中上浮的每个小气泡都相当于一个小真空室,气泡内 H_2、N_2 及 CO 等的分压接近于零,钢中的 [H]、[N] 以及碳氧反应产物 CO 将向小气泡中扩散并随之上浮排除。因此,吹氩对钢液具有"气洗"作用。例如,电弧炉冶炼不锈钢的返回吹氧法,在 1873K 下很难使 $w[C]$ 降至很低的数值;而在 AOD 法中,向钢液中吹入不断变换 Ar 与 O_2 比例的气体,可以降低碳氧反应中产生的 CO 分压,从而使钢液的碳含量达到超低碳水平。

Based on the current use of various molten steel refining methods, the vacuum degassing of steel can be divided into the following three categories:

综合目前各种钢液炉外精炼法的使用情况,钢的真空脱气可分为以下三类:

(1) Steel flow degassing. Steel flow degassing means that the falling steel flow is exposed to the vacuum and then collected into ingot mold, ladle or furnace, such as the vacuum pouring method (VC method).

(1) 钢流脱气。钢流脱气是指下落中的钢流被暴露在真空中,然后被收集到钢锭模、钢包或炉内的方法。如真空浇注法(VC 法)等。

(2) Ladle degassing. Ladle degassing means that the molten steel in the ladle is exposed to a vacuum and the molten steel is stirred by gas or electromagnetic means, such as VOD, VD, ASEA-SKF etc.

(2) 钢包脱气。钢包脱气是指钢包内钢水被暴露在真空中,并用气体或电磁搅拌钢水的方法,如 VOD、VD、ASEA-SKF 等方法。

(3) Cycle degassing. Cyclic degassing refers to that in the ladle, the molten steel is pressed into the vacuum chamber by the atmospheric pressure, exposed to the vacuum, and then flows out of the degassing chamber into the ladle, such as RH method.

(3) 循环脱气。循环脱气是指在钢包内,钢水由大气压力压入真空室内,暴露在真空中,然后流出脱气室进入钢包的方法,如 RH 法等。

2.5.1.4 Heating
2.5.1.4 加热

When the molten steel is refining outside the furnace, the temperature drops because of the heat loss. The heating function of the refining outside the furnace can avoid the high temperature steel production, ensure the normal pouring of steel liquid, and increase the flexibility of the refining process outside the furnace. The amount of refining agent, the final temperature and treatment time of the molten steel can be freely selected, so as to obtain the best refining effect.

钢液在进行炉外精炼时,由于有热量损失,造成温度下降。炉外精炼的加热功能可避免高温出钢和保证钢液正常浇注,增加炉外精炼工艺的灵活性,在精炼剂用量、钢液处理最终温度和处理时间方面均可自由选择,以获得最佳的精炼效果。

Common heating methods include electric heating (such as arc heating, induction heating and resistance heating), combustion heating of fuel (such as CO, heavy oil, natural gas, etc.), and chemical heating (exothermic chemical reaction, and Al commonly used as a heating agent). Among them, arc heating is the most important and the most effective and flexible heating method. Arc heating and chemical heating are described below.

常用的加热方法有电加热（包括电弧加热、感应加热和电阻加热）、燃料（如 CO、重油、天然气等）燃烧加热和化学加热（化学反应放热，目前常用 Al 作为发热剂）。其中，电弧加热是最重要也是效果最好、最灵活的加热方法。下面介绍电弧加热和化学加热。

Arc Heating

电弧加热

The principle of arc heating is similar to that of arc furnace. After the graphite electrode is electrified, an arc is generated between the electrode and the molten steel, and the molten steel is heated by the high temperature of the arc. Due to the high arc temperature, the arc length and the foaming slag should be controlled during the heating process to prevent the arc from corroding refractory materials. The basic composition of heating device includes furnace transformer, short net, electrode transverse arm, electrode gripper, electrode, electrode column and electrode regulator, etc. It can be three-electrode three-phase ac (arc) ladle furnace, single-electrode dc (arc) ladle furnace, or double-electrode dc (arc) ladle furnace. Arc heating can make the molten steel clean, but it may make the molten steel carbon.

电弧加热的原理与电弧炉相似，采用石墨电极通电后，在电极与钢液间产生电弧，依靠电弧的高温加热钢液。由于电弧温度高，在加热过程中需要控制电弧长度及造好发泡渣进行埋弧操作，以防止电弧对耐火材料产生高温侵蚀。加热装置的基本组成包括炉用变压器、短网、电极横臂、电极夹持器、电极、电极立柱和电极调节器等，其可以是三电极的三相交流（电弧）钢包炉、单电极的直流（电弧）钢包炉，还可以是双电极的直流（电弧）钢包炉。采用电弧加热对钢液无杂质污染，可保证钢水清洁，但可能使钢水增碳。

Ladle arc heating can achieve the following metallurgical purposes:

采用钢包电弧加热可以达到如下冶金目的:

(1) The molten steel can be produced at a lower temperature, thus improving the life of the resistant material of the primary smelting furnace.

(1) 钢水可以在较低的温度下出钢，从而提高了初炼炉耐材的寿命。

(2) The temperature, chemical composition and desulfurization and deoxidization of molten steel can be controlled more accurately.

(2) 可以更精确地控制钢水温度、化学成分和脱硫、脱氧操作。

(3) The ladle refining furnace with arc heating is used as a buffer to operate between the furnace and the caster.

(3) 将带电弧加热的钢包精炼炉作为一个在炼钢炉和连铸机之间运行的缓冲器。

(4) The desulfurization, deoxidation and alloying operations in the primary furnace can be moved to the refining furnace, thus greatly improving the productivity of the primary furnace, reducing the power consumption and electrode consumption of the primary furnace, and greatly improving the technical and economic indicators of the primary furnace.

(4) 可将初炼炉中的脱硫、脱氧及合金化操作任务移到精炼炉内，从而大大提高了初炼炉的生产率，降低了初炼炉的电耗、电极消耗，大大改善初炼炉的技术经济指标。

Ladle arc heating is used in vacuum arc degassing (VAD) and ladle furnace (LF).

炉外精炼工艺中，真空电弧脱气（VAD）、钢包炉（LF）等均采用钢包电弧加热。

Chemical Heating

化学加热

Common chemical heating methods are aluminum heating and silicon heating. Among them, aluminum heating is the most widely used method. It is the use of an air gun to blow oxygen dissolved in molten steel aluminum combustion, release a lot of heat to increase the steel temperature. The advantages of this method are as follows:

常用的化学加热方法有铝氧加热法和硅氧加热法。其中，铝氧加热法应用最为广泛。它是利用喷枪吹氧使钢水中的溶解铝燃烧，放出大量热能，使钢液升温。该法的优点是：

(1) As the spray gun is immersed in molten steel during oxygen blowing, it seldom produces flue gas.

(1) 由于吹氧时喷枪浸在钢水中，很少产生烟气。

(2) Oxygen is in direct contact with molten steel, which can predict the result of temperature rise accurately.

(2) 氧气全都与钢水直接接触，可以准确地预测升温结果。

(3) The method is no effect on ladle life.

(3) 对钢包寿命没有影响。

(4) The method is simple equipment, low investment cost.

(4) 设备简单，投资费用低。

However, if the operation is improper, the total oxide inclusion in steel is easy to increase.

但如果操作不当，易使钢中氧化物夹杂的总量升高。

It is important to note that in the use of chemical heating period, except for controling the amount of aluminum and oxygen blowing. The method also need to blow argon stirring to uniform temperature and composition, otherwise the hot steel will be concentrated in the top of the ladle.

需要注意的是，在使用化学加热期间，除了要控制加铝量和吹氧量外，还需要进行吹氩搅拌来均匀温度和成分，否则过热钢水会集中在钢包上部。

2.5.1.5 Spray and Feed Silk
2.5.1.5 喷吹和喂丝

Refining agents for metal liquids (molten iron or molten steel) in off-furnace refining are divided into two categories. One is powder or synthetic slag based on calcium compounds, and the other is alloying elements (such as Ca, Mg, Al, Si and rare earth elements). Adding these refining agents into the steel liquid can play the role of desulphurization, deoxidization, removal of inclusions, denaturation treatment of inclusions and adjustment of alloy composition.

炉外精炼中金属液（铁水或钢液）的精炼剂分为两类，一类为以钙化合物为基的粉剂或合成渣，另一类为合金元素（如 Ca、Mg、Al、Si 及稀土元素等）。将这些精炼剂加入钢液中，可起到脱硫、脱氧、去除夹杂物、进行夹杂物变性处理以及调整合金成分的作用。

The spray blowing method is to fluidize the refined powder with carrier gas (Ar) to form Gas-Solid two-phase flow, and directly send the refining agent into the molten steel through the spray gun. Because of the small particle size of the refining powder in the spray blowing method, the contact area between the powder and the molten steel is greatly increased after it enters the molten steel, so the refining effect can be significantly improved.

喷吹法是用载气（Ar）将精炼粉剂流态化，形成气—固两相流，通过喷枪直接将精炼剂送入钢液内部。由于在喷吹法中精炼粉剂粒度小，其进入钢液后与钢液的接触面积大大增加，因此可以显著提高精炼效果。

Feed wire method is to have oxidized, and the density of light alloy elements in the low carbon steel package in the core wire. Through the feed wire machine, if will be sent to the liquid steel. The advantages of feeding wire method are as follows:

喂丝法是将已氧化、密度轻的合金元素置于低碳钢包的芯线中，通过喂丝机将其送入钢液内部。喂丝法的优点是：

(1) It can prevent the oxidation of easily oxidized elements by the top slag on the air and steel surface, accurately control the amount of added alloy elements, improve and stabilize the utilization rate of alloy elements.

（1）可防止易氧化元素被空气和钢液面上的顶渣氧化，准确控制合金元素的添加数量，提高和稳定合金元素的利用率。

(2) There is no spillage during the adding process, which can avoid the reoxidation of molten steel.

（2）添加过程无喷溅，可避免钢液再氧化。

(3) There is small temperature drop in refining process, less equipment investment, and low processing costs.

（3）精炼过程温降小，设备投资少，处理成本低。

2.5.2 Main Methods of Off-furnace Refining
2.5.2 炉外精炼的主要方法

2.5.2.1 Vacuum Degassing
2.5.2.1 真空脱气法

In order to prevent hydrogen defects such as white spots in large steel castings and forgings, the primary purpose of vacuum refining was to remove hydrogen from liquid steel. The common vacuum degassing methods mainly include vacuum cycle degassing (RH) and ladle vacuum degassing (VD).

为了防止大型钢铸锻件产生白点等含氢缺陷，最初真空精炼的主要目的是脱除钢液中的氢，后来又增加了脱氮、真空碳脱氧、真空氧脱碳、改善钢液洁净度及合金化等功能。常见的真空脱气方法主要有真空循环脱气法（RH法）和钢包真空脱气法（VD法）等。

Method of RH
RH 法
- Operating Principle and Characteristics of RH Method
- RH 法的工作原理及特点

The working principle of RH method is shown in Figure 2-2. The device consists of a vacuum chamber and an exhaust device with two impregnated tubes for suction and discharge of molten steel. During the vacuum treatment, two impregnating tubes of the vacuum chamber are inserted into the molten steel of the ladle, and the molten steel is vented from the vacuum chamber. The molten steel rises from the two impregnating tubes to the differential pressure height (about 1.48m). At this time, the driving gas Ar is blown into the lower 1/3 of the ascending pipe, and the apparent density of the molten steel in the ascending pipe is smaller than that of the molten steel in the descending pipe. The molten steel circulates as shown in Figure 2-2 and degasses in the vacuum chamber.

RH 法的工作原理如图 2-2 所示。该法装置由具有吸入钢液和排出钢液的两根浸渍管的真空室及排气装置构成。在进行真空处理时，把真空室的两根浸渍管插入钢包的钢液中，从真空室排气，钢液从两根浸渍管上升到压差高度（约 1.48m）。这时在上升管下部约 1/3 处吹入驱动气体 Ar，则上升管钢液的表观密度比下降管钢液的密度小，钢液就像如图 2-2 所示那样循环运动，并在真空室内脱气。

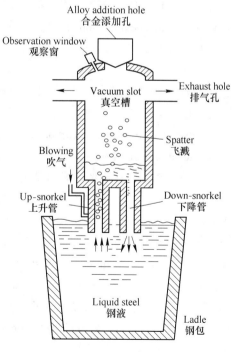

Figure 2-2　RH method
图 2-2　RH 法的工作原理

RH method has the characteristics of good degassing effect, fast treatment speed, small temperature drop in treatment process, large treatment capacity and wide application range. It is suitable for large quantities of liquid steel degassing treatment, flexible operation and reliable operation. The RH method has a wide range of applications, and developed rapidly during the converter development period. At the same time, it has formed a process system for mass production of special steel by matching with the new ultra high power large electric arc furnace.

RH 法具有脱气效果好、处理速度快、处理过程温降小、处理容量大和适用范围广等特点。其适用于大批量的钢液脱气处理，操作灵活，运转可靠。RH 法适用范围广，在转炉大发展时期获得迅速发展，同时与新兴的超高功率大型电弧炉相配套，形成了大批量生产特殊钢的工艺体系。

- Refining Effect of RH Process
- RH 法的精炼效果

The refining effect of RH process are as follows:

RH 法的精炼效果包括：

(1) Dehydrogenation. The dehydrogenation effect of RH method is obvious, dehydrogenation of deoxidized steel is about 65%, and dehydrogenation of undeoxidized steel is 70%.

(1) 脱氢。RH 法的脱氢效果明显，脱氧钢可脱氢约 65%，未脱氧钢可脱氢 70%。

(2) Nitrogen. As with other vacuum degassing methods, the effect of RH method is not obvious.

(2) 脱氮。与其他真空脱气法一样，RH 法的脱氮效果不明显。

(3) The DNA. Carbon has a certain deoxidization during cyclic treatment. Especially when the original oxygen content is high (such as the treatment of undeoxidized steel), this effect is more obvious.

(3) 脱氧。循环处理时碳有一定的脱氧作用，特别是当原始氧含量较高（如处理未脱氧的钢）时，这种作用就更明显。

(4) Decarburization. RH method has a strong decarbonization ability. After taking certain measures, it can achieve a better decarbonization effect in a short time.

(4) 脱碳。RH 法具有很强的脱碳能力，采取一定的措施后，可以在较短的时间内获得较好的脱碳效果。

(5) Quality of steel. With the reduction of hydrogen, oxygen, nitrogen and nonmetal inclusions, the longitudinal and transverse mechanical properties of steel are uniform. Elongation, reduction of area and impact toughness are improved, and the machining properties and mechanical properties of steel are significantly improved. The range of steel processed by RH process is very wide, including forging steel, high strength steel, all kinds of carbon steel, alloy structural steel, bearing steel, tool steel, stainless steel, electrical steel, deep drawing steel and other high value-added products.

(5) 钢的质量。钢液经处理后，由于其中氢、氧、氮及非金属夹杂物的减少，使钢的纵向和横向力学性能均匀，伸长率、断面收缩率和冲击韧性得以提高，钢的加工性能和力学性能得到显著改善。RH 法处理的钢种范围很广，包括锻造用钢、高强钢、各种碳素钢、合金结构钢、轴承钢、工具钢、不锈钢、电工钢、深冲钢等各种高附加值产品。

VD Method

VD 法

- Working Principle and Evolution Process of VD Method
- VD 法的工作原理及演变过程

The ladle vacuum degassing method is called VD method. It is a method of blowing argon into the molten steel in a ladle placed in a vacuum container for refining, and its principle is shown in Figure 2-3. This method is different from the RH method. The ladle that fully eliminates the slag in the process of steelmaking and only liquid steel is placed in the vacuum chamber. After the lid is closed and the gas is vented, the liquid steel is stirred by blowing argon through the permeable brick at the bottom of the ladle.

钢包真空脱气法称为 VD 法，它是向放置在真空容器中的钢包里的钢液吹氩进行精炼的一种方法，其原理如图 2-3 所示。此方法与 RH 法不同，将充分排除炼钢过程中的渣，只有钢液的钢包放置在真空室内，盖上盖子排气后，通过装在钢包底部的透气砖吹氩搅拌钢液。

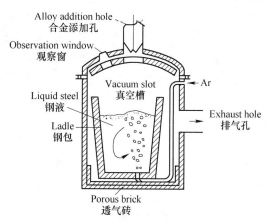

Figure 2-3　Working principle of VD method

图 2-3　VD 法的工作原理

The early ladle vacuum degassing equipment is composed of ladle, vacuum chamber and vacuum system. It is a static degassing device without argon and electromagnetic stirring system. Its main purpose is to degassing liquid steel and the effect is not obvious. Because the vacuum degassing equipment is no heating function, so when steel steel to make the liquid steel overheating, overheating temperature according to the furnace capacity and different. Larger stoves can overheat at lower temperatures, and smaller ones at higher temperatures. Now it seems that this kind of steel processing equipment is relatively simple, but it has brought a new era of steel making technology, which is the beginning of people master clean steel production technology.

早期的钢包真空脱气设备由钢包、真空室、真空系统组成，是一种静态脱气装置，没有氩气和电磁搅拌系统，其主要目的就是使钢液脱气，效果不明显。这种真空脱气设备因为没有加热功能，所以出钢时要使钢液过热，过热温度根据炉容量的不同而不同。较大的炉子过热温度可以小些，较小的炉子过热温度要大些。这种钢液处理设备比较简单，但却带来了一个炼钢技术的新时代，是人们掌握洁净钢生产技术的开始。

VD method is rarely used alone, often with the heating function of LF method and so on. Because VD refining equipment can effectively remove gas and inclusions, and construction investment and production costs are far lower than RH and DH methods, VD furnace has a more obvious advantages, widely used in small-scale electric furnace manufacturers for the refining of special steel.

VD法一般很少单独使用，往往与具有加热功能的LF法等双联。由于VD法精炼设备能有效地去除气体和夹杂物，而且建设投入和生产成本均远远低于RH法和DH法。因此，VD炉具有较明显的优势，广泛用于小规模电炉厂家等进行特殊钢的精炼。

- VD Refining Effect
- VD法的精炼效果

VD method is one of the main methods to reduce and control the gas content in molten steel. In the process of VD furnace smelting, the strong inert gas stirring and the molten pool reaction can ensure the full reaction between steel and slag and realize the desulphurization of steel liquid. Through feeding treatment, the sulfide inclusion can be denaturated.

VD法是减少和控制钢液中气体含量的主要手段之一，同时还具有脱氧、去除夹杂物、调整钢液成分和控制钢液温度的功能。在VD炉冶炼过程中，强烈的惰性气体搅拌和熔池反应可确保钢渣间充分反应，实现钢液脱硫。通过喂丝处理，还可以对硫化物夹杂做变性处理。

In a word, the quality of finished steel is much better than that of unfinished steel. Through refining, the content of gas and oxygen in steel was significantly reduced, and the inclusion rating was also significantly reduced.

总之，精炼钢的质量比未精炼钢的质量好得多。通过精炼，钢中气体、氧的含量都显著降低，夹杂物评级也都明显降低。

2.5.2.2 Ladle Refining with Heating Function
2.5.2.2 有加热功能的钢包精炼法

The prominent problem in the process of outside treatment is that the temperature of molten steel will inevitably decrease, which limits the refining time of molten steel and the amount of alloy added, and makes it difficult to control the pouring temperature, which affects the stable production and quality control of the subsequent continuous casting process. Compared with pure vacuum treatment, a prominent feature of ladle furnace refining method with heating function, can be carried out on the molten steel in ladle heating, for refining the task of heat absorption and heat loss in the refining process can be compensated by heat. In this way, the ladle furnace in liquid steel temperature and refining time is no longer dependent on early tapping temperature of the furnace, the amount and type of alloy to join also has greatly increased, and the steel varieties increased significantly.

在炉外处理过程中比较突出的问题是钢液温度不可避免地要降低，使钢水精炼时间和合金加入量都受到一定的限制，浇注温度难以控制，对后续连铸工序的稳定生产和质量控制产生影响。与单纯的真空处理相比，钢包炉精炼法的一个突出特点是具有加热功能，可以对钢包内钢液进行加热，为完成精炼任务的吸热以及在精炼过程中的散热损失均可通过加热得到补偿。这样，钢包炉在钢液温度和精炼时间方面不再依赖初炼炉的出钢温度，合金加入的种类和数量也大大增加，钢的品种显著增加。

There are three typical heating ladle furnace refining methods, which are vacuum refining furnace method of arc heating (ASEA—SKF method), vacuum arc degassing refining furnace method (VAD method) and arc heating ladle blowing argon furnace method (LF method).

典型的加热钢包炉精炼法有三种，即电弧加热的真空精炼炉法（ASEA—SKF 法）、真空电弧脱气精炼炉法（VAD 法）和电弧加热的钢包吹氩炉法（LF 法）。

ASEA—SKF method

ASEA—SKF 法

In order to further expand the refining function and overcome the inclusion problem caused by the mixing of steel slag from electric furnace during the smelting of bearing steel, ASEA-SKF ladle refining furnace was developed. It has the function of vacuum degassing, arc heating and electromagnetic stirring of steel liquid in ladle, and is a kind of all-purpose off-furnace refining method, which can degassing, deoxidizing, decarbonizing, desulfurizing, heating, removing inclusions, adjusting alloy composition and so on.

为了进一步扩大精炼功能，克服在冶炼轴承钢时采用电炉钢渣混冲出钢而产生的夹杂问题，ASEA—SKF 法在钢包精炼炉中被广泛使用。该方法具有在钢包内对钢液进行真空脱气、电弧加热、电磁搅拌的功能，是一种万能型的炉外精炼方法。同时也可以进行脱气、脱氧、脱碳、脱硫、加热、去除夹杂物、调整合金成分等操作。

ASEA-SKF furnace can be combined with electric arc furnace and converter, and almost complete all the tasks in the steelmaking process. Therefore, the structure of ASEA—SKF furnace is relatively complex. The equipment of ASEA—SKF furnace mainly includes ladle containing liquid steel, water-cooled electromagnetic induction stirrer and its frequency converter, arc heating system, vacuum sealing furnace cover and vacuum pumping system, alloy and slag feeding system, etc. Some ASEA—SKF furnaces are also equipped with argon blowing stirring system and oxygen blowing system. In order to ensure the objective control of molten steel composition and temperature, vacuum temperature measurement, sampling and vacuum feeding equipment are very necessary. But it is still difficult to realize these functions without failure. In addition, the advanced ASEA-SKF furnace uses the computer control system.

ASEA—SKF 炉可以与电弧炉和转炉配合，几乎能完成炼钢过程的所有任务，因此 ASEA—SKF 炉的结构较复杂。ASEA—SKF 炉的设备主要有盛装钢液的钢包、水冷电磁感应搅拌器及其变频器、电弧加热系统、真空密封炉盖和抽真空系统、合金及渣料加料系统等，有些 ASEA—SKF 炉还配有吹氩搅拌系统和吹氧系统。为了保证钢水成分和温度的目标控制，真空测温、取样以及真空加料设备是十分必要的，但这些功能的无故障实现仍有一定难度。另外，先进的 ASEA—SKF 炉都采用了计算机控制系统。

ASEA—SKF furnace layout is divided into two types: trolley mobile and rotary furnace cover. Trolley mobile is more common, and its structure is shown in Figure 2-4. It consists of a ladle placed on the trolley, a ladle cover for vacuum treatment connected with the vacuum equipment, and a ladle cover for heating with three-phase ac electrode. The processing process of

ASEA—SKF furnace with revolving furnace cover is similar to that of trolley mobile, except that the ladle is placed in a fixed induction agitator, and the heating furnace cover and vacuum degassing furnace cover can be rotated alternately. Its structure is shown in Figure 2-5.

ASEA—SKF 炉的布置形式分为台车移动式和炉盖旋转式两种。台车移动式较为常见，其结构如图 2-4 所示，是由放在台车上的一个钢包、与真空设备连接真空处理用的钢包盖和设置三相交流电极加热用的钢包盖构成。炉盖旋转式布置的 ASEA—SKF 炉，其处理过程与台车移动式相似，只是钢包放到固定的感应搅拌器内，加热炉盖和真空脱气炉盖能旋转交替使用，其结构如图 2-5 所示。

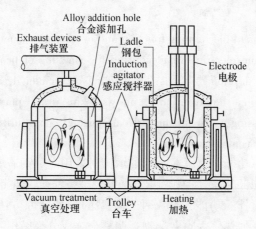

Figure 2-4　ASEA—SKF with mobile vehicle arrangement
图 2-4　台车移动式布置的 ASEA—SKF 炉

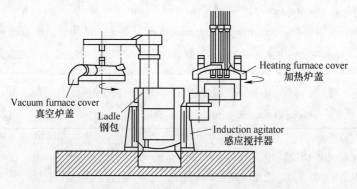

Figure 2-5　ASEA—SKF with rotary arrangement of furnace cover
图 2-5　炉盖旋转式布置的 ASEA—SKF 炉

It can be seen that ASEA—SKF furnace has the function of arc heating and low-frequency electromagnetic stirring, which is not available in general vacuum degassing equipment. Its main advantages are as follows:

由此可见，ASEA—SKF 炉具有电弧加热与低频电磁搅拌的功能，是一般真空脱气设备所不具备的。它的主要优点有：

(1) Liquid steel temperature can be quickly uniform, and conducive to the improvement of steel cleanliness. It can also reduce the consumption of refractory.

(1) 钢液温度能很快均匀，有利于钢洁净度的提高，并可减少耐火材料的消耗。

(2) It can make the added alloy melt quickly, and the composition is uniform and stable.

(2) 使加入的合金熔化快，成分均匀、稳定。

(3) Arc heating can improve the fluidity of molten slag, accelerate the reaction speed of steel—slag, and help deoxidize and remove inclusion.

(3) 电弧加热可提高熔渣的流动性，加快钢—渣反应速度，有利于脱氧、去除夹杂。

(4) Electromagnetic stirring can improve the efficiency of vacuum degassing.

(4) 电磁搅拌可提高真空脱气的效率。

VAD Method

VAD 法

VAD method is arc heating ladle degassing method, also called vacuum arc ladle degassing method. It is a ladle refining method which uses argon to stir steel liquid, and adds an arc heating device on the lid of the vacuum chamber. The method was developed by Finkle & Sons in 1967 in cooperation with Mohr company to solve the problem of the temperature drop of molten steel during ladle degassing. So it is also known as Finkle—Mohr method (or Finkle. VAD method). VAD method can also be considered as the addition of arc heating device on the basis of VD method (ladle degassing method). The VAD refining unit is mainly composed of ladle, vacuum system, arc heating system and bottom blowing argon gas system, as shown in Figure 2-6.

VAD 法即电弧加热钢包脱气法，或称真空电弧钢包脱气法，是用氩气搅拌钢液，并在真空室的盖子上增设电弧加热装置的钢包精炼法。该法是美国 Finkle & Sons 公司为解决钢包脱气过程中钢水温度下降的问题，与 Mohr 公司合作，于 1967 年开发成功的方法，因此也称为 Finkle—Mohr 法或 Finkle. VAD 法。VAD 法也可以认为是在 VD 法（钢包脱气法）的基础上增加了电弧加热装置。VAD 精炼装置主要由钢包、真空系统、电弧加热系统和底吹氩气系统等设备组成，其示意图如图 2-6 所示。

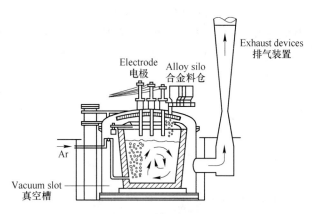

Figure 2-6 Schematic Diagram of VAD Equipment

图 2-6 VAD 设备示意图

VAD method has a variety of refining methods, such as arc heating, argon blowing stirring, vacuum degassing, ladle slagging and alloying. The method has the following characteristics：

VAD 法具有电弧加热、吹氩搅拌、真空脱气、钢包内造渣及合金化等多种精炼手段,能对钢液进行脱硫、脱氧、脱氢、脱氮、去夹杂处理。该法具有以下特点:

(1) As the heating is carried out under vacuum, a good reducing atmosphere can be formed to prevent the oxidation of molten steel during the heating process. And a good degassing effect can be obtained during the heating process.

(1) 由于加热是在真空下进行的,可形成良好的还原性气氛,防止钢液在加热过程中的氧化,在加热过程中还可以获得良好的脱气效果。

(2) The refining furnace is completely sealed, with little noise during heating and almost no smoke and dust during heating.

(2) 精炼炉完全密封,加热过程噪声较小,加热过程中几乎无烟尘。

(3) The pouring temperature can be accurately adjusted, and the ladle lining can fully store heat. The temperature drop is stable during pouring.

(3) 能够准确地调整浇注温度,而且钢包内衬充分蓄热,浇注时温降稳定。

(4) Due to sufficient stirring in the refining process, the composition of molten steel is uniform and stable.

(4) 由于精炼过程中搅拌充分,钢液成分均匀、稳定。

(5) The composition can be fine-tuned under vacuum conditions. It can also be added a large number of alloys, and smelt a wide range of carbon steel.

(5) 可以在真空条件下进行成分微调,可加入大量的合金,能冶炼范围很广的碳素钢。

(6) It can achieve a variety of refining purposes in a station, and be added to slag and other slag for desulfurization and decarburization.

(6) 可以在一个工位达到多种精炼目的,可以加入造渣剂和其他渣料进行脱硫和脱碳。

LF Method
LF 法

LF (Ladle Furnace) method is composed of argon stirring, graphite electrode arc heating under atmospheric pressure, and white slag refining technology. LF method has the functions with desulphurization and deoxidization with strong reducing slag, inclusion control, melting ferroalloys by electric arc heating, and adjusting composition and temperature, etc.

LF (Ladle Furnace) 法是采用氩气搅拌,在大气压力下用石墨电极埋弧加热,再与白渣精炼技术组合而成。LF 法的功能有:用强还原性渣脱硫、脱氧,进而进行夹杂物控制;用电弧加热熔化铁合金;调整成分、温度等。

- LF Method of Equipment Composition
- LF 法的设备组成

The main equipment of LF method includes furnace body, arc heating system, alloy and slag charging system, wire feeding system, bottom blowing argon system, furnace cover and cooling water system, etc., as shown in Figure 2-7. Due to its simple equipment, low investment cost, flexible operation and good refining effect, it has become a rising star in ladle refining, and has been widely used and developed in the metallurgical industry.

LF 法的主要设备包括炉体、电弧加热系统、合金及渣料加料系统、喂线系统、底吹氩系统、炉盖及冷却水系统等，其设备如图 2-7 所示。由于设备简单、投资费用低、操作灵活和精炼效果好，其成为钢包精炼的后起之秀，在冶金行业得到广泛的应用和发展，在我国的炉外精炼设备中已占据主导地位。

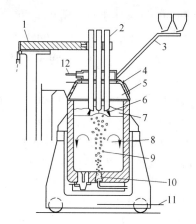

Figure 2-7 LF equipment schematic diagram
图 2-7 LF 设备示意图

1—One electrode transverse arm；2—One electrode；3—Feeding chute；4—A water-cooled furnace cover；5—An inert atmosphere in a furnace；6—An electric arc；7—One slag；8—Gas stirring；9—Liquid steel；10—One air plug；11—Ladle car；12—Water-cooled smoke hood and alloy steel
1—电极横臂；2—电极；3—加料溜槽；4—水冷炉盖；5—炉内惰性气氛；6—电弧；7—炉渣；8—气体搅拌；9—钢液；10—透气塞；11—钢包车；12—水冷烟罩和合金钢

- Refining Function and Characteristics of LF Method
- LF 法的精炼功能及特点

During LF refining, the operation can be simplified as submerged arc heating, inert gas stirring liquid steel, alkaline white slag and inert gas protection.

LF 精炼期间所进行的操作，可以简化为埋弧加热、惰性气体搅拌钢液、造碱性白渣和惰性气体保护。

LF itself generally does not have a vacuum system. Because the ladle and furnace cover seal, it can play the role of air isolation. When the graphite electrode is heated, the CO gas generated by the action of oxides (such as FeO, MnO, Cr_2O_3 and so on) in the slag and the argon from the stirred liquid steel increases the reducibility of the furnace gas, thus preventing the transfer of oxygen from the furnace gas to the metal and ensuring the reduction atmosphere in the furnace during refining. Refining the molten steel under the condition of reduction can further deoxidize, desulphurize and remove the nonmetallic inclusions, which is conducive to improving the quality of the molten steel.

LF 本身一般不具有真空系统，精炼时，由于钢包与炉盖密封，可起到隔离空气的作用。加热时石墨电极与渣中（FeO、MnO、Cr_2O_3 等）氧化物作用生成的 CO 气体及来自搅拌钢液的氩气，增加了炉气的还原性，这样就阻止了炉气中氧向金属的传递，保证了精炼时炉内的还原气氛。钢液在还原条件下精炼，可以进一步地脱氧、脱硫及去除非金属夹杂物，有利于钢液质量的提高。

Good argon stirring is another characteristic of LF refining. Argon stirring is beneficial to the chemical reaction between steel and slag, and it can accelerate the material transfer between steel and slag. Blowing argon agitator is also good for deoxidizing and desulphurizing of steel liquid, and can accelerate the reduction of oxide in slag. Another advantage of argon blowing agitation is that it can speed up the uniformity of temperature and composition in the molten steel and accurately adjust the complex chemical composition, which is essential for high quality steel. In addition, blowing argon agitation can also remove the nonmetallic inclusion in the molten steel. It is very difficult for the inclusions in molten steel to float up by nature. Strong stirring must be adopted to improve the dynamic conditions, and enough stirring time should be allowed to make the inclusions accumulate and float up, so as to remove the inclusions and reduce the oxygen content.

良好的氩气搅拌是 LF 精炼的又一特点。氩气搅拌有利于钢—渣之间的化学反应，它可以加速钢—渣之间的物质传递。吹氩搅拌还有利于钢液脱氧、脱硫反应的进行，可加速渣中氧化物的还原。吹氩搅拌的另一作用是可以加速钢液中温度与成分的均匀，能精确调整复杂的化学组成，而这对优质钢来说又是必不可少的。此外，吹氩搅拌还可以去除钢液中的非金属夹杂物。钢液中的夹杂物靠自然上浮是很困难的，必须采取比较强的搅拌，改善动力学条件，并且要有足够的搅拌时间使夹杂物聚集上浮，达到去除夹杂物、降低氧含量的目的。

LF arc heating is similar to the arc furnace smelting process, using three graphite electrodes for heating. When the electrode is inserted into the slag layer, the arc generated between the electrode and the molten steel is buried by the white slag.

LF 电弧加热类似于电弧炉冶炼过程，是采用三根石墨电极进行加热的。加热时电极插入渣层中，采用埋弧加热法，电极与钢液之间产生的电弧被白渣埋住。这种方法辐射热小，对炉衬有保护作用，与此同时加热的热效率也比较高，热利用率好。

The LF method uses white slag for refining, which is different from other refining methods which mainly rely on vacuum degassing. The white slag has a strong reducing property in LF, which is the result of the interaction between the good reducing atmosphere and argon stirring in LF. The general amount of slag is 2% ~ 8% of the amount of liquid steel. The content of oxygen, sulfur and inclusions in steel can be reduced by the refining of white slag. LF smelting can not add deoxidizer, but relying on the adsorption of white slag to oxide to achieve the purpose of deoxidization. The precondition for making good alkaline white slag is to control the composition, temperature and smelting process parameters of molten steel. The amount of oxidized slag in the early stage is small (no slag steel). The molten steel has been deoxidized, the lining of the ladle is alkaline refractory, slag should be easy to melt, and the total content of FeO and MnO in slag should be less than 1.0% (by mass).

LF 法是利用白渣进行精炼，它不同于主要靠真空脱气的其他精炼方法。白渣在 LF 内具有很强的还原性，这是 LF 内良好的还原气氛和氩气搅拌互相作用的结果。一般渣量为钢液量的 2%~8%。通过白渣的精炼作用，可以降低钢中的氧、硫及夹杂物含量。LF 冶炼时可以不加脱氧剂，而是靠白渣对氧化物的吸附来达到脱氧的目的。造好碱性白渣的前提条件是：控制好钢液成分、温度和熔炼过程参数，前期氧化渣量少（无渣出钢），钢液已经脱氧，钢包内衬为碱性耐火材料，渣应易熔化，渣中 FeO 与 MnO 总含量（质量分数）应低于 1.0%。

The four refining functions of LF influence depend on and promote each other. The reducing atmosphere in the furnace and the steel slag stirring under the condition of heating can improve the refining capacity of white slag and create an ideal steelmaking environment, so that the quality and productivity can be produced better than ordinary arc furnace steel.

LF 四大精炼功能是互相影响、互相依存与互相促进的。炉内的还原气氛以及有加热条件下的钢渣搅拌，提高了白渣的精炼能力，创造了一个理想的炼钢环境，从而能生产出质量和生产率优于普通电弧炉钢的钢种。

Task 2.6　Continuous Pouring of Steel
任务 2.6　钢的连续浇注

2.6.1　Overview of Steel Pouring
2.6.1　钢的浇注概述

Billet or ingot casting is the final forming process of steelmaking products, which is directly related to the output and quality of steelmaking production. In recent years, continuous steel casting has gradually replaced die casting as the main method of liquid steel pouring.

铸坯或铸锭是炼钢产品最终成形的工序，直接关系到炼钢生产的产量和质量。钢的浇注通常采用模铸和连铸两种方法。近年来，连续铸钢法已经逐渐取代模铸法，成为钢液浇注的主要方法。

2.6.2　Models and Characteristics of Continuous Casting Machine
2.6.2　连铸机的机型及特点

Continuous casting is a process in which molten steel is poured, condensed and cut by a continuous casting machine to obtain the billet directly. The type of continuous caster directly affects the output and quality of billet, capital construction investment and production cost. Since the industrialization of continuous casting in the 1950s, continuous casting machines have undergone more than 20 years of development.

连铸是把钢液用连铸机浇注、冷凝、切割，直接得到铸坯的工艺。连铸机的机型直接影响铸坯的产量和质量、基本建设投资及生产成本。从 20 世纪 50 年代连铸工业化以来，经历了 20 多年的发展，连铸机的机型基本上完成了一个由立式、立弯式到弧形的演变过程。

According to the structure shape, the continuous casting machine can be divided into vertical continuous casting machine, vertical bending continuous casting machine, multi-point bending vertical bending continuous casting machine, arc continuous casting machine (divided into straight mold and curved mold), multi-radius arc (oval) continuous casting machine and horizontal continuous casting machine. A brief diagram of these models used in industrial production is shown in Figure 2-8.

连铸机按结构外形,可分为立式连铸机、立弯式连铸机、多点弯曲的立弯式连铸机、弧形连铸机(分直形结晶器和弧形结晶器两种)、多半径弧形(即椭圆形)连铸机和水平式连铸机等。如图 2-8 所示为这几种用于工业生产的连铸机机型简图。

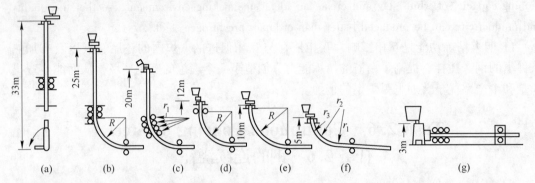

Figure 2-8 Schematic diagram of continuous casting machines
图 2-8 连铸机机型简图

(a) Vertical continuous casting machine; (b) Vertical bending continuous casting machine; (c) Straight mold multi-point bending continuous casting machine; (d) Straight mold arc continuous casting machine; (e) Arc continuous casting machine; (f) Multi-radius arc (oval) continuous casting machine; (g) Horizontal continuous casting machine
(a) 立式连铸机;(b) 立弯式连铸机;(c) 直形结晶器多点弯曲连铸机;(d) 直形结晶器弧形连铸机;
(e) 弧形连铸机;(f) 多半径弧形(椭圆形)连铸机;(g) 水平式连铸机

Vertical continuous caster mold, secondary cooling and solidification of slab shear equipment are set on a vertical line, thus it is advantageous to the inclusions in molten steel floating. Cooling condition of slab in each direction is uniform, and casting during the solidification process is not subjected to bending and straightening deformation. Even crack sensitivity high steel grade can be smoothly continuous casting. But its disadvantages are: high casting equipment, steel hydrostatic pressure; heavy equipment and maintenance is not convenient; the installation of vertical continuous casting machine requires a very high plant or pit, and infrastructure costs are also high.

立式连铸机的结晶器、二冷段和全凝固铸坯的剪切等设备均设置在一条垂直线上,因而有利于钢水中夹杂物的上浮。铸坯各方向的冷却条件较均匀,并且铸坯在整个凝固过程中不受弯曲、矫直等变形作用,即使裂纹敏感性高的钢种也能顺利地连铸。但其缺点是:铸机设备高,钢水静压力大;设备较笨重,维修也不方便;安装立式连铸机需要很高的厂房或地坑,基建费用也高。

Vertical bending continuous casting machine is a kind of transition model in the development of continuous casting. Its upper half is the same as the vertical continuous casting machine. But after the casting billet solidifies, it is bent 90, so that the casting billet along the horizontal direction. In this way, it not only has the advantages of good vertical continuous casting machine inclusions floating conditions, but also the height lower than the vertical continuous casting machine with the horizontal billet. Billet length is not limited, and casting billet transport is also more convenient. Vertical bending continuous casting machine is mainly used for casting billet with small section. For casting billet with large section, the metallurgical length has been very long and the advantage of reducing the height of equipment is not obvious. In addition, the stress inside the billet of this model is large at the top bending and straightening points, which is easy to produce internal cracks.

立弯式连铸机是连铸发展过程中的一种过渡机型。其上半部与立式连铸机相同，而在铸坯全凝固后将其顶弯90°，使铸坯沿水平方向出坯。这样，它既具有立式连铸机夹杂物上浮条件好的优点，又比立式连铸机的高度低，而且水平出坯，铸坯定尺长度不受限制，铸坯的运送也较方便。立弯式连铸机主要适用于小断面铸坯的浇注，对于大断面铸坯来说，全凝固后再顶弯，冶金长度已经很长了，其降低设备高度方面的优势已不明显。此外，该机型铸坯在顶弯和矫直点内部应力较大，容易产生内部裂纹。

Arc-shaped continuous casting machine is the most important types of continuous casting machine at home and abroad. The main features of arc continuous casting machine are as follows:

弧形连铸机是目前国内外最主要的连铸机机型，其又分为直形结晶器和弧形结晶器两种类型。弧形连铸机的主要特点是：

(1) Because the layout in the range of 1/4 arc, its height is lower than that of the vertical and vertical bending type continuous casting machine. It also makes the weight lighter, the investment cost is lower, and the equipment installation and maintenance is convenient. So it is widely used.

(1) 由于布置在1/4圆弧范围内，其高度低于立式与立弯式连铸机，这就使得它的设备重量较轻，投资费用较低，设备安装与维护方便，因而得到广泛应用。

(2) Due to the low height of the equipment, the steel hydrostatic pressure that the billet bears in the solidification process is relatively small, which can reduce the inner crack and segregation of the billet shell caused by the bulging deformation. It is also conducive to improve the billet quality and the drawing speed.

(2) 由于设备高度较低，铸坯在凝固过程中承受的钢水静压力相对较小，可减少坯壳因鼓肚变形而产生的内裂与偏析，有利于改善铸坯质量和提高拉速。

(3) The main problem of arc continuous casting machine is that nonmetallic inclusions tend to converge to the inner arc during the solidification process of molten steel, which easily leads to the uneven distribution of inclusions in the billet. In addition, the inner and outer arcs tend to produce uneven cooling, resulting in the center segregation and affecting the casting quality.

(3) 弧形连铸机的主要问题在于，钢水凝固过程中非金属夹杂物有向内弧聚集的倾向，易造成铸坯内部夹杂物分布不均匀。此外，内外弧易产生冷却不均，造成铸坯中心偏析而影响铸坯质量。

In order to further reduce the height of the caster, the elliptical caster was developed. It refers to the downward arc radius from the mold gradually become larger, the mold and the secondary cooling section clamp roller arrangement in a quarter of the elliptic arc, also known as ultra-low head continuous casting machine. The basic characteristics of this type are the same as that of the arc mould casting machine except that the arc area adopts multi-radius and the height is reduced. However, as the elliptical caster is multi-radius, its installation, arc adjustment are more complex. Radian inspection and the maintenance of the caster are also more difficult.

为了进一步降低连铸机高度,发展了椭圆形连铸机。椭圆形连铸机是指从结晶器向下圆弧半径逐渐变大,将结晶器和二冷段夹辊布置在 1/4 椭圆弧上,又称为超低头连铸机。这种机型除了弧形区采用多半径、高度有所降低外,其基本特点与弧形结晶器连铸机相同。但是由于椭圆形连铸机是多半径的,其安装、对弧调整均较复杂,弧度的检查和连铸机的维护也比较困难。

In order to minimize the height of the caster, the main equipment (tundish, mold, secondary cooling section, casting machine and cutting equipment) are arranged in a horizontal position. This caster is called horizontal caster. Its tunic is closely connected to the crystallizer and is connected with a separation ring. When casting, the mold does not vibrate, but through the casting machine to drive the casting billet to do pull, push, and stop different combinations of periodic movement to achieve. Compared with the curved caster, the horizontal caster has the following characteristics:

为了最大限度地降低连铸机高度,将其主要设备(中间包、结晶器、二冷段、拉坯机和切割设备)均布置在水平位置上,这种连铸机称为水平式连铸机。它的中间包与结晶器是紧密相连的,相连处装有分离环。拉坯时,结晶器不振动,而是通过拉坯机带动铸坯做拉、反推、停不同组合的周期性运动来实现。水平式连铸机与弧形连铸机相比,具有以下特点:

(1) It has low equipment height, low investment, fast construction, and suitable for the transformation of existing steelmaking workshops.

(1) 设备高度低,投资省,建设快,适合于现有炼钢车间的改造。

(2) The tunic is completely closed with the mould to realize non-oxidation casting. The casting billet is of good quality, and the liquid level detection and control system of the mould steel are not required.

(2) 中间包与结晶器全封闭,实现无氧化浇注,铸坯质量好,而且不需要结晶器钢液液面检测和控制系统。

(3) The casting billet has no bending and straightening during the solidification process. For alloy steel and special steel having difficulties in casting with arc continuous casting machine, the horizontal continuous casting machine can be used.

(3) 铸坯在凝固过程中无弯曲和矫直,对于用弧形连铸机浇注有困难的合金钢和特殊钢,可用水平式连铸机浇注。

(4) All equipment is installed on the ground, which is convenient for operation, accident treatment and maintenance.

(4) 所有设备均安装在地面上,操作、事故处理和维护都较方便。

The main problems of horizontal caster are as follows: the section of cast slab is small due to the limitation of inertia force during casting; the cost of graphite sleeve and separation ring of mold is higher; and the cost of casting billet is increased.

水平式连铸机存在的主要问题是:受拉坯时的惯性力限制,所浇注的铸坯断面较小;结晶器的石墨套和分离环价格较高,增加了铸坯成本。

2.6.3 Main Equipment of Continuous Casting Machine
2.6.3 连铸机的主要设备

2.6.3.1 Ladle Rotary Table
2.6.3.1 钢包回转台

Ladle turntable is usually located between the steel receiving span and the continuous casting casting span. A continuous casting machine is equipped with a ladle turntable. The turning radius of the rotary arm must be able to accept ladles from the steel receiving crane across the side, rotating 180°, stop at the middle of the pouring span above the car for pouring. The other end is stopped at the receiving end of the molten steel to replace the empty ladle. The rotating speed of ladle turntable is generally 0.1~1r/min, and the replacement time is 0.5~2min.

钢包回转台通常设在钢水接受跨和连铸浇注跨之间。一台连铸机配备一台钢包回转台。回转臂的回转半径必须能从钢水接受跨一侧的吊车接受钢包，旋转180°，停在浇注跨中间包车的上方进行浇注；而另一端则停在钢水接受跨，以更换空钢包。钢包回转台的回转速度一般为0.1~1r/min，更换钢包时间为0.5~2min。

Adopting ladle turntable, occupying a small area of pouring platform, is easy to locate with quick ladle replacement. When the accident or power failure can be used pneumatic or hydraulic motor quickly to the safe position of ladle, conducive to the realization of multiple furnace continuous pouring and pouring steel accident treatment.

采用钢包回转台，占用浇注平台的面积较小，易于定位，钢包更换迅速，发生事故或停电时可用气动或液压电动机迅速将钢包旋转到安全位置，有利于实现多炉连浇和浇钢事故的处理。

2.6.3.2 Tundish and Its Carrying Equipment
2.6.3.2 中间包及其载运设备

Tundish is a transition vessel device for pouring molten steel between ladle and crystallizer. Its function is to reduce the impact force (static pressure) of molten steel into the mold, stabilize the steel flow, promote the temperature of molten steel uniform, and conducive to the inclusion in the molten steel floating and separation. For the storage of molten steel and the separation of molten steel, the role connect the above and the next when multiple furnace continuous pouring. In addition, there are new processes such as tundish heating, spraying, filtering, alloy fine tuning, etc., which constitute tundish metallurgy.

中间包是位于钢包与结晶器之间用于钢液浇注的过渡容器装置。它的作用是减小钢水注入结晶器中的冲击力（静压力），稳定钢流，促使钢水温度均匀，有利于钢水中夹杂物的上浮与分离，用于储存钢水以及对钢水分流，多炉连浇时起到承上启下的作用。此外，还有中间包加热、喷吹、过滤、合金微调等新工艺，构成了中间包冶金。

The tundish consists of a body, a cover, a plug, a nozzle (or a sliding nozzle), etc. Tundish shell generally welded with 12~20mm thick steel plate, can also be used to cast steel to ensure that it in the high temperature environment pouring, slag, handling and rolling deformation. Tunic lining consists of permanent layer and working layer. There are two types of working layer materials: one with magnesium or magnesium calcium coating on the permanent layer, and the other with an insulating plate. The plug rod center of tunic is often cooled by compressed air or argon to extend its service life. For billet caster less than 120mm×120mm, a sizing nozzle is applied. In addition, retaining walls and dykes are often built in tundish to improve the flow field of molten steel, which is conducive to the temperature uniformity and the separation of inclusions floating up.

中间包由包体、包盖、塞棒和水口（或滑动水口）等组成。中间包外壳一般用12~20mm厚的钢板焊成，也可用铸钢件，要保证其在高温环境中浇注、清渣、搬运和翻包时不变形。中间包内衬由永久层和工作层组成。工作层的材质有两类：一类用镁质或镁钙质涂料喷涂在永久层上；另一类用绝热板。中间包的塞棒中心常通入压缩空气或氩气冷却，以提高其使用寿命。对于小于120mm×120mm的小方坯连铸机，应用定径水口。此外，中间包内常加砌挡墙和堤坝，以改善钢水的流场，有利于温度均匀和促使夹杂物上浮分离。

2.6.3.3 Mould and Its Vibration Device
2.6.3.3 结晶器及其振动装置

Mould is the most important part in continuous casting equipment, which is called the heart of continuous casting equipment. The mold requirements are: good thermal conductivity and rigidity, wear resistant inner surface, simple structure, small quality, easy to manufacture, installation, adjustment and maintenance, and low cost.

结晶器是连铸设备中最关键的部件，称为连铸设备的"心脏"，其主要作用是钢液在结晶器内冷却，初步凝固成一定坯壳厚度的铸坯外形，并被连续地从结晶器下口拉出而进入二冷区。对结晶器的要求是：具有良好的导热性和刚性，内表面耐磨，结构简单，质量小，易于制造、安装、调整和维修，造价低。

Structure of Crystallizer
结晶器的结构形式

The mold structure is generally composed of copper inner wall, water jacket and cooling water joint. In addition, there are inlet and outlet pipes and fixed frame.

结晶器的结构一般由铜内壁、水套和冷却水水缝三部分组成，此外，还有进出水管和固定框架等。

Crystallizers can be divided into straight and curved types. According to the section shape of casting billet, it can be divided into square billet, slab, round billet and shaped billet mold. According to its own structure, it can be divided into integral type, tube type, combined type and on-line width adjustment mold.

结晶器可分为直形和弧形两类。按铸坯断面形状，其可分为方坯、板坯、圆坯和异型坯结晶器；按本身结构，其可分为整体式、管式、组合式和在线调宽结晶器等。

Selections of Main Parameters of Mold
结晶器主要参数的选择

The Selections of main parameters of mold are as follows:

结晶器的主要参数包括：

(1) Section size of mold. Due to the factors such as shrinkage and deformation of the casting billet during condensation and straightening, it is required that the section size of the mould should be 2%~3% larger than the section size of the cold casting billet (the nominal section of the continuous casting billet). The thickness direction should be about 3% and the width direction should be about 2%.

(1) 结晶器的断面尺寸。由于铸坯在冷凝过程时收缩和矫直时变形等因素，要求结晶器的断面尺寸应比冷铸坯的断面（连铸坯公称断面）尺寸大 2%~3%（厚度方向约取 3%，宽度方向约取 2%）。

(2) The length of the mold. The length of the mold is a very important parameter. The longer the mold is, the thicker the shell of the mold is, and the better the casting safety is. However, if the mold is too long, the cooling efficiency will be reduced. The length of mold usually used is 700~900mm.

(2) 结晶器的长度。结晶器的长度是一个非常重要的参数。结晶器越长，在相同的拉速下，出结晶器的坯壳越厚，浇注安全性越好。然而，结晶器过于长的话，冷却效率就会降低。通常采用的结晶器长度为 700~900mm。

(3) Taper of mold. In order to reduce air gap, improve thermal conductivity and accelerate the formation of casting shell, the inner cavity of mould copper plate must be designed in a larger shape and a smaller shape.

(3) 结晶器的锥度。由于铸坯在结晶器内凝固的同时伴随着体积的收缩，结晶器铜板内腔必须设计成上大下小的形状，以减小气隙、提高导热性能、加速铸坯壳的生成。

Vibration Device of Mold
结晶器的振动装置

Mould vibration plays an important role in continuous casting. The up and down reciprocating operation of the mould actually plays the role of 'Stripping'. Because the adhesion between the blank shell and the copper plate is reduced due to the mould vibration, the excessive stress on the surface of the raw shell is prevented from causing the crack or more serious consequences. When the mold moves downward, due to the action of negative slipper (the downward speed of the mold is greater than the casting speed in the vibration process), the cracks on the surface of the shell can be healed and the ideal surface quality can be obtained.

结晶器振动在连铸过程中扮演着非常重要的角色。结晶器的上下往复运行实际上起到了"脱模"的作用。由于坯壳与铜板间的黏附力因结晶器振动而减小，防止了在初生坯壳表面产生过大应力而导致裂纹的产生或引起更严重的后果。当结晶器向下运动时，由于负滑脱（振动过程中结晶器下行速度大于拉坯速度）作用，可愈合坯壳表面裂痕，并有利于获得理想的表面质量。

2.6.3.4　Secondary Cooling Device
2.6.3.4　二次冷却装置

The secondary cooling device is composed of support guide system, water spray cooling system and installation base. The roles of the large cooling device are as follows:

二次冷却装置由支撑导向系统、喷水冷却系统和安装底座组成。其装置作用包括：

(1) The billet with liquid core is directly forced to be cooled by water (or gas-water), and solidification is accelerated to enter the tension correction area.

(1) 用水或气—水对带液芯的铸坯直接强制冷却，加速凝固，以进入拉矫区。

(2) Through the clamping roller and side guide roller, it guides and supports the movement of the billet and the ingot guide rod, so as to prevent the possible deformation of the billet such as bulge, diamond shape and bending.

(2) 通过夹辊和侧导辊，对铸坯和引锭杆的运动起导向和支撑作用，以防止铸坯可能产生的鼓肚、菱形、弯曲等变形。

(3) For the arc-shaped continuous casting machine with a straight mold, there is also a bending effect on the billet. For ultra-low head caster, this area is a section straightening area.

(3) 对于带直形结晶器的弧形连铸机，还有对铸坯的顶弯作用；对于超低头连铸机此区又是分段矫直区。

The process requirements of the cold zone are as follows:

冷区的工艺要求如下：

(1) The secondary cooling device should have sufficient strength and stiffness under the action of high temperature casting billet.

(1) 二冷装置在高温铸坯作用下应有足够的强度和刚度。

(2) Through the champing roller and side guide roller, it guides and supports the movement of the billet and the ingot guide rod, so as to prevent the possible deformation of the billet such as bulge, diamond shape and bending.

(2) 结构简单，对中准确，调整方便，能适应改变铸坯断面的要求，便于快速更换和维修。

(3) The amount of water spraying can be adjusted according to the requirements to adapt to the requirements of the casting section, steel type, casting temperature and casting speed.

(3) 能按要求调整喷水量，以适应铸坯断面、钢种、浇注温度和拉坯速度等变化的要求。

2.6.3.5　Casting Straightening Device
2.6.3.5　拉坯矫直装置

The function of the casting straightening device is to clamp and pull the casting billet, making it move continuously forward, and straightening the arc casting billet.

拉坯矫直装置的作用是夹持拉动铸坯，使之连续向前运动，并把弧形铸坯矫直，在浇注准备时还要把引锭杆送入结晶器下口。

The process requirements of the casting straightening device are as follows:

对拉坯矫直装置的工艺要求是：

(1) The resistance of mold and secondary cooling zone can be overcome in the pouring process, and the casting billet can be pulled out smoothly.

(1) 在浇铸过程中能克服结晶器和二冷区的阻力，把铸坯顺利拉出。

(2) It has good speed regulation performance to meet the requirements of changing steel type, section and ingot bar, etc. It can also realize closed-loop control on the automatic control liquid level casting system.

(2) 具有良好的调速性能，以适应改变钢种、断面和上引锭杆等的要求，对自动控制液面的拉坯系统能实现闭环控制。

(3) Under the premise of ensuring the quality of casting billet, complete solidification or straightening of casting billet with liquid core can be realized.

(3) 在保证铸坯质量的前提下，能实现完全凝固或带液芯铸坯的矫直。

(4) It has the simple structure, reliable work, and easy installation and adjustment.

(4) 结构简单，工作可靠，安装和调整方便。

2.6.3.6 Ingot Priming Device
2.6.3.6 引锭装置

Ingot priming device includes ingot head, ingot priming rod and ingot priming rod storage device. The function of the ingot drawing device is to block the under the mouth (usually the dummy head into crystallizer is about 200mm in the mouth, the tail in the interior of the machine in 500~1000mm length) of the mould at the start of casting and make the molten steel solidify at the head of the ingot. The casting billet is taken out with the drawing roller. After the casting billet enters the drawing levelling machine, the ingot rod is removed and enters the normal casting state. The ingot guide rod can be divided into flexible and rigid type according to its structure, and into two types of mounting and mounting according to its installation.

引锭装置包括引锭头、引锭杆和引锭杆存放装置。引锭装置的作用是在开浇时堵住结晶器的下口（通常引锭头伸入结晶器下口内约200mm，尾部在拉矫机内保持500~1000mm的长度），并使钢水在引锭头处凝固，通过拉辊把铸坯带出，在铸坯进入拉矫机后将引锭杆脱去，进入正常拉坯状态，引锭杆则送入存放装置待下次开浇时使用。引锭杆按结构形式可分为挠性和刚性两类，按安装方式又分为下装和上装两种。

2.6.3.7 Billet Cutting Device
2.6.3.7 铸坯切割装置

The function of the billet cutting device is to cut the billet into the required length in the process of casting, which has two forms of flame cutting and mechanical cutting.

铸坯切割装置的作用是在铸坯前进过程中将其切成所需的定尺长度，其有火焰切割和机械切割两种形式。

Generally in slab, green, round and multi-flow continuous casting machine use flame cutting. Its advantages are: the quality of the equipment is small, not limited by the casting temperature and the size of the section; the incision is more uniform; the size of the equipment is small. Its disadvantages are: The metal burn loss is large, for 1%~2%; There is pollution to the environment, it need to set up smoke and slag removal equipment; When cutting short fixed size, it need to add secondary cutting equipment. Mechanical cutting is used for billet caster. Its shear speed is fast, fixed size is adjustable, and metal loss is small. However, the equipment is of high quality and high power consumption.

一般在板坯、大方坯、圆坯和多流连铸机上多采用火焰切割。它的优点是：设备质量小，不受铸坯温度和断面大小限制，切口较齐，设备的外形尺寸较小。它的缺点是：金属烧损较大，为1%~2%；对环境有污染，需设置消烟和清渣设备；当切割短定尺时，需增加二次切割设备。机械切割多用于小方坯连铸机，其剪切速度快，定尺可调，金属损失小。但设备质量大，消耗功率较高，切口附近的铸坯部分易变形。

2.6.4 Solidification and Crystallization Theory of Molten Steel
2.6.4 钢液凝固结晶理论

2.6.4.1 Characteristics of Solidification and Crystallization of Molten Steel
2.6.4.1 钢液凝固结晶的特点

Crystallization Temperature Range
结晶温度范围

Steel is an alloy, containing elements such as C, Si, Mn, P, S, etc., and the solidification of steel actually belongs to non-equilibrium crystallization. So the crystallization of molten steel has different characteristics from that of pure metal. The molten steel contains various alloy elements, and its crystallization temperature is not one point, but within a temperature range, as shown in Figure 2-9. The molten steel starts to crystallize at T_1 and finishes at T_2. The difference between T_1 and T_2 is the crystallization temperature range in ΔT_0, given by:

钢是合金，含有 C、Si、Mn、P、S 等元素。钢的凝固实际上属于非平衡结晶，因此钢液的结晶具有不同于纯金属的特点。钢液中含有各种合金元素，它的结晶温度不是一点，而是在一个温度区间内，如图 2-9 所示。钢水在 T_1 时开始结晶，到达 T_2 时结晶完毕。T_1 与 T_2 的差值为结晶温度范围，用 ΔT_0 表示。其计算公式为：

$$\Delta T_0 = T_1 - T_2 \qquad (2-41)$$

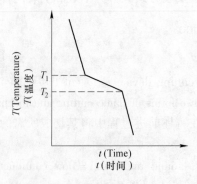

Figure 2-9 Curve of crystallization temperature of molten steel
图 2-9 钢水结晶温度变化曲线

Since the crystallization of molten steel is accomplished in a temperature range, the solid and liquid phases coexist in this temperature range. The actual crystallization state is shown in Figure 2-10. The molten steel completely solidifies on the left side of the S line, on the right side of the L line, and the solid-liquid phase coexistence between the S line and the L line (called the two-phase zone). The distance between the S line and the L line is called the width of the two-phase zone, Δx.

由于钢液结晶是在一个温度区间内完成的,在这个温度区间里固相与液相并存。实际的结晶状态如图 2-10 所示。钢液在 S 线左侧完全凝固,在 L 线右侧全部为液相,在 S 线与 L 线之间固—液相并存(称为两相区),S 线与 L 线之间的距离称为两相区宽度 Δx。

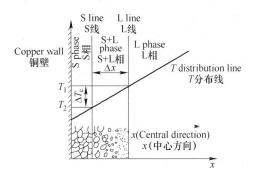

Figure 2-10 State diagram of two-phase region during molten steel crystallization
图 2-10 钢水结晶时两相区状态图

The effect of Δt on solidification structure can be seen from the relationship between crystallization temperature range and width of two-phase region. The grain size is larger when the cylinder is larger and smaller when the cylinder is larger. Large grain size means that the dendrites are developed, and the developed dendrites make the densification of the solidified tissue worse. The porosity is easy to form, and segregation is also more serious.

从结晶温度范围和两相区宽度的关系中可以看出 Δt 对凝固组织的影响。当缸较大时,晶粒度较大,反之则小。晶粒度大意味着树枝晶发达,发达的树枝晶使凝固组织的致密性变差,易形成气孔,偏析也较严重。

Chemical Segregation
化学成分偏析

During the crystallization of molten steel, due to the selective crystallization, the content of solute in the first solidified part is low. The solute is concentrated in the mother solution and the concentration gradually increases, so the content of solute in the last solidified part is high. Obviously, the distribution of solute concentration in the final solidified structure is not uniform. This phenomenon of uneven composition is called segregation.

钢液结晶时,由于选分结晶,最先凝固的部分溶质含量较低,溶质聚集于母液中,浓度逐渐增加,因而最后凝固的部分溶质含量则很高。显然,在最终凝固结构中溶质浓度的分布是不均匀的。这种成分不均匀的现象称为偏析。

When analyzing the composition distribution of a casting billet or a grain, it can be found that the solute concentration in the center of the casting billet is higher, while the solute concentration at the grain boundary of a grain is higher. The former is macroscopic segregation and the latter is microscopic segregation.

在分析一支铸坯或一个晶粒的成分分布时可发现，铸坯中心溶质的浓度较高，而一个晶粒的晶界处溶质的浓度较高。前者为宏观偏析，后者为显微偏析。

- Microsegregation
- 显微偏析

In the actual production, the molten steel crystallizes under the condition of rapid cooling, so it belongs to non-equilibrium crystallization, forming the inhomogeneity of solute concentration inside the grain, with low concentration at the central crystal axis and high concentration at the edge crystal. This dendritic segregation is called microsegregation or dendritic segregation.

实际生产中，钢液是在快速冷却的条件下结晶，因而属于非平衡结晶，形成了晶粒内部溶质浓度的不均匀性，中心晶轴处浓度低，边缘晶间处浓度高。这种呈树枝状分布的偏析称为显微偏析或树枝偏析。

- Macro Segregation
- 宏观偏析

During the solidification process, the liquid between dendritic dendrites is enriched with solute elements due to selective crystallization. And the flow of molten steel during the solidification process brings the liquid enriched with solute elements to the unsolidified area, so that the solute concentration in the final solidified part on the cross section of the casting slab is much higher than the original concentration. Many factors that cause steel flow, tear flow injection, temperature difference, the density difference, slab, deformation and solidification shrinkage and gas inclusions floatation, etc., can cause not solidification of liquid steel flow, resulting in the overall slab internal solute elements distribution inhomogeneity, called the macro segregation, also known as the macro segregation. The macroscopic segregation of billet can be demonstrated by chemical analysis or acid leaching.

钢液在凝固过程中，由于选分结晶，使树枝晶枝间的液体富集了溶质元素，再加上凝固过程中钢液的流动将富集了溶质元素的液体带到未凝固区域，使得铸坯横截面上最终凝固部分的溶质浓度远高于原始浓度。引起钢液流动的因素很多，如注流的注入、温度差、密度差、铸坯鼓肚变形、凝固收缩以及气体、夹杂物的上浮等，均能引起未凝固钢液的流动，从而导致整体铸坯内部溶质元素分布的不均匀性，此即宏观偏析，也称为低倍偏析。可通过化学分析或酸浸显示铸坯的宏观偏析。

- Solidification Shrinkage
- 凝固收缩

The phenomenon of thermal expansion and cold contraction is characterized by solidification shrinkage in the solidification process of molten steel. The volume of 1t liquid steel was reduced from $0.145m^3$ to $0.128m^3$ when cooled to room temperature solid steel. The shrinkage also varies with the composition and temperature drop.

热胀冷缩现象在钢液凝固过程中表现为凝固收缩。1t 的液态钢冷却到常温的固态钢，其体积由 $0.145m^3$ 缩小至 $0.128m^3$，收缩了近12%。随成分、温降的不同，其收缩量也有差异。

The contraction of molten steel with temperature drop and phase transition can be divided into the following three stages:

钢液的收缩随温降和相变可分为如下三个阶段：

(1) Liquid contraction: The shrinkage of molten steel from pouring temperature to liquidus temperature is called liquid shrinkage. At this stage the steel remains liquid and the shrinkage is 1%. The harm of liquid shrinkage is not big, especially for the continuous casting billet, the liquid shrinkage is filled by the continuous injection of molten steel, which has little influence on the solidified dimension and can be ignored.

(1) 液态收缩：钢液由浇注温度降至液相线温度过程中产生的收缩称为液态收缩，即过热度消失时的体积收缩。这个阶段钢保持液态，收缩量为1%。液态收缩危害并不大，尤其对于连铸坯而言，液态的收缩被连续注入的钢液所填补，对已凝固的外形尺寸影响极小，可以忽略。

(2) Solidification and contraction: The molten steel forms a solid phase with a temperature drop in the crystallization temperature range. The wider the crystallization temperature range, the greater the shrinkage, which is about 4% of the total. Because of the continuous addition of molten steel, it can also be considered that the shrinkage of solidification process has little effect on the structure of billet.

(2) 凝固收缩：钢液在结晶温度范围内形成固相并伴有温降，这两个因素均会对凝固收缩有影响。结晶温度范围越宽，则收缩量越大，其收缩量约是总量的4%。由于钢液的连续补充，也可认为凝固过程的收缩对铸坯的结构影响较小。

(3) Solid state contraction: Steel from the solid-phase line temperature to room temperature, steel in the solid state, the process of shrinkage is called solid shrinkage. It is also called linear shrinkage because the size of the billet is changed by shrinkage. The contraction of the solid state is 7% to 8% of the total. The shrinkage of the solid state is the largest, the thermal stress is generated in the process of temperature drop, and the tissue stress is generated in the process of phase change. Therefore, the solid shrinkage has a great influence on the casting quality.

(3) 固态收缩：钢由固相线温度降至室温，钢处于固态，此过程的收缩称为固态收缩。由于收缩使铸坯的尺寸发生变化，故其也称为线收缩。固态收缩的收缩量为总量的7%~8%。固态收缩的收缩量最大，在温降过程中产生热应力，在相变过程中产生组织应力，应力的产生是铸坯裂纹的根源。因此，固态收缩对铸坯质量影响甚大。

2.6.4.2 Solidification Structure of Continuous Casting Billet
2.6.4.2 连铸坯的凝固结构

The solidification of continuous casting billet is equivalent to that of ingot with a large ratio of height to width. In general, the continuous casting billet from the edge to the center is also composed of the chilled layer, columnar crystal zone and central equiaxial crystal zone (ingot core zone), which is not fundamentally different from ingot.

连铸坯凝固相当于高宽比特别大的钢锭凝固，且铸坯在连铸机内边运行边凝固，形成了很大的液相穴。一般情况下，连铸坯从边缘到中心是由激冷层、柱状晶区和中心等轴晶区（锭心区）组成，与钢锭无本质区别。

Chilling Layer

激冷层

The skin of the casting billet, called the chilled layer, is composed of fine equiaxed crystals. The width of the small equiaxed crystal zone is generally 2~5mm, which is formed under the condition of the highest cooling rate at the meniscus of the mould. Its thickness mainly depends on the superheat of the molten steel. The higher the pouring temperature is, the thinner the cooling layer will be. The lower the pouring temperature, the thicker the cooling layer.

铸坯表皮由细小等轴晶组成，称为激冷层。细小等轴晶区的宽度一般为2~5mm，它是在结晶器弯月面处冷却速度最高的条件下形成的。其厚度主要取决于钢水过热度，浇注温度越高，激冷层就越薄；浇注温度越低，激冷层就越厚。

Columnar Crystal Region

柱状晶区

The shrinkage during the forming process of the chilling layer of the billet results in an air gap on the wall of the mould at a depth of 100~150mm below the menisculum of the mould, which reduces the heat transfer rate. At the same time, the internal heat dissipation of molten steel increases the temperature of the cooling layer and no new crystal nuclei are generated. Under the condition of the development of directional heat transfer in molten steel, columnar crystal zone began to form. The columnar crystals near the chilled layer are fine and almost unbranched. From the vertical section, the columnar crystal is not completely perpendicular to the surface, but slopes upwards at an Angle (about 10°). From the outer edge to the center, the number of columnar crystals is less than variable (like a bamboo forest). The development of columnar crystals is irregular, and the transgranular structure may be formed through the center of the billet in some parts. For arc-shaped continuous casting machine, the low power structure of billet has asymmetry. Due to the action of gravity, the crystal sinks, inhibiting the growth of outer arc columnar crystal. So the columnar crystal on the inner arc side is longer than that on the outer arc side, and the cracks in the billet are often concentrated on the inner arc side.

铸坯激冷层形成过程中的收缩，使处于结晶器弯月面100~150mm以下的器壁产生气隙，从而降低了传热速度。同时，钢液内部向外散热使激冷层温度升高，不再产生新的晶核。在钢液定向传热得到发展的条件下，柱状晶区开始形成。靠近激冷层的柱状晶很细，基本上不分叉。从纵断面来看，柱状晶并不完全垂直于表面，而是向上倾斜一定角度（约10°），从外缘向中心，柱状晶的个数由多变少（呈竹林状）。柱状晶的发展是不规则的，在某些部位可能会贯穿铸坯中心而形成穿晶结构。对于弧形连铸机，铸坯低倍结构具有不对称性。由于重力作用，晶体下沉，抑制了外弧柱状晶的生长，故内弧侧的柱状晶比外弧侧要长些，且铸坯内裂纹也常常集中在内弧侧。

Central Isometric Zone

中心等轴晶区

With the advance of solidification front, the temperature gradient of solidification layer and solidification front gradually decreases, and the width of two-phase zone is continuously expanded. After the molten steel temperature of the casting slab core drops to the liquidus temperature, a large number of isometric crystals are generated and grow rapidly, forming an isometric crystal zone with irregular arrangement. The central isometric zone has visible, non-dense porosity and porosity, accompanied by elemental segregation. Compared with ingot, due to the development of columnar crystal in continuous casting billet, the central equiaxial crystal area is much narrower and the grain size is finer.

随着凝固前沿的推移，凝固层和凝固前沿的温度梯度逐渐减小，两相区宽度不断扩大，铸坯芯部钢水温度降至液相线温度后，大量等轴晶产生并迅速长大，形成无规则排列的等轴晶区。中心等轴晶区有可见的、不致密的疏松和缩孔，并伴随有元素的偏析。与钢锭相比，由于连铸坯柱状晶的发展，中心等轴晶区要窄得多，晶粒也细一些。

2.6.5 Determination of Drawing Speed of Continuous Casting
2.6.5 连铸拉速的确定

2.6.5.1 Throwing Speed
2.6.5.1 拉坯速度

Casting speed (referred to as drawing speed) refers to the length in m/min of casting billet pulled out by continuous casting machine per unit time, or the weight of casting billet pulled out per unit time in t/min. It is one of the important technological parameters of continuous casting machine. Its size determines the production capacity of continuous casting machine, and directly affects the solidification rate of molten steel, the metallurgical quality of billet and the safety of continuous casting process. If the drawing speed is too high, the thickness of the blank shell at the outlet of the mould will be insufficient to bear the casting force and the steel hydrostatic pressure, so that the blank shell will be cracked and steel leakage accident will occur. Even if there is no steel leakage, it will cause the slab to crack, when the stress generated by the hydrostatic steel pressure and casting force exceeds the critical stress for the steel to crack. Therefore, the drawing speed should be based on good metallurgical quality, continuous casting process safety and continuous casting machine production capacity. Generally, under certain technological conditions, the pulling speed has an optimal value, which is disadvantageous if it is too large or too small.

拉坯速度（简称拉速）是指连铸机每一流单位时间内拉出铸坯的长度（m/min），或每一流单位时间内拉出铸坯的质量（t/min）。它是连铸机重要的工艺参数之一，其大小决定了连铸机的生产能力，同时又直接影响着钢水的凝固速度、铸坯的冶金质量以及连铸过程的安全性。拉速过高会造成结晶器出口处坯壳厚度不足，从而不能承受拉坯力和钢水静压力，以致坯壳被拉裂而产生漏钢事故。当钢水静压力和拉坯力产生的应力超过钢产生裂纹的临界应力时，即使不漏钢，也会造成铸坯形成裂纹。因此，拉速应以获得良好的冶金质量、连铸过程安全性和连铸机生产能力为前提。通常在一定工艺条件下，拉速有一最佳值，过大或过小都是不利的。

2.6.5.2 Cooling Control in Continuous Casting Process
2.6.5.2 连铸过程冷却控制

The slab cooling control includes two parts: mould cooling (primary cooling) control and secondary cooling (secondary cooling) control.

铸坯冷却控制包括结晶器冷却(一次冷却)控制和二冷区冷却(二次冷却)控制两部分。

Mold Cooling Control
结晶器冷却控制

The cooling function of the mould is to ensure that the blank shell has enough thickness at the outlet of the mould to withstand the static pressure of the molten steel and prevent leakage. At the same time, the blank shell should be cooled evenly in the mould to prevent the appearance of surface defects. Whether the primary cooling can meet these requirements is mainly determined by the cooling capacity of the mold, heat flow distribution, parameters (length, taper, material, thickness, etc.) and cooling water quality, flow rate, flow rate and other factors.

结晶器冷却的作用是保证坯壳在结晶器出口处有足够的厚度,以承受钢水的静压力,防止拉漏,同时又要使坯壳在结晶器内冷却均匀,防止表面缺陷的产生。而一次冷却能否满足这些要求,主要是由结晶器的冷却能力、热流分布、参数(长度、锥度、材质、厚度等)以及冷却水的质量、流速、流量等因素来决定的。

The greater the flow of cooling water, the greater the cooling intensity. But when the flow of cooling water increases to a certain value, the cooling intensity is no longer increased.

冷却水的流量越大,冷却强度也越大。但当冷却水的流量增加到一定数值后,冷却强度就不再增加。

The pressure of mould cooling water is generally controlled within the range of 0.4~0.6MPa. In order to prevent the water gap of mould from producing intermittent boiling, the water pressure can be increased and the water gap can be narrowed to increase the water velocity and avoid the thermal deformation of mould copper pipe. In actual production operation, most of the water pressure should be maintained at about 0.5MPa, not less than 0.4MPa.

结晶器冷却水的压力一般控制在0.4~0.6MPa范围内。为防止结晶器水缝产生间断沸腾,可提高水压,缩小水缝,以增加水流速,避免结晶器铜管产生热变形。实际生产操作中大多数使水压保持在0.5MPa左右,不应低于0.4MPa。

In the process of production, the inlet and outlet temperature difference of mold cooling water should be kept stable to facilitate the uniform growth of the shell. The inlet and outlet temperature difference of the mold cooling water should be less than 10℃, generally controlled between 3℃ and 8℃.

生产中要保持结晶器冷却水进出温度差的稳定,从而使得坯壳均匀生长。结晶器冷却水进出温度差应小于10℃,一般控制在3~8℃之间。

Secondary Cooling Zone Cooling Control
二冷区冷却控制

After the casting billet is pulled out of the mold, its core is still liquid. In order for the casting billet to solidify completely before it enters the straightening point or before the cutting machine, it must be further cooled in the secondary cooling zone. Therefore, a billet cooling system is set up in the secondary cooling area of the continuous casting machine.

铸坯从结晶器拉出后,其芯部仍为液体,为使铸坯在进入矫直点之前或者在切割机之前完全凝固,就必须在二冷区进一步对铸坯进行冷却。为此,在连铸机的二冷区设有铸坯冷却系统。

The following principles shall be followed in determining the cooling strength:

确定冷却强度的原则。确定冷却强度时应遵循以下原则:

(1) When the casting billet pulled out of the mold enters the upper section of the secondary cooling zone, the internal liquid core quantity is large, the billet shell is thin, the thermal resistance is small, and the stress generated by solidification and shrinkage of the billet shell is also small. At this point, the increased cooling intensity can make the shell thicken rapidly, even at a higher pull speed will not leak. When the thickness of the billet increases to a certain degree, the cooling intensity should be gradually reduced with the increase of the thermal resistance of the billet, so as to avoid the excessive thermal stress on the surface of the billet causing cracks. Therefore, the whole secondary cooling zone should follow the principle of top-down cooling intensity from strong to weak.

(1) 由结晶器拉出的铸坯进入二冷区上段时,内部液芯量大,坯壳薄,热阻小,坯壳凝固收缩产生的应力也小。此时加大冷却强度可使坯壳迅速增厚,即使在较高的拉速下也不会拉漏。当坯壳厚度增加到一定程度以后,随着坯壳热阻的增加,应逐渐减小冷却强度,以免铸坯表面热应力过大而产生裂纹。因此,在整个二冷区应当遵循自上而下、冷却强度由强到弱的原则。

(2) In order to improve the productivity of continuous casting machine, it is necessary to adopt high drawing speed and high cooling efficiency. However, in order to improve the cooling efficiency, the surface temperature of the billet should be cooled uniformly both horizontally and longitudinally, so as to avoid the crack caused by the local drastic decrease of the surface temperature. Generally, the surface cooling rate of billet should be less than 200℃/m, and the surface temperature recovery speed should be less than 100℃/m. The larger the section, the lower the surface cooling speed and surface temperature recovery speed.

(2) 为了提高连铸机生产率,应当采取高拉速和高冷却效率。但在提高冷却效率的同时,要避免铸坯表面温度局部剧烈降低而产生裂纹,故应使铸坯表面横向及纵向都能均匀降温。通常铸坯表面冷却速度应小于200℃/m,铸坯表面温度回升速度应小于100℃/m。铸坯断面越大,其表面冷却速度及表面温度回升速度应越小。

(3) The secondary cooling water distribution should make the surface temperature of the casting billet avoid the brittle 'Pocket Area' when straightening, and control the temperature zone with the highest ductility of the steel with 700~900℃. The temperature range is the brittleness temperature zone of the billet. For low carbon steel, the surface temperature should be higher than 900℃ when straightening. For Nb bearing steel, the surface temperature of billet should be higher than 980℃ when straightening. In addition, in order to ensure the minimum

bulges formed between supporting rollers in the secondary cooling zone, the surface temperature of the billet should be limited in the whole secondary cooling zone, usually controlled at 1100℃. At the same time, the surface temperature of the billet after cutting should be controlled to be higher than 1000℃ when the billet is hot delivered and directly rolled.

（3）二冷配水应使矫直时铸坯表面温度避开脆性"口袋区"，控制在钢延性最高的温度区。700~900℃的温度范围是铸坯的脆性温度区。对于低碳钢，矫直时铸坯表面温度应高于900℃；对于含 Nb 的钢，矫直时铸坯表面温度应高于980℃。此外，为了保证铸坯在二冷区支承辊之间形成的鼓肚量最小，在整个二冷区应限定铸坯表面温度，通常控制在1100℃以下。同时，在铸坯进行热送和直接轧制时，还要控制切割后铸坯表面温度高于1000℃。

In determining the cooling strength must meet the needs of different types of steel, especially steel with strong crack sensitivity, which uses weak cooling.

在确定冷却强度时必须满足不同钢种的需要，特别是裂纹敏感性强的钢种，要采用弱冷。

Determination of Secondary Cooling Water Ratio
二冷比水量的确定

Secondary cooling intensity is often expressed as 'Specific Water Volume', which refers to the cooling water volume used by casting billet per unit weight through secondary cooling zone in L/kg. The water content of secondary cooling ratio varies with the steel type, the section size of casting billet, the type of continuous casting machine, the casting speed and other parameters. And it usually fluctuates between 0.3L/kg and 1.51L/kg. Table 2-1 shows the specific water quantities of different types of steel.

二次冷却强度常用"比水量"来表示，是指通过二冷区单位质量铸坯所使用的冷却水量，单位为 L/kg。二冷比水量随着钢种、铸坯断面尺寸、连铸机机型、拉坯速度等参数的不同而变化，通常在0.3~1.51L/kg 之间波动。表2-1列出了不同类别钢种的比水量。

Table 2-1　Specific water quantities of different types of steel
表 2-1　不同类别钢种的比水量　　　　　　　　　　　　（L/kg）

Types of steel 钢种类别	Water ratio 比水量	Types of steel 钢种类别	Water ratio 比水量
Common carbon steel and low alloy steel 普碳钢、低合金钢	1.0~1.2	Crack sensitive steel 裂纹敏感性强的钢	0.4~0.6
High carbon steel and alloy steel 中高碳钢、合金钢	0.6~0.8	High-speed steel 高速钢	0.1~0.3

When choosing the secondary cooling strength, the factors such as steel type, drawing speed, straightening temperature, heat transfer and direct rolling should also be considered. At present, the general use of 'Hot Line', also known as soft cooling. The surface temperature of the billet decreases slowly in the whole secondary cooling zone, and the billet temperature should be raised as much as possible to facilitate heat transfer or direct rolling under the condition that the billet does not bop up. The surface temperature before the billet straightening should not be lower than 900℃.

具体选择二冷强度时，除考虑钢种、拉速等因素外，还要考虑铸坯矫直温度以及是否热送、直接轧制等。目前，一般采用"热行"，也称软冷却。在整个二冷区铸坯表面温度缓慢下降，在保证铸坯不鼓肚的情况下应尽可能提高出坯温度，以便热送或直接轧制，至少也应做到铸坯矫直以前的表面温度不低于900℃。

2.6.5.3 Determination of Pouring Temperature
2.6.5.3 浇注温度的确定

The pouring temperature of continuous casting refers to the temperature of molten steel in the tundish. Generally, the temperature of molten steel in a furnace needs to be measured for 3 or 4 times in the tundish. The temperature should be measured at the beginning of pouring 5min, the middle stage of pouring process and 5min before the end of pouring. The average of the measured temperature is the average pouring temperature.

连铸浇注温度是指中间包内的钢水温度，通常一炉钢水需要在中间包内测3次或4次温度，即在开浇5min、浇铸过程中期和浇注结束前5min时均应测温，所测温度的平均值为平均浇注温度。

2.6.6 Quality of Continuous Casting Billet
2.6.6 连铸坯的质量

The types of quality of continuous casting billet are as follows:
连铸坯的质量包括：

(1) Surface quality of continuous casting billet. The surface quality of continuous casting billet determines whether the billet needs finishing before the hot processing. Surface defects mainly include various types of surface cracks (surface longitudinal cracks, surface transverse cracks, surface star-shaped cracks), deep vibration marks, surface slag, subcutaneous bubbles and pores, surface pits and heavy skin.

(1) 连铸坯表面质量。连铸坯表面质量的好坏决定了铸坯在热加工之前是否需要精整，是影响金属收得率和成本的重要因素，也是铸坯热送和直接轧制的前提条件。表面缺陷主要有各种类型的表面裂纹（表面纵裂纹、表面横裂纹、表面星状裂纹）、深振痕、表面夹渣、皮下气泡与气孔、表面凹坑和重皮。

(2) Internal quality of continuous casting billet. The internal quality of continuous casting billet refers to whether the billet has correct solidification structure, segregation degree, internal crack, inclusion content and distribution. Internal defects mainly include internal cracks (subcutaneous crack, intermediate crack, corner crack, straightening crack, central stellate crack), central segregation, central porosity, and central shrinkage cavity.

(2) 连铸坯内部质量。连铸坯内部质量是指铸坯是否具有正确的凝固结构、偏析程度、内部裂纹、夹杂物含量及分布状况等。内部缺陷主要有内部裂纹（皮下裂纹、中间裂纹、角部裂纹、矫直裂纹、中心星状裂纹）、中心偏析、中心疏松和中心缩孔。

(3) Shape defects of continuous casting billet. The shape defects of continuous casting billet include stripper (rhomboid) defects and billet bulge.

(3) 连铸坯形状缺陷。连铸坯形状缺陷包括脱方（菱形）缺陷和铸坯鼓肚。

Exercises

思考题

(1) What is the oxidizability of slag and how is the oxidizability of slag reflected in the process of steelmaking?

(1) 什么是炉渣的氧化性，在炼钢过程中熔渣的氧化性如何体现？

(2) What are the factors affecting dephosphorization in steelmaking process?

(2) 影响炼钢过程脱磷的因素有哪些？

(3) What are the factors affecting desulfurization in steelmaking process?

(3) 影响炼钢过程脱硫的因素有哪些？

(4) What kinds of deoxidizing methods do steel liquids have and what are their characteristics?

(4) 钢液的脱氧方式有哪几种，各有何特点？

(5) What are the characteristics of element oxidation, slag composition and temperature changes in oxygen top-blown converter steelmaking process?

(5) 氧气顶吹转炉炼钢过程中元素的氧化、炉渣成分和温度的变化体现出哪些特征？

(6) Describe the influence of gun position of oxygen gun on steelmaking and metallurgical process of converter.

(6) 简述氧枪的枪位对转炉炼钢冶金过程产生的影响。

(7) Describe the main tasks of arc furnace in melting period, oxidation period and reduction period.

(7) 简述电弧炉熔化期、氧化期及还原期的主要任务。

(8) What are the main preheating technologies for electric furnace steelmaking and scrap? Briefly describe its energy-saving effect.

(8) 电炉炼钢废钢预热技术主要有哪几种？简要说出其节能效果。

(9) Describe the advantages of dc arc furnace.

(9) 简述直流电弧炉的优越性。

(10) Describe the main equipment structure and functional characteristics of LF.

(10) 简述 LF 主要的设备构成及功能特点。

(11) Compare the similarities and differences between AOD and VOD furnace refining.

(11) 试比较 AOD 与 VOD 炉外精炼法之间的异同。

(12) Describe the main equipment composition of continuous casting machine.

(12) 简述连铸机的主要设备构成。

(13) Describe the common surface defects of continuous casting billet.

(13) 简述连铸坯常见的表面缺陷。

Project 3　Metal Pressure Machining
项目3　金属压力加工

Metal pressure processing refers to the process of plastic deformation of heated or unheated metal blanks to obtain raw materials, blanks (or parts) with certain shape, size and mechanical properties, also known as metal plastic molding, plastic processing or solid molding. Metal pressure processing plays an important role in metal forming process. Metal pressure processing is at the end of metallurgical process flow, which provides customers with needed products and services. This project mainly introduces the metal forming method, the basic principle of metal pressure processing and the basic concepts, basic theory and production process knowledge of steel rolling process and non-ferrous metal process.

金属压力加工是指借助外力作用使金属坯料（加热的或者不加热的）发生塑性变形，从而获得具有一定形状、尺寸和力学性能的原材料、毛坯（或零件）的成型工艺方法，也称为金属塑性成型、塑性加工或固态成型。金属压力加工在金属成型方法中占有重要的地位，金属压力加工环节处于冶金工艺流程中的终端，为用户提供所需要的产品与服务。本项目主要介绍金属成型方法、金属压力加工基本原理以及轧钢生产工艺与有色金属加工工艺等基本理论和生产工艺知识。

Task 3.1　Metal Forming Methods
任务3.1　金属成型方法

3.1.1　Casting
3.1.1　铸造

Casting is a kind of metal hot-working technology which has a history of about 6000 years. After the industrial revolution in the 18th century, the steam engine, textile machine and railway industries emerged, casting entered a new period of service for large industry, casting technology began to have a great development. In the 20th century, the development of casting is very fast. One of the important factors is that the progress of product technology requires the casting to have better mechanical and physical properties, as well as good mechanical properties. Another reason is that the development of the machinery industry itself and other industries such as chemical industry and instrument industry has created favorable material conditions for the foundry industry, such as the development of testing means, which has ensured the improvement and stability of the casting quality, the invention of electron microscope and so on helping people to go deep into the micro-world of metal, exploring the mystery of metal crystallization, studying the theory of metal solidification, and guiding the production of casting.

铸造是人类掌握比较早的一种金属热加工工艺，已有约 6000 年的历史。在 18 世纪的工业革命以后，蒸汽机、纺织机和铁路等工业兴起，铸件进入为大工业服务的新时期，铸造技术开始有了大的发展。进入 20 世纪，铸造的发展速度很快，其重要因素之一是产品技术的进步要求铸件具有更好的力学和物理性能，同时还应具有良好的机械加工性能。另一个原因是机械工业本身和其他工业（如化工、仪表等）的发展给铸造业创造了有利的物质条件，如检测手段的发展，保证了铸件质量的提高和稳定，并给铸造理论的发展提供了条件；电子显微镜等的发明，帮助人们深入到金属的微观世界，探查金属结晶的奥秘，研究金属凝固的理论，以指导铸造生产。

Casting is the process of melting solid metal into liquid and pouring it into mold with special shape. After solidification, the blank of metal parts with certain shape, size and performance can be obtained. Almost all metal materials can be cast into shape. Casting includes sand casting and special casting, and the material of sand mold casting is raw sand (such as quartz sand, magnesia, zircon sand, chromite sand, forsterite sand, kyanite sand, graphite sand, and iron sand), clay, water glass, resin and other auxiliary materials. Special casting mold includes pressure casting, investment casting, lost-foam casting, metal mold casting, ceramic mold casting and so on. Because the casting blank is near to forming, it can avoid machining or a little machining, and reduce the cost and the production time to a certain extent.

铸造是将固态金属熔化为液态并注入具有特定形状的铸型，凝固后获得具有一定形状、尺寸和性能的金属零件毛坯的成型方法。几乎所有金属材料都可铸造成型。铸造包括砂型铸造和特种铸造，砂型铸造的铸型的材料是原砂（原砂包括石英砂、镁砂、锆砂、铬铁矿砂、镁橄榄石砂、蓝晶石砂、石墨砂、铁砂等）、黏土、水玻璃、树脂及其他辅助材料；特种铸造的铸型包括压力铸造、熔模铸造、消失模铸造、金属型铸造、陶瓷型铸造等。铸造毛坯由于近乎成型，因此可达到免去机械加工或少量加工的目的，从而降低了成本，并在一定程度上减少了制作时间。

Casting is a relatively economical blank forming method, which can show its economy for complex shape parts, such as the cylinder block and head of automobile engine, marine propeller and exquisite artwork, etc. Some difficult-to-cut parts, such as nickel-based turbine parts, can not be molded without casting.

铸造是比较经济的毛坯成型方法，对于形状复杂的零件更能显示出它的经济性，如汽车发动机的缸体和缸盖、船舶螺旋桨以及精致的艺术品等。有些难以切削的零件，如燃气轮机的镍基合金零件，不用铸造方法则无法成型。

In addition, the range of casting parts size and weight is wide, almost unrestricted metal types; parts with General Mechanical Properties, but also with wear-resistant, corrosion-resistant, shock-absorbing and other comprehensive properties, this is the other metal forming methods such as forging, rolling, welding, punching and so on can not do. So far, in the machine-building industry, the number and tonnage of blank parts produced by casting is still the largest.

另外，铸造的零件尺寸和质量的适应范围很宽，金属种类几乎不受限制；零件在具有一般力学性能的同时，还具有耐磨、耐腐蚀、吸震等综合性能，这是其他金属成型方法（如锻、轧、焊、冲等）所做不到的。因此迄今为止，在机器制造业中用铸造方法生产的毛坯零件在数量和吨位上仍是最多的。

There are many kinds of casting, which can be divided into ordinary sand mold casting and special casting according to the molding method. Ordinary sand casting is also known as sand casting or sand, including green sand, dry sand and chemical hardening sand. Special casting molding materials can also be divided into natural mineral sand as the main molding materials of special casting (such as investment casting, mud casting, shell casting, negative pressure casting, full mold casting, ceramic mold casting, and lost-foam casting) and other special casting (such as metal mold casting, pressure casting, continuous casting, low pressure casting, and centrifugal casting).

铸造的种类很多，按造型方法习惯上分为普通砂型铸造和特种铸造。普通砂型铸造又称砂铸或翻砂，包括湿砂型、干砂型和化学硬化砂型三类。特种铸造按造型材料，又可分为以天然矿产砂石为主要造型材料的特种铸造（如熔模铸造、泥型铸造、壳型铸造、负压铸造、实型铸造、陶瓷型铸造、消失模铸造等）和以金属为主要造型材料的特种铸造（如金属型铸造、压力铸造、连续铸造、低压铸造、离心铸造等）两类。

According to the molding process, casting can be divided into gravity casting and die casting. Gravity casting divided into sand casting and hard mold casting, and it is to rely on gravity to melt liquid metal pouring cavity. Die casting is divided into low-pressure casting and high-pressure casting, and it is to rely on the additional pressure of molten metal, which will be immediately pressed into the casting cavity.

按照成型工艺，铸造可分为重力铸造和压力铸造。重力铸造分为砂铸和硬模铸造，是依靠重力将熔融金属液浇入型腔。压力铸造分为低压铸造和高压铸造，是依靠额外增加的压力将熔融金属液瞬间压入铸造型腔。

The casting process usually includes mold preparation, casting metal melting and pouring, and casting treatment and inspection, which are introduced as follows:

铸造工艺通常包括铸型准备、铸造金属熔化与浇注以及铸件处理和检验。现分别讨论如下：

(1) Preparation of a vessel (for casting liquid metal into a solid). Casting can be divided into sand mold, metal mold, ceramic mold, mud mold, graphite mold. According to the number of use, it can be divided into one-time type, semi-permanent type and permanent. The quality of mould preparation is the main factor that affects the quality of castings. Taking the most widely used sand casting as an example, mould preparation includes preparation of moulding materials, moulding and core-making. All kinds of raw materials used for molding and core-making in sand casting, such as foundry sand, molding sand binder and other auxiliary materials and molding sand, core sand and coating prepared by them, are collectively called molding materials. The task of molding material preparation is to select suitable raw sand, binder and auxiliary materials according to the requirements of casting and the properties of metal, and then mix them into molding sand and core sand with certain performance in certain proportion. The commonly used

sand mixing equipment are roller mixer, counter-flow mixer and continuous mixer. The latter is designed for mixing chemical self-hardening sand, continuous mixing, and mixing speed. Molding and coremaking are carried out according to the requirements of casting technology, on the basis of determining the molding method and preparing the molding materials. In many modern casting workshops, molding and coremaking have been mechanized or automated. There are high, medium and low pressure moulding machines, air impact moulding machines, no-box shooting moulding machines, cold box core making machines, hot box core making machines, and film coated sand core making machines.

(1) 铸型（使液态金属成为固态铸件的容器）准备。铸型按所用材料可分为砂型、金属型、陶瓷型、泥型、石墨型等；按使用次数可分为一次性型、半永久型和永久型。铸型准备的优劣是影响铸件质量好坏的主要因素。以应用最广泛的砂型铸造为例，铸型准备包括造型材料的准备和造型以及造芯两大项工作。砂型铸造中用来造型、造芯的各种原材料，如铸造原砂、型砂、黏结剂和其他辅料以及由它们配制成的型砂、芯砂、涂料等，统称为造型材料。造型材料准备的任务是按照铸件的要求和金属的性质，选择合适的原砂、黏结剂和辅料，然后按一定的比例把它们混合成具有一定性能的型砂和芯砂。常用的混砂设备有碾轮式混砂机、逆流式混砂机和连续式混砂机。后者是专为混合化学自硬砂而设计的，可连续混合，混砂速度快。造型、造芯是根据铸造工艺要求，在确定好造型方法、准备好造型材料的基础上进行的，铸件的精度和全部生产过程的经济效果主要取决于这道工序。在很多现代化的铸造车间里，造型、造芯都实现了机械化或自动化。常用的砂型造型、造芯设备有高、中、低压造型机，气冲造型机，无箱射压造型机，冷芯盒制芯机，热芯盒制芯机，以及覆膜砂制芯机等。

(2) Melting and pouring of casting metal. Casting metals are mainly cast iron, cast steel and cast non-ferrous metals and alloys. Metal smelting is not only the pure melting, but also the smelting process, so that the metal poured into the mold meets the expected requirements in terms of temperature, chemical composition and cleanliness. To this end, a variety of inspection tests for the purpose of quality control should be carried out during the smelting process, and liquid metal can be poured only after reaching the specified targets. Sometimes, in order to meet the higher requirements, the metal liquid has to be treated outside the furnace, such as desulphurization, vacuum degassing, refining, inoculation or metamorphism. Smelting metal commonly used equipment is cupola, arc furnace, induction furnace, resistance furnace, reflector furnace and so on.

(2) 铸造金属熔化与浇注。铸造金属（铸造合金）主要有各类铸铁、铸钢和铸造有色金属及合金。金属熔炼不仅仅是单纯的熔化，还包括冶炼过程，使浇进铸型的金属在温度、化学成分和洁净度方面都符合预期要求。为此，在熔炼过程中要进行以控制质量为目的的各种检查测试，液态金属在达到各项规定指标后方可允许浇注。有时为了达到更高的要求，金属液在出炉后还要经过炉外处理，如脱硫、真空脱气、炉外精炼、孕育或变质处理等。熔炼金属常用的设备有冲天炉、电弧炉、感应炉、电阻炉、反射炉等。

(3) Casting treatment and inspection. Casting treatment includes removal of foreign matter on core and casting surface, removal of spouts and risers, relieving burrs and slits, as well as heat treatment, shaping, anti-rust treatment and roughing. After the casting is removed from the self-

cooling mold, it has a gate, riser, metal burr and seam. The sand is adhered to the casting in the sand mold, so it must be cleaned. The equipment that carries out this kind of work has polisher, shot blasting machine, and gate riser cutter to wait. Sand cleaning of sand casting is a working procedure with poor working conditions, so it should be taken into account to create convenient conditions for sand cleaning when selecting molding methods. Some castings have to undergo post-treatment due to special requirements, such as heat treatment, shaping, rust prevention, rough machining, etc. Heat treatment can change or affect the structure and properties of cast iron. At the same time, it can also obtain higher strength, hardness, and improve its wear resistance. There are many types of heat treatment due to different purposes, and they can be divided into two main types. The first type is heat treatment whose structure will not change or should not change due to heat treatment, and the second type is heat treatment with a change in the basic structure. The first type of heat treatment program is mainly used to eliminate the internal stress, which is caused by the different cooling conditions during the casting process. For the second type of heat treatment, significant changes have taken place in the basic organization, which can be broadly divided into the following five categories:

（3）铸件处理和检验。铸件处理包括清除型芯和铸件表面上的异物，切除浇口和冒口，铲磨毛刺和披缝等凸出物，以及热处理、整形、防锈处理和粗加工等。铸件自浇注冷却的铸型中取出后，带有浇口、冒口、金属毛刺和披缝，砂型铸造的铸件还黏附着砂子，因此必须经过清理工序。进行这种工作的设备有磨光机、抛丸机、浇冒口切割机等。砂型铸件的落砂清理是劳动条件较差的一道工序，所以在选择造型方法时应尽量考虑到为落砂清理创造方便的条件。有些铸件因特殊要求，还要经过铸件后处理，如热处理、整形、防锈处理、粗加工等。借助热处理可以改变或影响铸铁的组织及性质，同时可以获得更高的强度和硬度，从而改善其磨耗抵抗能力等。由于目的不同，热处理的种类非常多，基本上可分成两大类：第一类是组织结构不会因经过热处理而发生变化或不应该发生改变的热处理；第二类则是基本组织结构发生变化的热处理。第一类热处理程序主要用于消除内应力，而此内应力是在铸造过程中由于冷却状况及条件不同而引起的，组织、强度及其他力学性质等不因热处理而发生明显变化。对于第二类热处理而言，基本组织发生了明显的改变，可大致分为以下五类：

(1) The purpose of softening annealing is to decompose carbides and reduce their hardness so as to improve the processability. For Ductile Iron, the aim is to obtain more ferrite structure.

（1）软化退火，其目的主要在于分解碳化物，将其硬度降低，从而提高加工性能。对于球墨铸铁而言，其目的在于获得更多的铁素体组织。

(2) The purpose of normalizing is to obtain pearlite and sorbite structure, and to improve the mechanical properties of the casting.

（2）正火，其主要目的是获得珠光体和索氏体组织，提高铸件的力学性能。

(3) The purpose of quenching is to obtain higher hardness or wear strength, and to get very high surface wear resistance.

（3）淬火，其目的主要是为了获得更高的硬度或磨耗强度，同时得到较高的表面耐磨特性。

(4) The purpose of surface hardening treatment is mainly to obtain the surface hardening

layer. At the same time it can get very high surface wear resistance.

（4）表面硬化处理，其目的主要是为了获得表面硬化层，同时得到甚高的表面耐磨特性。

(5) The purpose of precipitation hardening treatment is mainly to obtain high strength. But the elongation rate does not change drastically.

（5）析出硬化处理，其目的主要是为了获得高强度，而伸长率并不因此发生激烈的改变。

3.1.2 Pressure Processing
3.1.2 压力加工

In the plastic forming process, the external force acting on the metal blank mainly includes impact force and pressure. Hammer equipment produces impact force to defrom the metal, rolling mill and press to apply pressure to the metal blank to defrom the metal.

在塑性成型过程中，作用在金属坯料上的外力主要有冲击力和压力两种。锤类设备产生冲击力使金属变形，轧机与压力机对金属坯料施加压力使金属变形。

Steel and most non-ferrous metal and its alloys have a degree of plasticity that allows them to be pressed either hot or cold. Metal processing at a temperature above recrystallization temperature in a hot state is called hot-working hot deformation, while metal processing at a temperature below the recovery temperature in a cold state is called cold-working cold deformation. Metal pressure processing and metal casting, cutting, welding and other processing methods have the following characteristics：

钢和大多数有色金属及其合金都具有一定的塑性，因此可以在热态或冷态下进行压力加工。金属在热态下发生再结晶温度以上的温度条件下加工，称为热加工（热变形）；金属在冷态下发生回复温度以下的温度条件下加工，称为冷加工（冷变形）。金属压力加工与金属铸造、切削、焊接等加工方法相比，具有以下特点：

(1) The metal pressure processing is on the premise of keeping the metal integrity. Depending on the plastic deformation of material transfer to achieve the workpiece shape and size changes, it will not produce chips, so the material utilization rate is much higher. Besides the change of size and shape, the structure and properties of metal can also be improved during plastic processing.

（1）金属压力加工是在保持金属整体性的前提下，依靠塑性变形发生物质转移来实现工件形状和尺寸的变化，不会产生切屑，因而材料的利用率高得多。

(2) Especially for the casting billet, after plastic processing, it can make its structure compact, coarse grain crushing refinement and uniformity, so as to improve its performance. In addition, the flow line produced by plastic flow can also improve its performance.

（2）塑性加工过程中，除尺寸和形状发生改变外，金属的组织、性能也能得到改善和提高。尤其对于铸造坯而言，经过塑性加工可使其结构致密、粗晶破碎细化和均匀，从而使其性能提高。此外，塑性流动所产生的流线也能使其性能得到改善。

(3) The plastic processing process is convenient to realize the continuous and automatic production process, and it is suitable for mass production, such as rolling and drawing processing. So the labor productivity is high.

（3）塑性加工过程便于实现生产过程的连续化和自动化,适于大批量生产,如轧制、拉拔加工等,因而劳动生产率高。

(4) The dimension precision and surface quality of the plastic processed products are high.

（4）塑性加工产品的尺寸精度和表面质量高。

(5) The equipment is large, with high energy consumption.

（5）设备较庞大,能耗较高。

Due to the above characteristics, metal pressure processing can not only reduce the consumption of raw materials, high production efficiency, and stable product quality, but also effectively improve the organization and performance of metal. These technical and economic advantages are unique, so that it becomes one of the most important means of metal processing, and therefore occupies a very important position in the national economy. For example, in the production of iron and steel materials, in addition to a small number of parts made directly by casting, more than 90% of the total output of steel and more than 70% of the total output of non-ferrous metal, and plastic processing are required to meet the needs of machinery manufacturing, transportation, power and telecommunications, chemical, building materials, instrumentation, aerospace, defense industry, civil hardware and household appliances. Moreover, plastic processing itself is an important processing method often used by many of the above-mentioned departments to directly manufacture parts, such as automobile manufacturing, shipbuilding, aerospace, civil hardware and other parts of many departments needing to be plastic processing manufacturing.

金属压力加工由于具有上述特点,不仅原材料消耗少,生产效率高,产品质量稳定,而且还能有效地改善金属的组织和性能。这些技术上和经济上的独到之处和优势,使它成为金属加工中极其重要的手段之一,因而在国民经济中占有十分重要的地位。如在钢铁材料生产中,除了少部分采用铸造方法直接制成零件外,钢总产量的90%以上和有色金属总产量的70%以上,均需经过塑性加工成材才能满足机械制造、交通运输、电力电讯、化工、建材、仪器仪表、航空航天、国防军工、民用五金和家用电器等部门的需要。而塑性加工本身也是上述许多部门直接制造零件而经常采用的重要加工方法,如汽车制造、船舶制造、航空航天、民用五金等部门的许多零件都需经塑性加工制造。

Metal pressure processing methods are generally divided into five categories：

金属压力加工方法一般分为以下五种:

(1) Forging. Forging is a plastic processing method which uses the moving hammer of the forging hammer or compresses the workpiece with the indenter of the press. And it is used to manufacture various parts, profiles or blanks. It mainly includes two basic methods, such as free forging and model forging, which are introduced as follows：

（1）锻造。锻造是利用锻锤的运动锤击或用压力机的压头压缩工件的塑性加工方法,用于制造各种零件、型材或毛坯。其主要包括两种基本方式,即自由锻造和模型锻造。介绍如下:

1) Free forging, abbreviated as free forging, as shown in Figure 3-1(a). Free forging is the deformation of a heated metal blank between an anvil and an anvil by an impact free forging hammer or a press. It is used to make parts of relatively simple shapes.

1) 自由锻造，简称自由锻，如图3-1(a)所示。自由锻造是使已加热的金属坯料在上下砧之间承受冲击力（自由锻锤）或压力（压力机）而变形，可用于制造各种形状比较简单的零件毛坯。

2) Model forging, referred to as die forging, as shown in Figure 3-1(b). Model forging is to make the heated metal billet deformed by bearing the impact force (free forging hammer) or pressure (press machine) in the forging die with the preformed cavity. It is used to make the blank of various parts with simple shapes.

2) 模型锻造，简称模锻，如图3-1(b)所示。模型锻造是使已加热的金属坯料在已经预先制好型腔的锻模间承受冲击力（自由锻锤）或压力（压力机）而变形，可用于制造各种形状比较简单的零件毛坯。

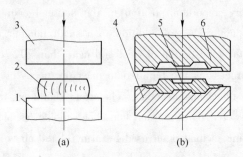

Figure 3-1 Forging
图 3-1 锻造
(a) Free forging; (b) Die forging
(a) 自由锻; (b) 模锻

1—Supporting body; 2, 5—Forging; 3—Forging hammer; 4—Lower die;
6—Upper die becoming the blank of the part consistent with the shape of the mold cavity
1—支撑体; 2, 5—锻件; 3—锻锤; 4—下模; 6—上模成为与型腔形状一致的零件毛坯

(2) Rolling (or calendering), as shown in Figure 3-2. Rolling is a process in which a metal blank is rolled through a gap between a pair of rotary rollers, resulting in a reduction in cross-section, a change in shape and an increase in length. Rolling includes non-heating of cold-rolled metal billet and heating of hot-rolled metal billet, which is used for manufacturing sheet, Bar, profile, tube, etc.

(2) 轧制（或称压延加工），其产品截面形状图如图3-2所示。轧制是使金属坯料通过一对回转轧辊之间的空隙而受到压延，使其断面减小、形状改变、长度增长的过程。轧制包括冷轧（金属坯料不加热）和热轧（金属坯料加热），用于制造板材、棒材、型材、管材等。

The main rolling equipment is called rolling mill. It is the equipment to realize the rolling process of metal, generally referring to the equipment to complete the whole rolling process, including the main equipment, auxiliary equipment, lifting transport equipment and ancillary equipment. However, the mill is usually only referred to as the main equipment, as shown in Figure 3-3. The trend of modern rolling mill development is continuous, automation, specialization, high quality products, and low consumption. Since the 1960s, great progress has been made in the design, research and manufacture of rolling mills, which has improved the

Figure 3-2 Section shape of rolled product
图 3-2 轧制产品截面形状图

performance of cold and hot strip mills, heavy plate mills, high-speed wire rod mills, H-section mills and tube mills. A series of advanced equipments such as wire mill with rolling speed up to 115m/s, fully continuous cold strip mill, 5500mm wide and heavy plate mill, and continuous H-beam mill have been developed. Hydraulic AGC, shape control, computer program control and test methods are becoming more and more perfect, and rolling varieties are expanding. All kinds of special structure rolling mills are developing to meet the requirements of new product quality and improve economic benefit.

轧制用主体设备称为轧机。它是实现金属轧制过程的设备，泛指完成轧材生产全过程的装备，包括主要设备、辅助设备、起重运输设备和附属设备等。一般所说的轧机往往仅指主要设备。如图 3-3 所示为轧机的传动与组成。现代轧机发展的趋向是连续化、自动化和专业化，产品质量高，消耗低。20 世纪 60 年代以来，轧机在设计、研究和制造方面取得了很大的进展，使带材冷热轧机、厚板轧机、高速线材轧机、H 型材轧机和连轧管机组等的性能更加完善，并出现了轧制速度高达 115m/s 的线材轧机、全连续式带材冷轧机、5500mm 宽厚板轧机和连续式 H 型钢轧机等一系列先进设备。轧机用的原料单重增大，液压 AGC、板形控制、电子计算机程序控制及测试手段越来越完善，轧制品种不断扩大。一些适用于连续铸轧、控制轧制等的新轧制方法，以及适应新的产品质量要求和提高经济效益的各种特殊结构轧机都在发展中。

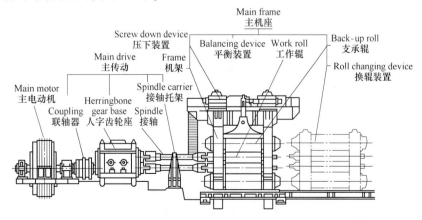

Figure 3-3 Transmission and composition of the mill
图 3-3 轧机的传动与组成

(3) Extrude, as shown in Figure 3-4. Extrusion is the process of extruding the metal blank which is placed in the cavity of the die from the die hole and forming it into a part, including cold extrusion, hot extrusion and compound extrusion. A variety of complex cross-section profiles or parts can be obtained by extrusion, which is suitable for the processing of low carbon steels, non-ferrous metal and their alloys, and alloy steels and refractory alloys, if appropriate process measures are taken. The extruded product is shown in Figure 3-5.

(3) 挤压，如图3-4所示。挤压是把放置在模具容腔内的金属坯料从模孔中挤出来，使其成型为零件的过程，包括冷挤压、热挤压和复合挤压。挤压可获得各种复杂截面的型材或零件，适用于低碳钢、有色金属及其合金的加工，如采取适当的工艺措施，还可对合金钢和难熔合金进行加工。挤压产品如图3-5所示。

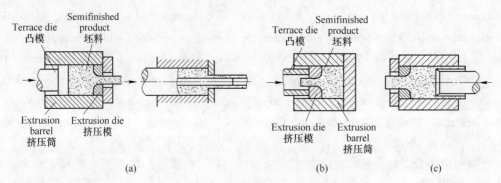

Figure 3-4　Extrusion
图 3-4　挤压
(a) Positive extrusion; (b) Reverse extrusion; (c) Composite extrusion
(a) 正挤压；(b) 反挤压；(c) 复合挤压

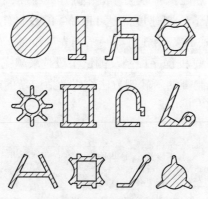

Figure 3-5　Extruded product
图 3-5　挤压产品

(4) Punching, as shown in Figure 3-6. Stamping is the process of forming metal slab under the impact force or pressure in the stamping die, which is divided into cold stamping and hot stamping. Sheet metal stamping is a process in which the metal sheet is placed in a stamping die and applied a force to cause the sheet to be sheared or deformed.

(4) 冲压，如图3-6所示。冲压是金属板坯在冲压模的冲击力或压力下成形的过程，分为冷冲压和热冲压。

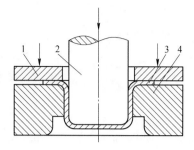

Figure 3-6　Punching
图 3-6　冲压

1—Pressure block; 2—Punch; 3—Stamping piece; 4—Concave die
1—压块；2—冲头；3—冲压件；4—凹模

(5) Drawing, as shown in Figure 3-7. Drawing is the process of drawing the metal blank through the die hole to reduce its cross-section, which is divided into cold drawing and hot drawing. The drawing process is mainly used to make all kinds of thin wires such as cables, thin-walled tubes and special geometric shapes as shown in Figure 3-8. Therefore, it is often used for the re-processing of rolled parts to improve product quality. Low carbon steels, and most non-ferrous metal and their alloys can be drawn.

(5) 拉拔，如图 3-7 所示。拉拔是将金属坯料拉过模孔以缩小其横截面的过程，可分为冷拉拔和热拉拔。拉拔工艺主要用于制造各种细线材（如电缆等）、薄壁管和特殊几何形状的型材（见图 3-8）。多数情况下拉拔是在冷态下进行的，所得到的产品精度高，故常用于轧制件的再加工，以提高产品质量。低碳钢和大多数有色金属及其合金都可以经拉拔成型。

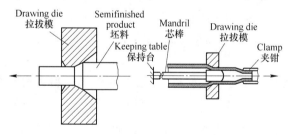

Figure 3-7　Drawing
图 3-7　拉拔

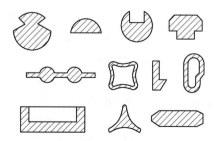

Figure 3-8　Drawing cross section
图 3-8　拉拔横截面

3.1.3 New Material Forming Technology
3.1.3 材料成型新技术

3.1.3.1 Powder Metallurgy
3.1.3.1 粉末冶金

Powder metallurgy is a process in which metal or a mixture of metal, and nonmetal powders is used as a raw material for forming and sintering metal materials, composites, and various types of articles. The powder metallurgy method is similar to the method used to produce ceramics, so a series of new powder metallurgy techniques can also be used to prepare ceramic materials. Because of the advantages of powder metallurgy technology, it has become the key to solve the problem of new materials and plays an important role in the development of new materials. It can also be used to make all kinds of iron and copper base parts such as automobiles, motorcycles, textile machinery, industrial sewing machines, electric tools, hardware tools, electrical appliances and construction machinery, etc.

粉末冶金是用金属粉末或金属粉末与非金属粉末的混合物作为原料，经过成型和烧结，制造金属材料、复合材料以及各种类型制品的工艺技术。粉末冶金法与生产陶瓷的方法有相似的地方，因此，一系列粉末冶金新技术也可用于陶瓷材料的制备。粉末冶金技术的优点已成为解决新材料问题的关键，在新材料的发展中起着举足轻重的作用。其可用于制作汽车、摩托车、纺织机械、工业缝纫机、电动工具、五金工具、电器及工程机械等各种粉末冶金（铁铜基）零件。

Powder metallurgy materials are divided into powder metallurgy porous materials, powder metallurgy friction materials, powder metallurgy structural parts, powder metallurgy working mold materials, powder metallurgy electromagnetic materials and powder metallurgy high-temperature materials, etc.

粉末冶金材料分为粉末冶金多孔材料、粉末冶金减摩材料、粉末冶金摩擦材料、粉末冶金结构零件、粉末冶金工模具材料、粉末冶金电磁材料和粉末冶金高温材料等。

The production processes of the powder metallurgy are as follows:
粉末冶金的生产过程为：

(1) Production of powders: The production processes of powder include the preparation of powder, mixing of powder materials and other steps. To improve the formability and plasticity of powders, plasticizers such as gasoline, rubber or paraffin wax are usually added.

(1) 生产粉末：粉末的生产过程包括粉末的制取、粉料的混合等步骤。为改善粉末的成型性和可塑性，通常加入汽油、橡胶或石蜡等增塑剂。

(2) Press Molding: The powder is pressed into the desired shape at a pressure of 500~600MPa.

(2) 压制成型：粉末在 500~600MPa 的压力下，被压成所需形状。

(3) Sintering: Sintering is carried out in a high temperature furnace or a vacuum furnace with a protective atmosphere. Sintering differs from metal melting in that at least one element

remains solid during sintering. During the sintering process, the powder particles are dispersed, recrystallized, melted, combined and dissolved to from the metallurgical products with certain porosity.

（3）烧结：烧结在保护气氛的高温炉或真空炉中进行。烧结不同于金属熔化，烧结时至少有一种元素仍处于固态。烧结过程中，粉末颗粒间通过扩散、再结晶、熔焊、化合、溶解等一系列的物理化学过程，成为具有一定孔隙率的冶金产品。

（4）Post processing：In general, the sintered parts can be used directly. However, for some parts which require high dimensional accuracy, high hardness and high wear resistance, sintering post - processing is needed. The post - treatment includes fine pressing, rolling, extrusion, quenching, surface quenching, oil impregnation and infiltration.

（4）后处理：一般情况下，烧结好的制件可直接使用。但对于某些要求尺寸精度高且具有高硬度、高耐磨性的制件，还要进行烧结后处理。后处理包括精压、滚压、挤压、淬火、表面淬火、浸油及熔渗等。

3.1.3.2 New Technologies for Pressure Processing
3.1.3.2 压力加工新技术

In recent years, many new processes and technologies have appeared in pressure processing, such as extrusion, rolling, precision die forging, multi-direction die forging, liquid die forging, superplastic forming, high - energy and high - speed forming. These new pressure processing technologies make the shape of the forging part close to the shape of the part as far as possible to achieve the goal of less or no cutting processing, thus saving raw materials and cutting processing workload, with higher productivity. Meanwhile, it can obtain reasonable fiber structure, improve the mechanical properties and use performance of parts.

近年来在压力加工生产中出现了许多新工艺和新技术，如零件的挤压成型、零件的轧制成型、精密模锻、多向模锻、液态模锻、超塑性成型以及高能高速成型等。这些压力加工新技术可尽量使锻压件的形状接近零件的形状，达到少或无切削加工的目的，从而可以节省原材料和切削加工工作量，具有更高的生产率。同时，还可以得到合理的纤维组织，提高零件的力学性能和使用性能。

Extrusion of Parts
零件的挤压成型

Extrusion is the process of applying strong pressure on the die to force the metal blank placed in the die to produce directional plastic deformation and extrude from the die hole to obtain the required parts or semi-finished products. It is characterized by：When extruding, the metal blank is deformed in three-dimensional compression state, so the plasticity of the metal blank can be improved, and the parts with complicated, deep hole, thin wall and special cross-section can be extruded；The parts have high precision and low surface roughness；The mechanical properties of the parts are improved and the raw materials are saved. The types of part extrusion molding are as follows：

挤压是施加强大压力作用于模具，迫使放在模具内的金属坯料产生定向塑性变形并从模孔中挤出,从而获得所需零件或半成品的加工方法。其特点是:挤压时金属坯料在三向受压状态下变形,因此可提高金属坯料的塑性,可以挤压出各种复杂形状、深孔、薄壁、异形截面的零件;零件精度高,表面粗糙度低;提高了零件的力学性能;节约了原材料。零件挤压成型的类型具体如下:

(1) Classification by metal flow direction and punch movement direction, it can be divided as follows:

(1) 按金属流动方向和凸模运动方向分类,可分为:

1) Positive extrusion. Forward extrusion is the process in which the metal flows in the same direction as the punch moves.

1) 正挤压。正挤压是指金属流动方向与凸模运动方向相同的挤压工艺。

2) Back extrusion. Backward extrusion is the extrusion process in which the metal flow direction is opposite to the punch movement direction.

2) 反挤压。反挤压是指金属流动方向与凸模运动方向相反的挤压工艺。

3) Compound extrusion. In the compound extrusion process, the flow direction of one part of metal is same as that of punch, and the flow direction of the other part of metal is opposite to that of punch.

3) 复合挤压。复合挤压过程中,坯料上一部分金属的流动方向与凸模运动方向相同,而另一部分金属的流动方向与凸模运动方向相反。

4) Radial extrusion. The radial extrusion is the extrusion process of the direction of 90° in the metal movement and the direction of the punch movement.

4) 径向挤压。径向挤压是指金属运动方向与凸模运动方向成90°的挤压工艺。

(2) According to the temperature of the metal blank, it can be divided as follows:

(2) 按金属坯料所具有的温度分类,可分为:

1) Hot extrusion. Hot extrusion is a process in which the deformation temperature of the billet is higher than the recrystallization temperature (same as the forging temperature). Hot extrusion of the deformation resistance is small, allowing a greater degree of deformation each time, but the product surface roughness.

1) 热挤压。热挤压是指坯料变形温度高于材料再结晶温度(与锻造温度相同)的挤压工艺。热挤压的变形抗力小,允许每次变形程度较大,但产品的表面粗糙。

2) Cold extrusion. Cold extrusion is a process in which the deformation temperature of the billet is often lower than the recrystallization temperature of the material at room temperature. The deformation resistance of cold extrusion is much higher than that of hot extrusion. But the surface of the product is smooth, and the inner structure of the product is processed hardening, which improves the strength of the product.

2) 冷挤压。冷挤压是指坯料变形温度低于材料再结晶温度(经常在室温下)的挤压工艺。冷挤压的变形抗力比热挤压高得多,但产品表面光洁,且产品内部组织为加工硬化组织,从而提高了产品的强度。

3) Warm extrusion. Warm extrusion is an extrusion process in which the billet temperature is between hot extrusion and cold extrusion. The metal is heated to an appropriate temperature below

the recrystallization temperature of 100~800℃ for the extrusion. The precision and mechanical properties of warm extrusion parts are slightly lower than those of cold extrusion parts.

3) 温挤压。温挤压为坯料温度介于热挤压和冷挤压之间的挤压工艺,即将金属加热到再结晶温度以下的某个合适温度(100~800℃)进行挤压。温挤压零件的精度和力学性能略低于冷挤压件。

In addition to the above extrusion method, there is a hydrostatic extrusion method, as shown in Figure 3-9. When hydrostatic extrusion, punch and blank do not directly contact, and the liquid pressure is up to 304MPa, and then liquid to the blank for the metal through the die and forming. Hydrostatic extrusion can increase the deformation of single extrusion, and the extrusion force is 10%~15% less than that of other extrusion methods.

除了上述挤压方法外,还有一种静液挤压方法,如图 3-9 所示。静液挤压时凸模与坯料不直接接触,而是给液体施加压力(压力可达 304MPa),再经液体传给坯料,使金属通过凹模而成型。静液挤压在坯料侧面无通常挤压时存在摩擦,变形较均匀,可提高一次挤压的变形量,挤压力也较其他挤压方法小 10%~15%。

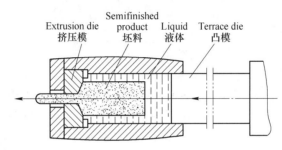

Figure 3-9 Hydrostatic extrusion
图 3-9 静液挤压

Rolling Forming of Parts
零件的轧制成型

The rolling method can produce not only profile but also all kinds of parts. It is more and more widely used in mechanical manufacturing. The rolling of the parts has a continuous static pressure process without shock and vibration. According to the difference between roll axis and billet axis, rolling can be divided into vertical rolling, cross rolling and cross rolling, introduced as follows:

轧制方法除了可生产型材外,还可生产各种零件,在机械制造业中得到了越来越广泛的应用。零件的轧制有一个连续静压过程,没有冲击和振动。根据轧辊轴线与坯料轴线的不同,轧制分为纵轧、横轧、斜轧等几种。介绍如下:

(1) Vertical Rolling. Vertical Rolling refers to the rolling method in which the axis of the roll and the axis of the billet are perpendicular to each other, including rolling of various profiles and plates, forging rolling, ring rolling and so on. Roll forging is a new process of applying rolling technology to forging production. It is used to compress the billet through a pair of relatively rotating rollers with circular modules, as shown in Figure 3-10. At present, forming roll forging is suitable for producing three types of forgings(long bar with flat cross section), such as wrench,

movable wrench, chain ring, etc. It is also forging with undeformed head and decreasing cross section area along the length direction, such as blade, etc.

（1）纵轧。纵轧是指轧辊轴线与坯料轴线互相垂直的轧制方法，包括各种型材与板带材轧制、辊锻轧制、辗环轧制等。辊锻轧制是把轧制工艺运用到锻造生产中的一种新工艺，它是使坯料在通过装有圆弧形模块的一对相对旋转的轧辊时受压而变形，如图 3-10 所示。目前，成型辊锻适用于生产三种类型的锻件（即扁截面的长杆件），如扳手、活动扳手、链环等，带有不变形头部而沿长度方向横截面面积递减的锻件，如叶片等。

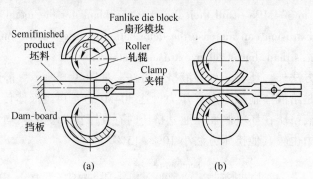

Figure 3-10　Roll forging rolling
图 3-10　辊锻轧制

（2）Cross Rolling. Cross rolling refers to a rolling method in which the roll axis is parallel to the workpiece axis and the roller rotates relative to the workpiece. It includes gear cross rolling, spiral cross rolling and wedge cross rolling, introduced as follows：

（2）横轧。横轧是指轧辊轴线与轧件轴线平行，且轧辊与轧件做相对转动的轧制方法。它包括齿轮横轧、螺旋横轧与楔横轧。介绍如下：

1) Gear cross rolling. Gear cross rolling is to make the roll with tooth shape and round blank to realize local continuous forming and rolling into gear. The cross rolling deformation is mainly in the radial direction, and the axial deformation is very small. There are both hot and cold rolling for gear cross rolling. The method can also be used for rolling sprockets, spline shafts, etc.

1) 齿轮横轧。齿轮横轧是使带齿形的轧辊与圆形坯料在对滚中实现局部连续成型，轧制成齿轮。这种横轧的变形主要在径向进行，轴向变形很小，齿轮横轧既有热轧也有冷轧。此方法还可以轧制链轮、花键轴等。

2) Spiral Cross Rolling. Spiral Cross rolling, also known as thread rolling, is two rollers with thread rotating in the same direction, driving round billet rotation. One of the rollers is radial feed, the billet rolling into thread. The cross rolling deformation is mainly in the radial direction.

2) 螺旋横轧。螺旋横轧又称螺纹滚压，是将两个带螺纹的轧辊（滚轮）以相同的方向旋转，带动圆形坯料旋转，其中一个轧辊径向进给，将坯料轧制成螺纹。这种横轧的变形主要在径向进行。

3) Cross wedge rolling. Cross wedge rolling is to take two rollers with wedge-shaped die to rotate in the same direction, driving round billet to rotate, and billet in the role of wedge-shaped roll rolled into a variety of shapes of step shaft. The deformation of the cross rolling is mainly radial

compression and axial extension. Figure 3-11 is a cross wedge rolling schematic diagram. Cross wedge rolling technology is mainly applied to the production of shaft parts with rotating body, such as all kinds of gear shafts in gearbox, camshaft and ball head pin in engine, etc. It can not only replace the roughing process to produce all kinds of shaft parts, but also can provide precision die forging blank for all kinds of die forging parts.

3) 楔横轧。楔横轧是将两个带楔形模具的轧辊以相同的方向旋转,带动圆形坯料旋转,坯料在楔形轧辊的作用下被轧制成各种形状的台阶轴。这种横轧的变形主要为径向压缩和轴向延伸。图 3-11 为楔横轧原理图。楔横轧工艺主要适用于带旋转体的轴类零件的生产,如汽车、拖拉机、摩托车、内燃机等变速箱中的各种齿轮轴,发动机中的凸轮轴、球头销等。它不仅可以代替粗车工艺来生产各种轴类零件,而且可以为各种模锻零件提供精密的模锻毛坯。

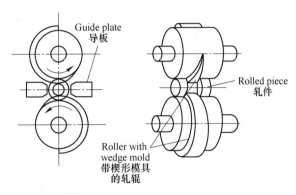

Figure 3-11 Principle of cross wedge rolling
图 3-11 楔横轧原理

(3) Skew rolling. Skew rolling is a rolling method in which the rollers are arranged obliquely and rotated in the same direction, and the workpiece is rotated in the opposite direction under the action of the rollers. At the same time, the roller moves axially (i.e., spirally).

(3) 斜轧。斜轧是指轧辊相互倾斜配置、以相同方向旋转,轧件在轧辊的作用下反向旋转,同时还做轴向运动(即螺旋运动)的轧制方法。

Precision Die Forging
精密模锻

Precision die forging is a kind of die forging technology which produces complicated shape and high precision forgings on the die forging equipment. For example, precision die-forged bevel gears whose tooth shape can be forged directly without further cutting. The main technological features of precision die forging are as follows:

精密模锻是在模锻设备上锻造出形状复杂、锻件精度高的锻件的模锻工艺。比如精密模锻伞齿轮,其齿形部分可直接锻出而不必再经切削加工。精密模锻的主要工艺特点有:

(1) It is necessary to calculate the size of the original blank accurately and to cut the material according to the quality of the blank strictly. Otherwise the size tolerance of the forgings will be increased and the precision will be reduced.

（1）需要精确计算原始坯料的尺寸，严格按坯料质量下料，否则会增大锻件尺寸公差，降低精度。

(2) The precision of precision forging depends on the machining precision of forging die to a great extent. So the precision of precision forging die chamber must be very high.

（2）精密模锻的锻件精度在很大程度上取决于锻模的加工精度，因此，精锻模膛的精度必须很高。

(3) Precision die forging is usually carried out on die forging equipment with high rigidity and precision.

（3）精密模锻一般都在刚度大、精度高的模锻设备上进行。

Multi-direction Die Forging

多向模锻

Multi-direction die forging is a die forging process in which the blank is placed in the forging die and pressed simultaneously or successively by several punches in different directions to obtain a precision forging with complicated shape.

多向模锻是将坯料放于锻模内，用几个冲头从不同方向同时或先后对坯料加压，以获得形状复杂的精密锻件的模锻工艺。

Liquid Die Forging

液态模锻

The essence of liquid die forging is to pour liquid metal directly into the metal die, acting on the liquid or semi-liquid metal with a certain pressure in a certain period of time to form it, and then crystallizing and producing local plastic deformation under this pressure. It is an advanced process developed on the basis of pressure casting, which is similar to squeeze casting.

液态模锻的实质是把金属液直接浇到金属模内，然后在一定时间内以一定的压力作用于液态（或半液态）金属上成型，并在此压力下结晶和产生局部塑性变形。液态模锻是在压力铸造的基础上逐渐发展起来，类似于挤压铸造的一种先进工艺。

Superplastic Forming

超塑性成型

Superplasticity is the property that the relative elongation of a metal or alloy is more than 100% when the deformation rate is low, the deformation temperature is about 1/2 of the melting point and the grain size is uniform. There is no necking phenomenon in the process of tensile deformation of superplastic metal, and the deformation stress is only one-tenth to one-tenth of that of normal metal. Therefore, the metal is very easy to shape and it can be made into complex parts by various technological methods. At present, the common superplastic forming materials are mainly zinc alloy, aluminum alloy, titanium alloy and some superalloys.

超塑性是指金属或合金在特定条件，即低的变形速率、一定的变形温度（约为熔点的1/2）和均匀的细晶粒度下，其相对伸长率超过100%以上的特性。超塑性状态下的金属在拉伸变形过程中不产生缩颈现象，变形应力仅为常态下金属变形应力的几分之一至几十分之一。因此，该种金属极易成型，可采用多种工艺方法制出复杂零件。目前常用的超塑性成型材料主要是锌合金、铝合金、钛合金和某些高温合金。

High-energy High-speed Forming

高能高速成型

High-energy high-speed forming is a forming method which releases high energy and deforms metal in very short time. The history of high-energy high-speed forming can be traced back to more than 100 years ago. At that time, due to the high cost and the limitation of industrial development, the process was not applied. With the development of aviation and missile technology, high-energy high-speed forming method has been put into production practice. High-energy high-speed forming mainly includes high-speed forming of kinetic energy by using high-pressure gas to make piston move at high speed, explosive forming of chemical energy produced by explosive, electro-hydraulic forming of electric energy and electromagnetic forming of magnetic force. Here is the main introduction to explosive molding. Explosive forming is a high-energy and high-speed forming process, using explosive substance to release huge chemical energy at the moment of explosion. The explosive forming process not only has the common characteristics of the high-energy and high-speed forming method. The die is simple, the part precision is high, the surface quality is high, the plastic deformation ability of the material can be improved, and the composite process is favorable. The utility model also has the characteristics of simple equipment and it is suitable for forming large parts. The explosive forming is mainly used for drawing, bulging and straightening of sheet metal, and it is often used for explosive welding, surface strengthening, pipe structure assembly, powder pressing and so on.

高能高速成型是一种在极短时间内释放高能量而使金属变形的成型方法。高能高速成型的历史可追溯到一百多年前，但当时由于成本太高及工业发展的局限，该工艺并未得到应用。随着航空及导弹技术的发展，高能高速成型方法才进入生产实践中。高能高速成型主要包括利用高压气体使活塞高速运动来生产动能的高能成型、利用火药爆炸产生化学能的爆炸成型、利用电能的电液成型以及利用磁场力的电磁成型。这里主要介绍爆炸成型。爆炸成型是利用爆炸物质在爆炸瞬间释放出巨大的化学能，对金属坯料进行加工的高能高速成型工艺。爆炸成型不仅具有高能高速成型方法共有的特点（即模具简单，零件精度高、表面品质高，可提高材料的塑性变形能力，利于采用复合工艺），还具有设备简单、适于大型零件成型的特点。爆炸成型主要用于板材的拉深、胀形和校形，还常用于爆炸焊接、表面强化、管件结构装配、粉末压制等。

Task 3.2　Fundamentals of Metal Pressure Processing
任务3.2　金属压力加工的基本原理

3.2.1　Physical Mature of Plastic Deformation of Metals
3.2.1　金属塑性变形的物理本质

Metal and alloy materials are made up of polycrystals, which are composed of many grains with different crystal directions. Each grain can be regarded as a single crystal with a grain boundary of rather small thickness between the grains. The plastic deformation of metal is the deformation of each grain and the deformation between grains. The deformation of each grain is equivalent to that of a single crystal. Under the action of external force, through the movement of dislocations, the crystals slip and twinning, and the plastic deformation of metals are realized. In addition to the deformation within the crystal, the deformation of the grain boundary also occurs during polycrystalline deformation, which is not only related to the dislocation movement, but also plays an important role in the diffusion process.

金属及合金材料由多晶体构成，多晶体是由许多结晶方向不同的晶粒所组成。每个晶粒可看成是一个单晶体，晶粒之间存在厚度相当小的晶界。金属塑性变形是指每个晶粒的变形和晶间变形。每个晶粒的变形相当于单晶体变形，在外力作用下，通过位错的移动，晶体发生滑移和孪生，从而实现金属的塑性变形。而多晶体变形时，除晶内变形外，晶界也发生变形，这类变形不仅与位错运动有关，而且扩散过程也起着很重要的作用。

Slip is the relative movement of one part of a crystal along a certain plane and in a certain direction on that plane under the action of an external force, as shown in Figure 3-12. This crystal plane is called the slip plane, and this crystal direction is called the slip direction. The slip plane and the slip direction always occur along the crystal plane and the crystal direction with the highest atomic density because the atomic distance between the crystal planes with the highest atomic density is small and the binding force between the atoms is strong. That is to say, the bonding force between the crystal plane and the slip direction (shown in Figure 3-12) is weak. The product of the slip surface and the value of the slip direction are called the slip system, as shown in Table 3-1.

滑移是指晶体在外力的作用下，其中一部分沿着一定晶面和在这个晶面上的一定晶向，对另一部分产生的相对移动（见图3-12）。此晶面称为滑移面，此晶向称为滑移方向。滑移面和滑移方向总是沿着原子密度最大的晶面和晶向发生，这是因为原子密度最大的晶面原子间距小、原子间的结合力强，同时其晶面间的距离较大，即晶面与晶面间的结合力较弱。滑移面与滑移方向数值的乘积称为滑移系（见表3-1）。

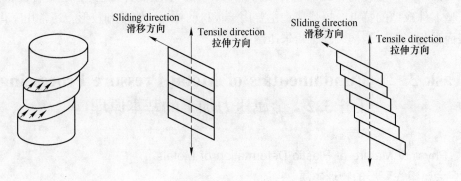

Figure 3-12　Slip
图 3-12　滑移

Table 3-1 Main slip plane, direction and system of metal
表 3-1 金属的主要滑移面、滑移方向和滑移系

Lattice 晶格	Body centered cubic lattice 体心立方晶格		Face centered cubic lattice 面心立方晶格		Close packed hexagonal lattice 密排六方晶格	
Slip planes 滑移面	{110}×6	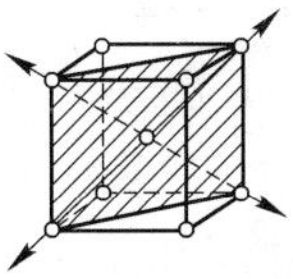	{111}×4	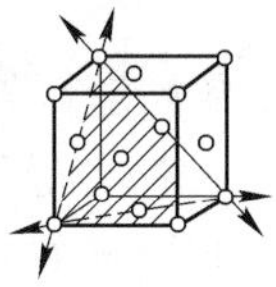	{0001}×1	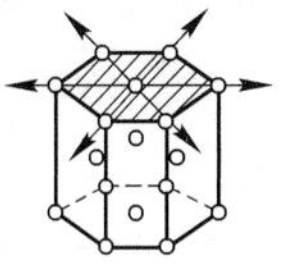
Slip direction 滑移方向	<111>×2		<110>×3		<100>×3	
Slip systems 滑移系	6×2=12		4×3=12		1×3=3	

In the slip process, the slip is first generated in the local region and gradually expanded until the whole slip surface is completed, instead of all the atoms along the slip surface producing rigid relative slip simultaneously. The reason for the local slip in this region is that there are dislocations and there is a large stress concentration. Although the stress acting on the slip surface is low, the stress is large enough to cause the object to slip in this local region. In the sliding process, when a dislocation moves along the sliding surface, a dislocation of the size of atom spacing is produced in the crystal. For a crystal to have a slip band displacement, thousands of dislocations would have to be displaced. At the same time, when the dislocation moves to the surface of the crystal causing a displacement of the size of the interatomic spacing, the dislocation disappears, as shown in Figure 3-13. But in the plastic deformation process, in order to ensure the continuous plastic deformation, there must be a large number of new dislocation. Therefore, it can be considered that the essence of crystal slip process is the process of dislocation migration and multiplication.

滑移过程是在其局部区域首先产生滑移并逐步扩大，直至最后整个滑移面上都完成了滑移，而不是沿着滑移面上所有原子同时产生刚性的相对滑移。此局部区域之所以首先产生滑移，是因为在该处存在着位错，并引起很大的应力集中，虽然在整个滑移面上作用的应力较低，但在此局部区域内应力已大到足够引起物体滑移。在滑移过程中，当一个位错沿滑移面移动过后，晶体产生一个原子间距大小的位错。若使晶体产生一个滑移带的位移量，则需上千个位错产生移动。同时，当位错移至晶体表面产生一个原子间距大小的位移后，位错便消失，如图 3-13 所示。但在塑性变形过程中，为保证塑性变形的不断进行，必须有大量新的位错出现，这些新的位错的产生是指位错理论中位错的增殖。因此可认为，晶体滑移过程的实质是位错的移动和增殖的过程。

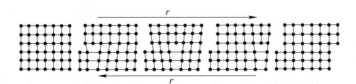

Figure 3-13 Slip mechanism
图 3-13 滑移机理

Besides the slip mode, twinning is one of the important ways in metal plastic deformation. Twinning is a part of a crystal moving in a relative direction along a certain plane and direction under the action of shear stress. The result is that a part of the crystal is in a symmetrical position with the orientation of the original crystal. Its crystal plane and crystal direction are twin crystal plane and crystal direction respectively, as shown in Figure 3-14.

金属的塑性变形除以滑移方式进行外，孪生也是其重要方式之一。孪生是指晶体在切应力的作用下，其一部分沿某一定晶面和晶向，按一定的关系发生相对的位向移动，使晶体的一部分与原晶体的位向处于相互对称的位置。其晶面和晶向分别为孪生晶面和晶向，如图3-14所示。

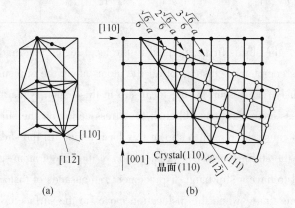

Figure 3-14 Twinning of a face-centered cubic crystal
图 3-14 面心立方晶体的孪生过程
(a) Atomic position before shear; (b) Atomic position after shear
(a) 切变前原子位置；(b) 切变后原子位置

3.2.2 Effects of Metal Plastic Deformation on Microstructure and Properties of Materials
3.2.2 金属塑性变形对材料组织性能的影响

3.2.2.1 Changes in Microstructure and Properties of Cold Deformed Metals and Alloys
3.2.2.1 冷变形金属与合金中组织与性能的变化

Microstructure Changes in Cold-deformed Metals and Alloys
冷变形金属与合金中组织的变化

- Change in Grain Shape
- 晶粒形状变化

During the cold deformation, the shape of the inner grain changes with the change of the shape. The tendency of grain change is to be elongated, elongated or flattened along the direction of maximum principal deformation. If the deformation is large, the grain presents a fibrous stripe, as shown in Figure 3-15. At the same time as the grain is elongated, the inclusions and the second phase in the metal are also elongated or stretched in the extension direction, forming a chain-like arrangement. The greater degree of deformation, the more obvious the fibrous tissue. The transverse mechanical properties of the deformed metal, which are generally perpendicular to the fiber direction, are decreased due to the existence of fiber structure.

金属冷变形中，随外形改变，内部晶粒形状也发生相应变化。晶粒变化趋势是沿最大主变形的方向被拉长、拉细或压扁。若变形程度很大，则晶粒呈现出一片纤维状条纹，如图 3-15 所示。在晶粒被拉长的同时，金属中的夹杂物和第二相也在延伸方向上拉长或拉伸，呈链状排列，这种组织称为纤维组织。变形程度越大，纤维组织越明显。纤维组织的存在使变形后的金属横向与纵向不同，一般垂直于纤维方向的横向力学性能降低。

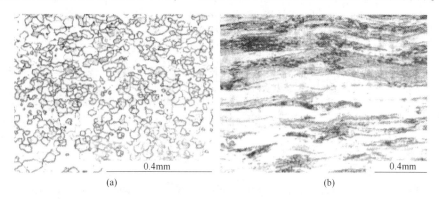

Figure 3-15　Changes of metal grain shape before and after cold rolling

图 3-15　冷轧前后金属晶粒形状的变化

(a) Annealing structure before deformation; (b) Cold-rolled deformed tissue after deformation

(a) 变形前退火组织；(b) 变形后冷轧变形组织

- Substructure Appearing in Grain
- 晶粒内出现亚结构

During plastic deformation of polycrystals, the grains deform due to different orientations, which hinder and promote each other. Generally, at the beginning of plastic deformation, they begin multi-system sliding and form disordered dislocation tangles, and the dislocation density and lattice distortion in these regions are very low. Each small region is called a unit cell, and the boundary of adjacent units is called a cell wall. The dislocation density is very high. The cell walls are parallel to the low-index crystal plane. The larger the deformation, the smaller the cell size. These small cells are called subcrystals, and this structure is called a substructure, as shown in Figure 3-16.

多晶体塑性变形时，各晶粒由于取向不同而变形，相互阻碍又相互促进。一般刚开始塑性变形时就开始多系滑移，形成分布杂乱的位错缠结，在这些缠结区域的内部位错密度很低，晶格的畸变很小。每个小区域称为晶胞，相邻晶胞的边界称为胞壁，其位错密度很大。胞壁是平行于低指数晶面排列的。变形量越大，晶胞的尺寸越小。这些小晶胞称为亚晶，这种组织称为亚结构，如图 3-16 所示。

This dislocation distribution is the main form of energy storage. During cold deformation, the process of increasing dislocation density is called work hardening, as shown in Figure 3-17.

这种位错分布是储存能的主要形式。冷变形过程中，位错密度增加的过程即为加工硬化过程，如图 3-17 所示。

Figure 3-16　Plastic deformation structure
图 3-16　塑性变形结构

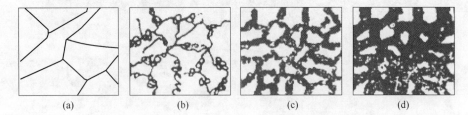

Figure 3-17　Changes in dislocation density and distribution during cold deformation
图 3-17　冷变形过程中位错密度和分布的变化

(a) No deformation; (b) 10 per cent deformation; (c) 50 per cent deformation; (d) Degree of deformation 200%
(a) 无变形；(b) 变形程度10%；(c) 变形程度50%；(d) 变形程度200%

In order to continue to deform in the process of metal deformation, the stress must be increased. The property of a metal that increases in strength and decreases in plasticity as a result of deformation is called work hardening. Work Hardening can cause a metal to undergo cold deformation with uniform cross-section because where there is deformation, there is hardening. It causes the deformation to spread to other places where it has not been deformed for a while. The result of this alternation is that the deformation of the cross section of the product tends to be uniform. Work hardening can improve the properties of metal materials, especially for those metal materials which can not be strengthened by conventional heat treatment. But the work hardening also has disadvantages in the cold working process, due to the increase of deformation resistance and plastic decline, it is often difficult to continue the processing, and it need to increase the annealing process in the process, such as cold rolling, cold drawing.

金属在变形过程中为了继续形变，必须增加应力。这种金属因形变而使其强度升高、塑性降低的性质，称为加工硬化。加工硬化可以使金属得到截面均匀一致的冷变形，这是因为哪里有变形，哪里就有硬化，从而使变形分布到其他暂时没有变形的部位上去。这样反复交替的结果就是使产品截面的变形趋于均匀。加工硬化可以改善金属材料性能，特别是对那些用一般热处理手段无法使其强化的无相变的金属材料，形变硬化是更加重要的强化手段。但加工硬化也有缺点，在冷加工过程中，由于变形抗力的升高和塑性的下降，往往使继续加工发生困难。因此需在工艺过程中增加退火工序，如冷轧、冷拔等。

- Grain Orientation Change
- 晶粒位向改变

As shown in Figure 3-18, the polycrystal of the metal is composed of a number of grains arranged irregularly. However, in the process of processing deformation, when the deformation reaches a certain degree, the orientation of the crystal lattice in each grain rotates, so that the specific crystal plane and the crystal tend to line up in a certain direction. Thus, the grains with disordered original orientation are ordered and have strict orientation relationship. The ordered structure of a metal formed under cold deformation is called deformed texture. When the deformation direction is the same, the greater the deformation degree is, the more obvious the orientation is.

如图 3-18 所示，金属的多晶体是由许多排列不规则的晶粒所组成。但在加工变形过程中，当达到一定的变形程度以后，由于在各晶粒内晶格取向发生了转动，特定的晶面和晶向趋于排成一定方向，从而使原来位向紊乱的晶粒出现有序化，并有严格的位向关系。金属在冷变形条件下所形成的有序排列的组织结构称为变形织构。变形方向一致时，变形程度越大，位向表现得越明显。

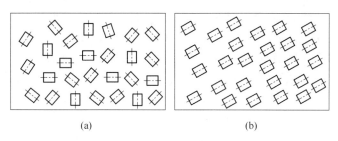

Figure 3-18 Arrangement of polycrystalline grains

图 3-18 多晶体晶粒的排列

(a) Disordered arrangement of grains; (b) Neat arrangement of grains

(a) 晶粒的紊乱排列；(b) 晶粒的整齐排列

Changes in Properties of Cold-deformed Metals and Alloys
冷变形金属与合金中性能的变化

- Change in Mechanical Properties
- 力学性能改变

In the process of deformation, lattice distortion, grain elongation and refinement, substructure and non-uniform deformation occurred made the resistance index (proportional limit, elastic limit, yield limit, strength limit, hardness limit, etc.) of metal increase with the increase of deformation degree. Due to the intergranular and intergranular failure and non-uniform deformation, the elongation, reduction of section and other metal plasticity indexes decrease with the increase of deformation degree.

变形中产生晶格畸变、晶粒的拉长和细化、出现亚结构以及产生不均匀变形等，使金属的变形抗力指标（比例极限、弹性极限、屈服极限、强度极限、硬度等）随变形程度的增加而升高。由于变形中产生晶内和晶间的破坏以及不均匀变形等情况，使延伸率、断面收缩率等金属塑性指标随变形程度的增加而降低。

- Changes of Physical Chemistry
- 物理化学性质的改变

In the process of cold deformation, a large number of micro-cracks and voids are produced in the deformed metal due to the breakage of intergranular and intragranular materials, which reduces the density of the deformed metal. For example, the density of annealed steel is $7.8659/cm^3$, but it is reduced to $7.789/cm^3$ after cold deformation.

在冷变形过程中，由于晶内和晶间物质的破碎，在变形金属内产生大量的微小裂纹和空隙，使变形金属的密度降低。例如，退火状态钢的密度为 $7.8659/cm^3$，而经冷变形后则降低至 $7.789/cm^3$。

The electrical conductivity of metals generally varies with the degree of deformation, especially when the degree of deformation is not significant. For example, the unit resistance of copper increases by 1.5% when the copper is stretched by 4%, by 2% when the copper is stretched by 40%, and by very little when the copper is further deformed by 85%.

金属的导电性一般是随变形程度的改变而变化的，特别是当变形程度不大时尤为显著。例如，赤铜的拉伸程度为4%时，其单位电阻增加1.5%；当拉伸变形程度达40%时，单位电阻增加2%；继续增大变形程度至85%时，此数值变化甚小。

The cold deformation reduces the thermal conductivity of the metal. For example, after the cold deformation, the thermal of the copper crystal conductivity reduces by 78%.

冷变形使金属导热性降低，如铜的晶体在冷变形后，其热导率降低了78%。

Cold deformation can change the magnetism of metal. The magnetic saturation is almost constant, the coercivity and hysteresis increase 2~3 times due to cold deformation, while the maximum permeability of the metal decreases. For some diamagnetic metals, such as copper, silver, lead and brass, cold deformation can increase their susceptibility to magnetization, and copper and brass can even change from diamagnetic state to paramagnetic state. Cold deformation will reduce its susceptibility to magnetization. While for some metals such as gold, zinc, tungsten, molybdenum, white steel, the magnetism is virtually unaffected by cold deformation.

冷变形可改变金属的磁性。磁饱和基本上不变，矫顽力和磁滞可因冷变形而增加2~3倍，而金属的最大磁导率则降低。对于某些抗磁性金属，如铜、银、铅及黄铜等，冷变形可提高其对磁化的敏感性，这时铜及黄铜甚至可由抗磁状态转变为顺磁状态；对于顺磁金属，冷变形将降低其对磁化的敏感性；而对于像金、锌、钨、钼、白钢这样一些金属的磁性，实际上不受冷变形的影响。

Cold deformation will increase the solubility and decrease the corrosion resistance of the metal. For example, when brass is cold-deformed, it is eroded faster by ammonia gas in the air. Some people think that the corrosion resistance is reduced because of the influence of residual stress. The higher the residual stress is, the greater the solubility of the metal is, and the worse the corrosion resistance is. It is due to the atom in a distorted state, and the cause of the increase in atomic potential energy.

冷变形会使金属的溶解性增加，耐蚀性降低。例如，黄铜经冷变形后，其在空气中被阿摩尼亚气体侵蚀的速度加快。关于耐蚀性降低的原因，有的认为是由于残余应力的影响，残余应力越大，则金属的溶解性越大，耐蚀性越差；有的认为溶解性越大，耐蚀性越小，是由于原子处于畸变状态、原子势能增加的缘故。

The fiber structure and texture of metal and alloy after cold deformation will make the deformed metal and alloy produce anisotropy. That is to say, the material has different properties in different directions.

金属与合金经冷变形后所出现的纤维组织及织构,均会使变形后的金属与合金产生各向异性,即材料在不同方向上具有不同的性能。

3.2.2.2 Changes in Microstructure and Properties of Cold-formed Metals During Heating
3.2.2.2 冷变形金属在加热时的组织与性能变化

When the metal is heated after cold plastic deformation, it is usually changed in three stages: recovery, recrystallization and grain growth. The three stages are not necessarily separate but often overlap.

冷塑性变形后的金属加热时,通常是依次发生回复、再结晶和晶粒长大三个阶段的变化。这三个阶段不是决然分开的,常有部分重叠。

Recovery is the process of changing the substructure and properties of the cold plastic deformed metal before the change of the optical microstructure (i.e. before the formation of recrystallized grains).

回复是指经冷塑性变形的金属在加热时,在光学显微组织发生改变前(即在再结晶晶粒形成前)所产生的某些亚结构和性能的变化过程。

After a large amount of cold deformation is heated to a temperature of about $0.5\%T$ (T is the melting point of the metal). After a certain period of time, new equiaxed grains with much lower density of crystal defects will nucleate and grow in the cold deformation Matrix. This process is called recrystallization until the cold deformed grains are completely exhausted. When the recrystallization process is completed, these new grains will merge and grow at a slower rate, which is called grain growth process.

将经过大量冷形变的金属加热到大约 $0.5\%T$(T 为金属熔点)的温度,经过一定时间后,会有晶体缺陷密度降低的新等轴晶粒在冷形变的基体内形核并长大,直到冷形变晶粒完全耗尽为止,这个过程称为再结晶。再结晶过程完成后,这些新晶粒将以较慢的速度合并并长大,这就是晶粒长大过程。

When the metal is heated after cold plastic deformation, the most remarkable changes of its microstructure and properties take place in the recrystallization stage. Recrystallization is an important softening method to eliminate work hardening and to control the size, shape and uniformity of grains and to obtain or avoid preferential orientation of grains. Controlling the recrystallization process by various factors will have a great influence on the strength, toughness, thermal strength, stamping property and electromagnetic property of metal materials.

冷塑性变形后的金属加热时,其组织和性能最显著的变化是在再结晶阶段发生。再结晶是消除加工硬化的重要软化手段,也是控制晶粒大小、形态、均匀程度,获得或避免晶粒的择优取向的重要手段。通过各种影响因素对再结晶过程进行控制,将对金属材料的强韧性、热强性、冲压性和电磁性等产生重大的影响。

3.2.2.3 Recovery and Recrystallization of Metals During Hot Deformation
3.2.2.3 金属在热变形过程中的回复及再结晶

Recovery and recrystallization also occur during hot working or after the end of hot deformation. In the process of plastic deformation, the metal is generally accompanied with work hardening. The metal with work hardening will recover or recrystallize at high temperature. In the hot working process, the deformation temperature is higher than the recrystallization temperature, so the work hardening and the recovery or recrystallization softening always exist at the same time. The state of recovery or recrystallization can be divided into five forms: static recovery, static recrystallization, dynamic recovery, dynamic recrystallization and sub-dynamic recrystallization.

金属在热加工过程中或热变形终止后也会发生回复和再结晶。金属在塑性变形过程中，一般都伴随有加工硬化现象，有加工硬化的金属在高温下就会发生回复或再结晶。就热加工过程而言，变形温度高于再结晶温度，因此在变形体内，加工硬化与回复或再结晶软化过程总是同时存在的。就回复或再结晶发生的状态来看，其可分为五种形态，即静态回复、静态再结晶、动态回复、动态再结晶和亚动态再结晶。

The static recovery or static recrystallization after hot working is carried out by using the residual heat after the end of plastic deformation of hot working. Dynamic recovery and dynamic recrystallization occur during plastic deformation, not after deformation stops. Subdynamic recrystallization refers to the process of thermal deformation interrupted. At this time, dynamic recrystallization is not complete, and the remaining tissue will continue to occur no pregnancy recrystallization. Figure 3-19 is a schematic diagram of the dynamic and static concepts of recovery and recrystallization. In the nature of recovery and recrystallization, dynamic recovery and dynamic recrystallization are no different from static state. Both dynamic recovery and dynamic recrystallization during hot working can soften hot deformed metal.

热加工后的静态回复或静态再结晶是在塑性变形终止后，利用热加工后的余热进行的。动态回复和动态再结晶是在塑性变形过程中发生的，而不是在变形停止之后。亚动态再结晶是指在热变形的过程中中断热变形，此时动态再结晶还未完成，遗留下来的组织将继续发生无孕期的再结晶。图 3-19 为回复和再结晶动静态概念示意图。从回复和再结晶的本质来讲，动态回复和动态再结晶与静态没有什么不同，热加工过程中的动态回复和动态再结晶都能使热变形金属软化。

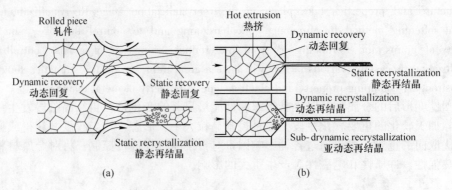

Figure 3-19 Dynamic and static concepts of recovery and recrystallization

图 3-19 回复和再结晶动静态概念示意图

(a) Deformation rate of 50%; (b) Deformation rate of 99%

(a) 变形率为 50%；(b) 变形率为 99%

Organizational Changes in Dynamic Response
动态回复时的组织变化
- Variation of Dislocation Density and Distribution
- 位错密度及分布的变化

In the first stage, the dislocation density increased from $10^{10} \sim 10^{11} \text{mm}^{-2}$ to $10^{11} \sim 10^{12} \text{mm}^{-2}$. In the second stage, the dislocation density increases from $10^{11} \sim 10^{12} \text{mm}^{-2}$ to $10^{14} \sim 10^{15} \text{mm}^{-2}$. At this stage, the dislocation entanglement appears and the substructure begins to form. In the third stage, because of the dynamic recovery, the velocity of dislocation is a function of the strain rate and temperature, while the velocity independent of the deformation is a function of the dislocation density and the degree of difficulty in the recovery mechanism. So the dislocation density remains constant.

第一阶段中，位错密度由 $10^{10} \sim 10^{11} \text{mm}^{-2}$ 增至 $10^{11} \sim 10^{12} \text{mm}^{-2}$。在第二阶段，位错密度由 $10^{11} \sim 10^{12} \text{mm}^{-2}$ 增至 $10^{14} \sim 10^{15} \text{mm}^{-2}$，这一阶段出现位错缠结，开始形成亚结构。第三阶段中，由于动态回复的缘故，产生位错的速度（是应变速度及温度的函数，而与形变量无关）与位错相消的速度（是位错密度及回复机制发生难易程度的函数）相等，因此位错密度保持不变。

- Changes of Subcrystalline
- 亚晶的变化

The increase of dislocation density leads to the recovery process. The rate of dislocation disappearance increases with the increase of strain, and finally reaches the steady-state rheological stage where the propagation and disappearance of dislocations reach a balance and no work hardening occurs. At this stage, the main characteristics of the subcrystals, such as the dislocation density between cell walls, the dislocation density between cell walls, the average distance between dislocation density and the orientation difference between cell-like substructures, remain unchanged. In addition, although the shape of the grains changes with the shape of the workpiece, the subgrains remain equiaxed, even if the variables are large.

位错密度的增大导致回复过程发生，位错消失的速率随应变的增大而不断增大，最后终于达到位错增殖与消失达到平衡、不再发生加工硬化的稳态流变阶段。在这个阶段，亚晶的一些主要特征，如胞壁之间的位错密度、胞壁的位错密度、位错密度之间的平均距离、胞状亚结构之间的取向差始终保持不变。此外，虽然晶粒的形状随工件外形的改变而改变，亚晶粒却始终保持为等轴状，即使变量很大也是如此。

- Elongated Grain with the Increase of Deformation
- 晶粒随变形量的增加而不断被拉长

The dynamic recovery can not be regarded as the static recovery of cold working because it avoids the accumulation of cold working effect and the deformation metal can not reach the dislocation density. Dynamic recovery produces a sub-crystal, which can not be obtained by cold-working static recovery. Figure 3-20 shows the dynamic recovery stress-strain curve and changes of grains and subgrains at each stage.

热加工动态回复避免了冷加工效应积累，因而形变金属达不到冷加工的位错密度，动态回复不能看成冷加工静态回复。动态回复产生亚晶,不能靠冷加工静态回复得到。动态回复的应力-应变曲线及各阶段晶粒和亚晶的变化如图 3-20 所示。

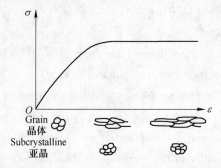

Figure 3-20　Dynamic recovery stress-strain curve and changes of grains and subgrains at each stage
图 3-20　动态回复的应力-应变曲线及各阶段晶粒和亚晶的变化

Dynamic Recrystallization
动态再结晶

- Stress-strain Curve When Dynamic Recrystallization
- 发生动态再结晶时的应力-应变曲线

Similarly, the dynamic recrystallization temperature is higher than the dynamic recovery temperature. Dynamic recrystallization, as shown in Figure 3-21, begins with a change in stress-strain curve stress at completion. At the beginning of dynamic recrystallization, the flow stress decreases gradually. And at the end of dynamic recrystallization, the new work hardening occurs and the flow stress rises again.

与静态下的情况相似,动态再结晶温度比动态回复温度高。如图 3-21 所示为动态再结晶开始与完成时应力-应变曲线流变应力的变化。当动态再结晶开始时,流变应力逐渐下降;当动态再结晶结束时,新的加工硬化开始出现,流变应力重新上升。

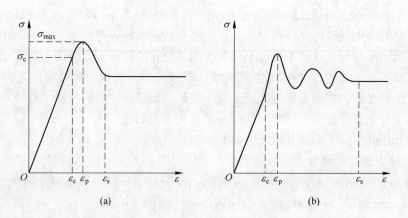

Figure 3-21　Dynamic recrystallization process reflected on the stress-strain curve
图 3-21　动态再结晶过程在应力-应变曲线上的反映
(a) Continuous dynamic recrystallization; (b) Intermittent dynamic recrystallization
(a) 连续动态再结晶;(b) 间断动态再结晶

- Changes of Microstructure during Dynamic Recrystallization
- 动态再结晶时组织结构的变化

According to the morphology of the dynamic recrystallization stress-strain curve, it can be divided into three stages, including the working hardening stage ($0<\varepsilon<\varepsilon_c$), the initial phase of dynamic recrystallization ($\varepsilon_c \leqslant \varepsilon<\varepsilon_n$) and the steady rheological phase ($\varepsilon \geqslant \varepsilon_n$), where ε_p is the strain of σ_{max} corresponding to the peak stress, ε_c is the critical strain at the beginning of dynamic recrystallization, and ε_n is the strain 95% from the beginning of dynamic recrystallization to the completion of dynamic recrystallization. Steady state rheological stress σ_n is less than peak stress σ_p and higher than yield stress σ_s.

根据动态再结晶应力-应变曲线的形态,可将其分为三个阶段,即加工硬化阶段($0<\varepsilon<\varepsilon_c$)、动态再结晶初始阶段($\varepsilon_c \leqslant \varepsilon<\varepsilon_n$)和稳态流变阶段($\varepsilon \geqslant \varepsilon_n$)。其中,$\varepsilon_p$为对应峰值应力$\sigma_{max}$的应变;$\varepsilon_c$为开始动态再结晶的临界应变,$\varepsilon_c=(0.6\sim0.8)\varepsilon_p$;$\varepsilon_n$为从动态再结晶开始到动态再结晶完成95%的应变。稳态流变应力σ_n小于峰值应力σ_p,高于屈服应力σ_s。

The microstructure characteristics of dynamic recrystallization are as follows:

动态再结晶的组织特点如下:

(1) The grains remain equiaxed.

(1) 晶粒保持为等轴状。

(2) The grain size is very uneven.

(2) 晶粒大小很不均匀。

(3) The grains are irregular concave and convex.

(3) 晶粒呈现不规则的凹凸状。

(4) Even the metals which are easy to form annealing twins, annealing twins after dynamic recrystallization are also rare.

(4) 即使是易于形成退火孪晶的金属,动态再结晶后退火孪晶也很少见。

- Subdynamic Recrystallization Diagram
- 亚动态再结晶图

The process of nucleation during deformation and recrystallization after deformation is called subdynamic recrystallization. Metadynamic recrystallization also causes softening of the metal. Because such recrystallization has formed nuclei during thermal deformation and has no incubation period, it proceeds very rapidly after deformation stops, an order of magnitude faster than conventional static recrystallization. Figure 3-22 shows the dynamic recrystallization stress-strain curve.

形变中形核在形变结束后再长大的再结晶过程,称为亚动态再结晶。亚动态再结晶同样会引起金属的软化。因为这类再结晶在热变形中已形成晶核且没有孕育期,所以变形停止后进行得非常迅速,比传统的静态再结晶要快一个数量级。动态再结晶应力-应变曲线如图3-22所示。

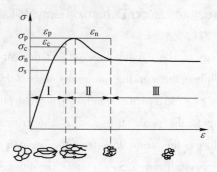

Figure 3-22　Dynamic recrystallization stress-strain curve
图 3-22　动态再结晶应力-应变曲线

Ⅰ—Work hardening stage；Ⅱ—Dynamic recrystallization initial stage；Ⅲ—A steady-state creep stage
Ⅰ—加工硬化阶段；Ⅱ—动态再结晶初始阶段；Ⅲ—稳态流变阶段

3.2.2.4　Changes in Microstructure and Properties of Metals during Hot Deformation
3.2.2.4　金属在热变形过程中的组织与性能变化

　　Hot-working deformation can be considered as the overlapping of work hardening and recrystallization. During this process, the following changes have taken place in the structure and properties of the metal：
　　热加工变形可认为是加工硬化和再结晶两个过程的相互重叠。在此过程中，金属的组织与性能发生以下变化：
　　(1) Shrinkage cavity, porosity, voids and air bubbles in as-cast metal structure are compacted or welded. The densification of metal during deformation due to work hardening also recovers with recrystallization.
　　(1) 铸态金属组织中的缩孔、疏松、空隙、气泡等缺陷得到压密或焊合。金属在变形中由于加工硬化所造成的不致密现象，也随着再结晶的进行而恢复。
　　(2) Grain refinement and inclusion breakage can be achieved during hot-working deformation. In as-cast metals, columnar grains and coarse equiaxed grains can be transformed into fine equiaxed grains by forging or rolling and recrystallization.
　　(2) 在热加工变形中可使晶粒细化和夹杂物破碎。铸态金属中，柱状晶和粗大的等轴晶粒经锻造或轧制等热加工变形后，加上再结晶的同时作用，可变成较细小的等轴晶粒。
　　(3) The formation of fibrous tissue is also an important feature of hot working deformation. The fibrous structure formed in hot working deformation of as-cast metal is different from that formed in cold working deformation due to the elongated grains. The former is caused by the elongation of the insoluble material at the grain boundary in the as-cast microstructure. There are coarse primary crystalline grains in as-cast metals, and there are thin layers of non-metallic inclusions on the boundaries of the grains. During the deformation, the large grains are broken and elongated in the direction of maximum metal flow. At the same time, the crystal layer containing non-metallic inclusions is also elongated in this direction. When the deformation is large enough, the inclusions can be drawn into thin strips. As a result of complete recrystallization, the elongated

grain becomes a lot of fine equiaxed grains, while the insoluble material at grain boundary and in the grain remains in an elongated state, to form fibrous tissue. Due to the appearance of fiber structure, deformed metals have different mechanical properties in longitudinal and transverse directions.

(3) 形成纤维组织也是热加工变形的一个重要特征。铸态金属在热加工变形中所形成的纤维组织，与金属在冷加工变形中由于晶粒被拉长所形成的纤维组织不同。前者是由于铸态组织中晶界上的非溶物质的拉长所造成。在铸态金属中存在粗大的一次结晶的晶粒，在其边界上分布有非金属夹杂物的薄层。在变形过程中这些极大的晶粒遭到破碎，并在金属流动最大的方向上被拉长。与此同时，含有非金属夹杂物的晶间薄层在此方向上也被拉长。当变形程度足够大时，这些夹杂物可被拉成细条状。在变形过程中由于完全再结晶的结果，被拉长的晶粒变成许多细小的等轴晶粒，而位于晶界和晶内的非溶物质却不因再结晶而改变，仍处于拉长状态，形成纤维状的组织。由于纤维组织的出现，变形金属在纵向和横向具有不同的力学性能。

(4) The banded structure is formed in the process of hot deformation. The zonal structure can be divided into two forms: grain zonal structure and inclusion zonal structure. As shown in Figure 3-23, pearlite sometimes appears in a ribbon-like arrangement during hot deformation of V-N-TI microalloyed steel, because the inclusions are arranged in a fibrous pattern during hot working, and after slow cooling, the ferrite first precipitates around the inclusions and forms a row, and the pearlite then precipitating in a row, and forming a band structure. The carbide particles which are broken after hot working are arranged along the extension direction of the steel to form carbide banded structure. The banded structure in the steel also affects the mechanical properties of the steel.

(4) 金属在热变形过程中产生带状组织。这种带状组织可表现为晶粒带状和夹杂物带状两种形式。如图3-23所示，钒氮钛微合金化钢在热变形中有时会出现珠光体，呈带状排列。这是因为热加工时夹杂物排列成纤维状，缓慢冷却后，铁素体首先在夹杂物的周围析出而排列成行，珠光体也随之成行析出，形成带状组织。热加工后被破碎了的碳化物颗粒沿钢材的延伸方向排列，从而形成碳化物带状组织，钢材中出现的带状组织也会影响钢材的力学性能。

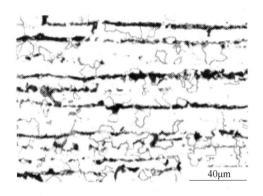

Figure 3-23　Strip structure in vanadium nitride titanium microalloyed steel
图 3-23　钒氮钛微合金化钢中的带状组织

3.2.3 Plastic Flow Law of Metal
3.2.3 金属塑性流动规律

3.2.3.1 Volume Invariance Law
3.2.3.1 体积不变定律

In metal plastic deformation, it is generally considered that the volume of the material before and after deformation remains constant when the density of the material is neglected. Figure 3-24 shows the dimensions of the metal before and after rolling deformation.

在金属塑性变形时,当忽略材料在加工过程中的密度变化时,一般认为材料变形前后的体积保持不变。金属材料轧制变形前后的尺寸如图3-24所示。

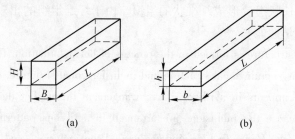

Figure 3-24 Dimensions of metal material before and after rolling deformation
图 3-24 金属材料轧制变形前后的尺寸
(a) Before rolling; (b) After rolling
(a) 轧前; (b) 轧后

3.2.3.2 Minimum Resistance Law
3.2.3.2 最小阻力定律

In plastic forming, when a metal particle has the possibility to move in several directions, it moves in the direction of the least resistance. This is actually a general principle of mechanics, which can be used to determine qualitatively the direction of flow of metal particles.

在塑性成型中,当金属质点有向几个方向移动的可能时,它向阻力最小的方向移动。这实际上是力学的普遍原理,它可以用来定性地确定金属质点的流动方向。

When there is friction on the contact surface, the flow model of the prism upsetting is shown in Figure 3-25. The friction force acting on the blank end face is τ. Because the frictional resistance of a particle flowing toward a free surface on a contact surface is proportional to the distance from the free surface, the closer the free boundary is, the less the resistance is, and the metal particle must flow in this direction. Thus, four flow regions are formed, bounded by the bisector of the four corners and the midline of the length direction, all of which have the shortest distance from each particle to the boundary line. As a result of this flow, less metal flows out in the width direction than in the length direction, so that the cross-section becomes elliptical after upsetting. It can be imagined that the section under upsetting tends to be equal to the frictional resistance in each direction until it is a circular shape. Therefore, the law of minimum resistance is also known as the law of minimum circumference in upsetting.

当接触表面存在摩擦时，棱柱体镦粗时的流动模型如图 3-25 所示。压板作用于坯料端面的摩擦力为 τ。因为接触面上质点向自由表面流动的摩擦阻力与质点离自由表面的距离成正比，所以距离自由边界越近，阻力越小，金属质点必然沿这个方向流动。这样就形成了四个流动区域，以四个角的二等分线和长度方向的中线为分界线，这四个区域内的质点到各自边界线的距离都是最短距离。这样流动的结果是，宽度方向流出的金属少于长度方向，因此镦粗后的断面成椭圆形。可以想象，不断镦粗必然趋于达到各向摩擦阻力均相等的断面——圆形为止。因此，最小阻力定律在镦粗中也称为最小周边定则。

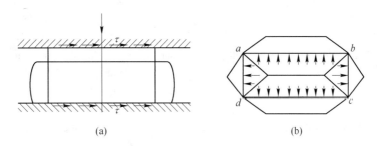

Figure 3-25　Flow model of prism upsetting
图 3-25　棱柱体镦粗时的流动模型
τ—Linking piece on the end of the billet tau friction
τ—压板作用于坯料端面的摩擦力

3.2.3.3　Uneven Deformation
3.2.3.3　不均匀变形

If the deformation states of metals in the deformation zone are the same at all particle points (or in all tiny volumes), that the occurrence, development direction and magnitude of the deformation are all the same in their respective axial directions, the deformation in this volume can be considered as uniform and the object is assumed to be in a uniform state of deformation. Otherwise, it is called non-uniform deformation. In plastic forming, the deformation is actually non-uniform due to the non-uniformity of metal properties, the influence of friction and tool shape, and the mutual restriction of different deformation zones. As shown in Figure 3-26, the velocity distribution of the metal particles along the width of the strip is not uniform in the deformation zone, which leads to non-uniform deformation.

若变形区金属各质点处（或各微小体积内）的变形状态相同，即在它们相应的各个轴向向上，变形的发生情况、发展方向及变形量的大小都相同，则这个体积内的变形可视为均匀的，并且认为该物体所处变形状态是均匀的，否则统称为不均匀变形。塑性成型时，由于金属本身性质的不均匀，摩擦和工具形状的影响，以及不同变形区之间的相互制约，实际上都是不均匀变形。如图 3-26 所示，板带材轧制时，在变形区内，沿轧件宽度上金属质点的运动速度分布是不均匀的，从而引起不均匀变形。

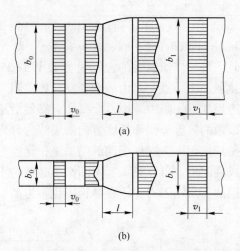

Figure 3-26　Velocity distribution diagram of metal particles moving along the strip width

图 3-26　沿带材宽度金属质点运动的速度分布图

(a) Wide plate; (b) A narrow board

(a) 宽板; (b) 窄板

b_0, b_1—Width before and after the rolling deformation; v_0, v_1—Velocity of metal particles of strip rolling before and after deformation; l—Length of contact arc

b_0, b_1—带材轧制变形前、后的宽度; v_0, v_1—带材轧制变形前、后金属质点的运动速度; l—接触弧长度

When the diameter-to-height ratio of the blank is larger than the cylinder or the width-to-thickness ratio of the blank larger than the rectangle, the contact surface friction force is smaller, the deformation degree is very small, and the uneven deformation of the double-drum shape is often produced. At this point, the metal near the contact surface layer produces obvious plastic deformation, while the central layer deforms little or no, forming a very prominent surface deformation, as shown in Figure 3-27.

金属塑性加工时，当坯料径高比（圆柱试件）或宽厚比（矩形试件）大，接触表面摩擦力较小，变形程度甚小时，常易产生双鼓形的高向上明显的不均匀变形。这时接触表面层附近的金属产生明显的塑性变形，而中心层变形很小甚至不变形，形成很突出的表面变形，如图 3-27 所示。

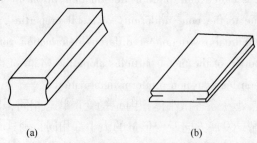

Figure 3-27　Double drum and fold shapes on the side of the rolled piece during rolling

图 3-27　轧制时轧件侧面的双鼓形及折叠形

(a) Double drum; (b) Folded shape

(a) 双鼓形; (b) 折叠形

Task 3.3　Rolling Process
任务 3.3　轧钢生产工艺

3.3.1　Classification of Steel Products
3.3.1　钢材品种的分类

Steel products can be divided into carbon steel, low alloy steel and alloy steel, According to the section shape of products it can also be divided into strip steel, section steel and steel pipe.

钢材品种按钢种可分为碳素钢、低合金钢与合金钢,按产品断面形状可分为板带钢、型钢与钢管等。

3.3.1.1　Classification of Strip Steel
3.3.1.1　板带钢的分类

The shape of the strip steel products is flat, the cross-section is rectangular, and the width-thickness ratio is large. When classifying strip steel products, the general supply of leaflets is known as the steel plate, and the roll will be called supply strip.

板带钢产品的外形扁平,断面呈矩形,宽厚比大。板带钢产品分类时,一般将单张供应的称为钢板,将成卷供应的称为带钢。

According to product size specification, plate and strip steel can be generally divided into thick plate (including medium plate and extra-thick plate), thin plate and extremely thin strip (foil). In the thick plate, the thickness of the medium plate is 3.0~20mm, the thickness of the thick plate is 20~60mm, the thickness of the extra-thick plate is more than 60mm, and the maximum thickness can be 500mm. The thickness of the strip is 0.2~3.0mm, that of the conventional hot rolled strip is 1.2mm, and that of the ultra thin strip is 0.8mm. The thickness of the extremely thin strip is less than 0.2mm, and the current foil can be as thin as 0.001mm.

按产品尺寸规格,板带钢一般可分为厚板(包括中板和特厚板)、薄板和极薄带材(箔材)三类。厚板中,中板的厚度为 3.0~20mm,厚板的厚度为 20~60mm,特厚板的厚度大于 60mm,最厚可达 500mm。薄板的厚度为 0.2~3.0mm,常规热轧薄板的厚度为 1.2mm,超薄带钢则生产到 0.8mm。极薄带材的厚度小于 0.2mm,目前箔材最薄可达 0.001mm。

According to usage, plate and strip steel can be divided into shipbuilding plate, boiler plate, bridge plate, pressure vessel plate, automobile plate, plating plate tin plate, galvanized plate etc.

按用途,板带钢可分为造船板、锅炉板、桥梁板、压力容器板、汽车板、镀层板(镀锡板、镀锌板等)、电工钢板、屋面板、深冲板、焊管坯、复合板以及不锈、耐酸、耐热等特殊用途钢板等。

3.3.1.2 Classification of Section Steel
3.3.1.2 型钢的分类

According to section shape, section steel can be divided into simple section and complex section. The shape of simple section steel is simple, including square steel, round steel, flat steel and hexagon steel. Simple section steel is also called bar. Complex section or special-shaped steel has obvious convex and concave parts. Forming is more difficult, including channel steel, I-beam and other special-shaped steel. The section shape and size of periodic section steel change periodically along the longitudinal direction of the steel, which can be produced by longitudinal rolling, cross rolling, cross rolling or cross wedge rolling.

按断面形状，型钢可分为简单断面和复杂断面两类。简单断面型钢没有明显的凸凹分肢部分，外形比较简单，包括方钢、圆钢、扁钢及六角钢等。简单断面型钢又称为棒材。复杂断面（或异形）型钢有明显的凸凹分肢部分，成型比较困难，包括槽钢、工字钢及其他异形钢等。周期断面型钢的断面形状和尺寸呈周期性沿钢材纵轴方向变化，可用纵轧、斜轧、横轧或楔横轧方法生产。

According to the use department, the section steel can be divided into railway section steel (rails, fishplates, rails for switches, and wheels, hoops), automobile section steel (tyres, retainers and locks), shipbuilding section steel (L-shaped steel, bulb flat steel, Z-shaped steel, and marine window frame steel), structure and construction section steel (H-beam, joist steel, channel, angle, crane rail, window frame and door frame steel, steel sheet pile, etc), mining section steel (U-shaped steel, bamboo steel, trough steel, mining I-beam, scraper steel), machinery manufacturing special-shaped steel and so on.

按使用部门，型钢可分为铁路用型钢（钢轨、鱼尾板、道岔用轨、车轮、轮箍）、汽车用型钢（轮箍、轮胎挡圈和锁圈）、造船用型钢（L型钢、球扁钢、Z字钢、船用窗框钢）、结构和建筑用型钢（H型钢、工字钢、槽钢、角钢、吊车钢轨、窗框和门框用钢、钢板桩等）、矿山用钢（U型钢、竹型钢、槽帮钢、矿用工字钢、刮板钢）和机械制造用异型钢材等。

According to section size and unit length, section steel can be divided into rail, steel beam, large section, medium and small section.

按断面尺寸和单位长度，型钢可分为钢轨、钢梁、大型材、中小型材。

3.3.1.3 Classification of steel Tubes
3.3.1.3 钢管的分类

Steel pipe is a kind of steel with two ends open and hollow section, and its length is larger than the section girth. Steel pipe accounts for 8%~15% of the total steel, and it is widely used in the national economy.

钢管是指两端开口并具有中空断面，而且其长度与断面周长之比较大的钢材。钢管占全部钢材总量的 8%~15%，在国民经济中应用范围极为广泛。

According to the section shape, the steel pipe can be divided into circular pipe and special-shaped pipe.

按断面形状,钢管可分为圆形管与异形管。

According to the production method, the steel pipe can be divided into welded pipe and seamless pipe. Its products are mainly used in petroleum industry, natural gas transmission, city gas transmission, power and communication pipe network, engineering construction, automobile, machinery and other manufacturing industries.

按生产方法,钢管可分为焊管和无缝管,其产品主要用于石油工业、天然气输送、城市输气、电力和通信管网、工程建筑和汽车、机械等制造业。

As a raw material manufacturing industry, the industry of iron and steel is not only the foundation of manufacturing industry, but also an important pillar industry of the country.

钢铁工业作为原材料制造工业,既是制造业的基础,又是国家重要支柱产业,钢铁工业的进步是衡量国民经济和社会发展的重要标志。

3.3.2 Production Process of Strip Steel
3.3.2 板带钢生产工艺

According to the different production methods and product varieties, the production process of plate and strip can be divided into plate production, hot strip production and cold strip production.

板带钢生产工艺按生产方法及产品品种的不同,分为中厚板生产、热轧带钢生产与冷轧带钢生产。

3.3.2.1 Production Process of Medium Plate
3.3.2.1 中厚板生产工艺

Medium plate is an important steel material on which the development of national economy depends. It is an indispensable steel variety in the course of industrialization and development. It is mainly used as steel for mechanical structures, buildings, vehicles, pressure vessels, bridges, shipbuilding and pipelines.

中厚板是一个国家国民经济发展所依赖的重要钢铁材料,是工业化进程和发展过程中不可缺少的钢铁品种。它主要用作机械结构、建筑、车辆、压力容器、桥梁、造船、输送管道用钢。

The technological processes of plate production are as follows:

中厚板生产的工艺流程具体如下:

(1) Preparation of the slab. According to the order specified in the rolling schedule, the slab is hoisted to the loading roller by crane. Then the loading personnel is responsible for checking the slab according to the plan, and the slab is identified by the corresponding instructions.

(1) 板坯的准备。板坯上料是根据轧制计划表中所规定的顺序,由起重机吊到上料辊道上,再由上料人员负责根据计划核对板坯,并通过相应的指令完成对板坯的识别。

(2) Heating of Slab. After the slab is transported to the furnace by the raw material conveying roller, the heating furnace is pushed by the pusher. The purpose of heating the raw

material is to make the raw material have good plasticity and low deformation resistance during rolling. There are three kinds of heating furnace in common use in plate production: continuous heating furnace, chamber heating furnace and soaking pit furnace.

（2）板坯的加热。板坯由原料输送辊道输送到炉后，由推钢机推进加热炉。原料加热的目的是使原料在轧制时有好的塑性和低的变形抗力。中厚板生产常用的加热炉有三种，即连续式加热炉、室式加热炉和均热炉。

（3）Descaling. The billet produced by the heating furnace is descaled by conveying the descaling box on the roller table. The purpose of descaling is to remove the oxide scale produced by the billet during heating by the action of high-pressure water, so as to avoid pressing into the surface of the steel plate and causing surface defects.

（3）除鳞。由加热炉出来的坯料，通过输送辊道送入除鳞箱进行除鳞。除鳞的目的是将坯料在加热时产生的氧化铁皮通过高压水的作用去除干净，以免压入钢板表面而造成表面的缺陷。

（4）Roughing. The Slab is rolled repeatedly through a reversible roughing mill. The main task of rough rolling is to widen the slab to the required width and to get the intermediate slab thickness for finishing mill. In the rough rolling stage, large pressure should be used as far as possible to refine the grain and improve the product performance.

（4）粗轧。板坯经过可逆式粗轧机进行反复轧制。粗轧阶段的主要任务是将板坯展宽到所需的宽度和得到精轧机所需要的中间坯厚度，粗轧后中间坯的宽度和厚度由测宽仪和测厚仪来测得。粗轧阶段，在满足轧机的强度条件和咬入条件的情况下应尽量采用大压下，以此来细化晶粒，提高产品的性能。

（5）Water curtain cooling of intermediate billet. For some plates with special properties, the opening and finishing temperature of finishing rolling should be controlled strictly, and the temperature of intermediate billet should be lowered by water curtain cooling.

（5）中间坯水幕冷却。对于一些有特殊性能要求的板材，需要严格控制其精轧的开、终轧温度，此时需要用水幕冷却对中间坯进行降温。

（6）Finish Rolling. After the secondary scale is removed by high pressure water, the medium billet is rolled in finishing mill. The finishing mill generally adopts the trapezoidal speed system of low speed biting, high speed rolling and low speed throwing. Thickness gauge and temperature gauge are installed at the exit of finishing mill to control the quality of products accurately. The main task of finish rolling stage is quality control, including thickness, shape, surface quality and performance control.

（6）精轧。中间坯经过高压水除去二次氧化铁皮后，进入精轧机进行轧制。精轧机一般采用低速咬入、高速轧制、低速抛出的梯形速度制度。精轧机出口处设有测厚仪和测温仪，以便精确控制产品的质量。精轧阶段的主要任务是质量控制，包括厚度、板形、表面质量和性能控制。

（7）Straightening. Hot Straightening is an indispensable process for flatness and straightness and surface quality of plate. The modern medium and heavy plate mill adopts the four-layer 9~11-roller powerful leveler, and the levelling temperature is usually between 600~750℃. If the straightening temperature is too high, the plate may warp when it is cooled in the cooling bed. If the straightening temperature is too low, the straightening effect is not good, and the residual stress on the surface of the plate is high after straightening, which reduces the performance of the plate.

（7）矫直。热矫直是使板形平直，保证板材表面质量不可缺少的工序。现代中厚板厂都采用四重式 9~11 辊强力矫直机，矫直终了温度一般在 600~750℃。若矫直温度过高，则矫直后钢板在冷床上冷却时可能会发生翘曲；若矫直温度过低，则矫直效果不好，矫直后钢板表面的残余应力高，降低了钢板的性能。

（8）Cooling. Cooling after rolling can be divided into process cooling and natural cooling. Process cooling (forced cooling), through laminar cooling, water curtain or mist to reduce the temperature of the steel plate. Natural cooling is the natural cooling of steel plates in the air on a cooling bed.

（8）冷却。钢板轧后冷却可分为工艺冷却和自然冷却。工艺冷却（即强制冷却）是通过层流冷却、水幕式或气雾的方式来降低钢板的温度。自然冷却是使钢板在冷床上于空气中自然冷却。

（9）Neat. Finishing processes include surface quality inspection, marking, cutting and printing of steel plates. The cutting of steel plate is carried out by means of double-sided shears or circular shears, while the cutting head, tail and fixed length are carried out by means of fixed length shears.

（9）精整。精整工序包括钢板的表面质量检查、划线、切割、打印等。钢板的切割通过双边剪或圆盘剪切边，切头尾与定尺可通过定尺剪来实现。

（10）Heat treatment, Most products through rolling or on-line control rolling and cooling after rolling control can meet the performance requirements. When there are special requirements for the properties of medium and heavy plate or the properties after rolling can not meet the requirements of customers, the finished plate usually needs to be treated by roller-hearth normalizing furnace to improve the comprehensive mechanical properties of the products.

（10）热处理。当然，大多数产品可通过轧制或在线控制轧制与轧后控制冷却达到性能要求。当对中厚板的性能有特殊要求，或中厚板轧后的有关性能达不到用户要求时，通常需要将成品钢板装入辊底式常化炉进行处理，以提高产品的综合力学性能等。

3.3.2.2　Hot Rolled Strip Production Process
3.3.2.2　热轧带钢生产工艺

At the end of 1990s, thin slab continuous casting and rolling (TSCR) technology appeared again, which made steel-making and continuous casting and rolling technology merge into one, and a short-flow production mode appeared, which made the traditional metallurgical process face new challenges. The new production process is more energy-saving, more cost-effective, and if can produce thinner hot-rolled strip products. The hot rolled strip steel production workshop is shown in Figure 3-28.

20 世纪 90 年代末又出现了薄板坯连铸连轧生产技术，使炼钢连铸与轧制技术融为一体，出现了短流程的生产模式，这使传统的冶金工艺流程面临新的挑战。新的生产工艺更节能，更具成本优势，可生产出更薄的热轧带钢产品品种。热轧带钢生产车间的平面布置图如图 3-28 所示。

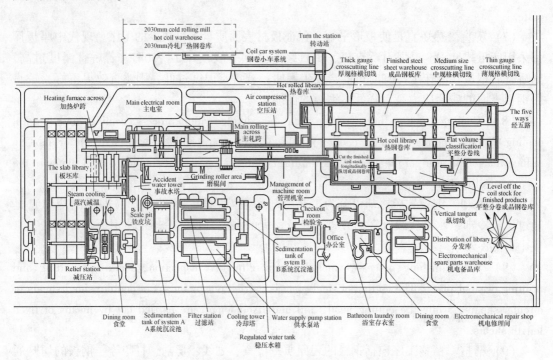

Figure 3-28　Layout of a 2050mm hot rolled strip steel production workshop
图 3-28　某 2050mm 热轧带钢生产车间的平面布置图

The traditional hot strip production process includes raw material preparation, heating, rough rolling, finishing rolling, cooling and coiling, introduced as follows:

传统的热轧带钢生产工艺过程，主要包括原料准备、加热、粗轧、精轧、冷却及卷取等。其介绍如下：

(1) Preparation and heating of raw materials. The raw materials used in hot strip rolling are generally continuous cast slabs, ground or hot loaded into a continuous walking beam furnace for heating. The continuous casting slab is larger in size, with 150~250mm in thickness and even 12000~15000mm in length. The yield and yield can be increased by increasing slab weight. At present, the weight of hot rolled strip per unit width is more than 30kg/mm. In order to improve the heating intensity, the furnace adopts multi-point heating mode of 6~8 points. In order to ensure the heating quality of the billet, the continuous walking beam furnace is adopted.

(1) 原料准备与加热。热轧带钢生产所用原料一般采用连铸板坯，经修磨或热装进入连续步进式加热炉加热。连铸板坯的尺寸较大，厚度多为 150~250mm，长度甚至达到 12000~15000mm，增大坯重可提高产量与成材率。目前热轧带钢单位宽度的板卷质量达到 30kg/mm 以上。为提高加热强度，炉子采用多点（6~8 点）供热方式。为保证坯料加热质量，采用连续步进式加热炉。

(2) Roughing. High pressure water descaling should be carried out on the raw material surface before rough rolling to improve the surface quality of the steel plate and prevent the iron oxide sheet from pressing in. The formation of roughing mill determines the form of hot strip mill. The roughing mill of a semi-continuous tropical mill, consisting of one or two reversing mills, is the same as that of a medium plate mill; the roughing mill of the 3/4 continuous tropical mill is the addition of a set of continuous mills (two) to the roughing mill of the semi-continuous tropical mill. At present, most of the tropical mills are semi-continuous. Roughing mill in addition to the horizontal roller also has a vertical roller frame, and constitute a universal mill. The purpose of vertical rolling is to control the width of slab and the accuracy of strip width.

（2）粗轧。粗轧前原料表面要进行高压水除鳞，以提高钢板表面质量，防止氧化铁皮压入。粗轧机的构成决定了带钢热连轧机组的形式。半连续式热带轧机的粗轧机组，由一架或两架可逆式轧机组成，与中厚板轧机的构成相同；3/4连续式热带轧机的粗轧机组，是在半连续式热带轧机的粗轧机组上增加一组连轧机组（两架）。目前新建的热带轧机多为半连续式布置。粗轧机上除水平辊外还设有立辊机架，构成万能轧机。立轧的目的是为了控制板坯的宽展及带钢宽度精度。

(3) Finish Rolling. Before finishing rolling, there are rotary drum flying shears, descaling box and other equipment. The aim of the flying shear strip head is to make the strip feed into the finishing mill correctly and the cold head is not easy to scratch the roll surface, so that the strip tail after coiling does not hinder the transportation on the transportation chain because of the flash. The finishing mill unit is usually composed of 6 or 7 continuous mill stands. During rolling, continuous rolling relationship is formed between each stand, and the strip is connected with the underground coiler after entering the hot output roller.

（3）精轧。精轧前设有转筒式飞剪与除鳞箱等设备。飞剪剪切带坯头部的目的是使带坯正确喂入精轧机组，且冷头不易划伤轧辊表面；飞剪剪切带坯尾部的目的是使卷取后的带钢尾部不会因出现飞边而妨碍在运输链上的运输。精轧机组一般由6架或7架连轧机架组成。轧制时各架之间形成连轧关系，带钢进入热输出辊道后与地下卷取机相连。

(4) Cooling and coiling. The strip temperature should be controlled after leaving the finishing mill so that the strip head can be coiled before entering the coiler to ensure the comprehensive mechanical properties of the strip. Laminar cooling device is often used for temperature control of strip steel. Coiler coiling operation must be synchronized with the hot output roll table and finishing mill stand to ensure high-speed and stable rolling and coiling. After coiling, the hot rolled strip is transported to the intermediate warehouse for cooling. Besides being used as raw material for cold rolling mill, it is also processed into commodity plate or coil through finishing line.

（4）冷却及卷取。带钢出精轧机组后，需要对带钢温度进行控制，以使带钢头部在进入卷取机前完成组织转变而进行卷取，保证带钢的综合力学性能。带钢温度控制多采用层流冷却装置以实现精确控温。卷取机的卷取操作必须与热输出辊道及精轧机架同步运行，以保证高速稳定地轧制与卷取。卷取后的热轧带卷通过运输传送至中间库冷却，除供冷轧带钢厂做原料外，还通过精整作业线加工成商品板或卷。

3.3.2.3 Cold Rolled Strip Production Process
3.3.2.3 冷轧带钢生产工艺

Cold-rolled strip steel is used as raw material to produce products with thinner dimension and higher dimension precision after cold-working deformation at room temperature. Compared with hot-rolled strip, the cold-rolled strip has thinner thickness, higher dimensional accuracy and better shape quality. The uniformity of product properties is better because it is not affected by the processing temperature, and it has been widely used because of its high deep-drawing property and surface treatment to improve the anti-corrosion property.

冷轧带钢生产是利用热轧带钢做原料，在室温条件下经冷加工变形生产出尺寸更薄、尺寸精度更高的产品。与热轧带钢产品相比，冷轧带钢产品的厚度更薄，尺寸精度和板形质量更高，产品性能的均匀性因不受加工变形温度的影响而更好，其由于具有高深冲性能以及可通过表面处理提高抗腐性能而获得更广泛的应用。

The production process of cold rolled strip includes pickling, cold rolling, annealing, leveling and finishing, introduced as follows:

冷轧带钢生产工艺流程主要包括酸洗、冷轧、退火、平整及精整等工序。其介绍如下：

(1) Pickling. The process of removing oxide from the surface of cold rolled strip is called descaling. Descaling methods include pickling, alkali washing and mechanical descaling. Acid pickling is often used, and alkaline washing is often used for descaling of special steel grades.

(1) 酸洗。冷轧带钢表面氧化物的去除过程称为除磷。除磷的方法有酸洗、碱洗和机械除磷等。采用较多的是酸洗方法，碱洗常用于特殊钢种的除磷。

(2) Cold rolled. The descaled slab is rolled to the finished product thickness on a cold rolling mill, usually without intermediate annealing. Cold rolling is divided into single-piece rolling and coil rolling.

(2) 冷轧。除磷后的板带坯在冷轧机上轧制到成品的厚度，一般不经中间退火。冷轧分为单片轧制和成卷轧制。

(3) Annealing. The aim of annealing is to eliminate the cold rolling work hardening and to make the steel sheet recrystallize and soften, so that it has good plasticity.

(3) 退火。退火的目的在于消除冷轧加工硬化，使钢板再结晶软化，从而具有良好的塑性。

(4) Flatness. Flatness is defined as light cold rolling with a reduction of 0.5% to 4%. The purpose of the flattening is to prevent the strip from drawing to a significant yield step, to obtain the necessary mechanical properties, to improve the strip shape, and to achieve the required surface roughness.

(4) 平整。平整是指以0.5%~4%的压下率轻微冷轧。平整的目的是：防止带钢拉伸发生明显的屈服台阶，并得到必要的力学性能；改善带钢的板形，达到所要求的表面粗糙度。

(5) Neat. The cold rolled strip is generally cut by the shearing machine after leveling. Slitting

is used for cutting edges or strips according to the required width, and cross cutting is used to cut strips into sheets according to the required length. The finished product strip after cutting is inspected and classified (or on-line automatic sorting and packaging), coated with anti-rust oil packaging factory. Typical cold-rolled strip production process, as shown in Figure 3-29.

（5）精整。一般冷轧板带经平整后送剪切机组剪切。纵剪用于剪边或按需要的宽度分条，横剪是将板带按需要的长度切成单张板。剪切后的成品板带经检验分类后（或在线自动化分选包装），涂防锈油包装出厂。典型的冷轧板带生产工艺流程如图3-29所示。

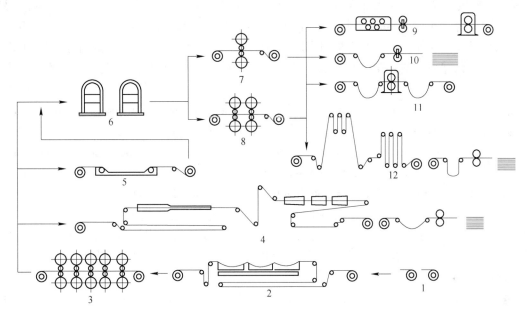

Figure 3-29 Typical production process of cold rolled strip steel
图3-29 典型的冷轧带钢生产工艺流程

1—Hot rolled strip coil; 2—Continuous pickling; 3—Cold continuous rolling; 4—Continuous hot-dip galvanizing; 5—Electrolytic cleaning; 6—One-cover annealing furnace; 7—Single frame flat; 8—A pair of racks are flat; 9—volume; 10—Cross shear; 11—Slitting shear; 12—Continuous electroplating of tin

1—热轧带卷；2—连续酸洗；3—冷连轧；4—连续热镀锌；5—电解清洗；6—罩式退火炉；7—单机架平整；8—双机架平整；9—重卷；10—横剪；11—纵剪；12—连续电镀锡

3.3.3 Section Steel Production Process
3.3.3 型钢生产工艺

Section steel productions have the following characteristics:
型钢生产具有以下特点：

（1）The cross section of the product is rather complicated. In addition to square steel, round steel, flat steel and other simple cross-section products, most of the profiled steel products have a great impact on rolling production. Due to the variety of cross-section shape of rolled piece, serious non-uniform deformation will occur when rolling profile, especially for special-shaped steel, which will bring corresponding bad results. It is more difficult to calculate forward slip, width spread, force and energy parameters in section steel production because of the difference of temperature, deformation degree and roll diameter. In addition, the serious non-uniform deformation has adverse effects on the quality of rolled products, energy consumption, roll consumption, guide and guard design and installation, pass adjustment, mill output, etc. It is also difficult to produce and straighten products after continuous rolling.

（1）产品的断面比较复杂。除方钢、圆钢、扁钢等简单断面产品外，型钢大多数为异形断面产品，这就给轧制生产带来了很大的影响。由于轧件断面形状多样，轧制型材，特别是异形钢时，必然产生严重的不均匀变形，因而带来相应的不良后果。轧件各部的温度、变形程度、轧辊直径的不同，使型钢生产中前滑、宽展、力能参数的计算要比钢板生产中困难得多。此外，严重的不均匀变形，对轧制产品的质量、能耗、轧辊消耗、导卫设计与安装、孔型的调整、轧机的产量等都有不利影响，组织连轧生产、轧后产品的矫直也具有较大困难。

（2）A wide variety of products. In addition to a few specialized section mills, most section mills produce a variety of products and specifications, resulting in a large variety of billet specifications, roll reserves, the number of guide and guard devices, so that the production management is very complex. And the frequency of roll change, rolling mill installation adjustment technical requirements are higher, which greatly affects the effective production time of the mill.

（2）产品的品种多。除少数专业化型钢轧机外，大多数型钢轧机生产的产品品种和规格繁杂多样，因而造成坯料的品种规格多、轧辊储备量大、导卫装置数量多，使生产管理工作极为复杂；且换辊次数频繁，轧机安装调整技术要求较高，从而大大影响了轧机有效生产时间。

（3）There are many types and structures of rolling mills. Section steel varieties and specifications, sizes vary greatly, coupled with their different production requirements, making the section steel mill structure and many types, including a variety of mill types and layout. There are two-high mill, three-high mill, four-high universal pass mill, multi-roll pass mill, y-type mill, rolling mill of 45° and cantilever mill, etc. In the arrangement of rolling mill, there are transverse mill, tandem mill, checkerboard mill, semi-continuous mill and fully continuous mill, etc.

（3）轧机的结构和类别多。型钢的品种和规格很多，尺寸相差很大，加上各自生产要求不同，使得型钢轧机的结构和类别很多，包括各种轧机类型和布置形式。在轧机结构形式上，有二辊式轧机、三辊式轧机、四辊万能孔型轧机、多辊孔型轧机、Y型轧机、45°轧机和悬臂式轧机等；在轧机布置形式上，有横列式轧机、顺列式轧机、棋盘式轧机、半连续式轧机和全连续式轧机等。

The section steel production processes are introduced as follows. Rail is an important part of railway running track, and it is a special-shaped section steel with only y axial symmetry. Its cross section can be divided into rail head, rail waist and rail bottom three parts. The rail head is the part in contact with the wheel, and the rail bottom is the part in contact with the sleeper. Different countries have different requirements on the technical conditions of rail, but the cross-section shape of rail is the same, as shown in Figure 3-30.

下面以重轨为例，介绍型钢生产工艺流程。钢轨是铁路运行轨道的重要组成部分，是仅 Y 轴对称的异形断面钢材。其横截面可分为轨头、轨腰和轨底三部分。轨头是与车轮相接触的部分，轨底是接触轨枕的部分。世界各国对钢轨的技术条件有不同的要求，但钢轨的横截面形状都是一致的，如图 3-30 所示。

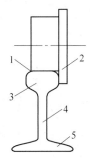

Figure 3-30　Cross section shape of rail
图 3-30　钢轨的横截面形状
1—Tread；2—Wheel；3—Rail head；4—One-track waist；5—Rail
1—踏面；2—车轮；3—轨头；4—轨腰；5—轨底

The gauge of the rail is expressed in terms of weight per meter. The weight range of ordinary rail is 5~78kg/m, and the weight of crane rail can reach 120kg/m. The common specifications of rail are 9kg/m, 12kg/m, 15kg/m, 22kg/m, 24kg/m, 30kg/m, 38kg/m, 43kg/m, 50kg/m, 60kg/m, and 75kg/m. Generally, the rails below 30kg/m are called light rail, and the rails above 30kg/m are called heavy rail. Light rail is mainly used for short-distance, light-load and low-speed special line railway in forest, mine and salt field.

钢轨的规格以每米长的质量表示。普通钢轨的质量范围为 5~78kg/m，起重机轨的质量可达 120kg/m。钢轨常用的规格有 9kg/m、12kg/m、15kg/m、22kg/m、24kg/m、30kg/m、38kg/m、43kg/m、50kg/m、60kg/m、75kg/m。通常将 30kg/m 以下的钢轨称为轻轨，将 30kg/m 以上的钢轨称为重轨。轻轨主要用于森林、矿山、盐场等工矿内部的短途、轻载、低速的专线铁路；重轨主要用于长途、重载、钢轨用于工业结构件。

Heavy rail production process is shown in Figure 3-31, and pass system for rolling heavy rail is shown in Figure 3-32.

重轨生产工艺流程如图 3-31 所示，轧制重轨的孔型系统如图 3-32 所示。

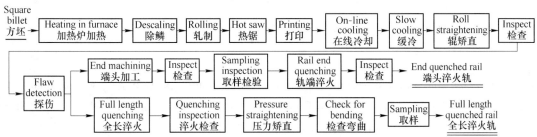

Figure 3-31　Heavy rail production process
图 3-31　重轨生产工艺流程

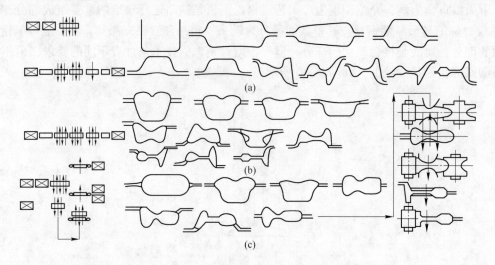

Figure 3-32　Pass system for rolling heavy rail
图 3-32　轧制重轨的孔型系统
(a) Cross rolled pass system; (b) Straight rolled pass system; (c) Universal pass system
(a) 斜轧孔型系统; (b) 直轧孔型系统; (c) 万能孔型系统

3.3.4　Hot Rolled Seamless Steel Tube Production Process
3.3.4　热轧无缝钢管生产工艺

Hot-rolled seamless steel tube is a kind of steel tube which is made by piercing and rolling the solid billet to meet the product standard. The production process includes the basic processes of preparation before rolling, heating, piercing, tube rolling, tube sizing and reducing (including stretch reducing, Tube Cooling, tube cutting, segment, straightening, flaw detection, manual inspection), spray mark printing, packing, etc. There are three main deformation processes: tube piercing, tube rolling, and tube sizing and reducing.

热轧无缝钢管是将实心管坯穿孔并轧制成符合产品标准的钢管。其生产工艺流程包括管坯轧前准备、管坯加热、管坯穿孔、轧管、钢管定径与减径（包括张力减径）、钢管冷却、钢管切头尾、分段、矫直、探伤、人工检查、喷标打印、打捆包装等基本工序，主要有管坯穿孔、轧管以及钢管定径与减径三个变形工序。

3.3.4.1　Pre Rolling Preparation of Tube Blank
3.3.4.1　管坯轧前准备

The tube blank has the section shape of circle, square, polygon and so on. Square or polygon billets with wavy edges are used for pressure piercing, while circular billets are used for skew piercing due to deformation conditions.

管坯有圆形、方形、多边形等断面形状。压力穿孔选用方形、带波浪边的方形或多边形管坯；斜轧穿孔则受变形条件限制，需选用圆形管坯。

3.3.4.2 Billet Heating
3.3.4.2 管坯加热

The purpose of tube blank heating is to improve the plasticity of tube blank, to reduce the resistance to deformation, to facilitate the plastic deformation, to reduce the energy consumption of processing, and to improve the microstructure and properties of steel by dissolving carbides and diffusing non-metallic phases. The defects of oxidation, decarburization, carbonization, overheat and overheat of tube blank should be prevented when the tube blank is heated. The heating temperature should be accurate and uniform, and the burning loss should be reduced. The tube-billet reheating furnace has ring furnace, step furnace, inclined bottom furnace and induction furnace and so on. The ring-shaped heating furnace is used in most of the modern hot-rolled seamless steel tube mills.

管坯加热的目的是为了提高管坯塑性，降低变形抗力，有利于塑性变形和降低加工能耗；使碳化物溶解和非金属相扩散，改善钢的组织性能。管坯加热时要防止管坯表面氧化、脱碳、增碳、过热和过烧等缺陷，还需保证加热温度准确且均匀、烧损少等基本要求。管坯加热炉的形式有环形炉、步进炉、斜底炉和感应炉等。现代热轧无缝钢管机组大多采用环形加热炉。

3.3.4.3 Billet Piercing
3.3.4.3 管坯穿孔

Tube piercing is the process of making hollow capillaries from solid billet. It is the most important deformation process in hot rolling seamless steel tube production. There are three common methods for piercing of tube blank (such as cross-rolling piercing with two-roll, vertical large guide disc with Disher, cone-shaped and three-roll piercing), pressure piercing and push rod piercing with PPM, as shown in Figure 3-33.

管坯穿孔是将实心管坯穿制成空心毛管的工艺过程，是热轧无缝钢管生产中最重要的变形工序。常见的管坯穿孔方法有斜轧穿孔（二辊式穿孔、立式大导盘（狄舍尔）穿孔、锥形辊（菌式）穿孔和三辊式穿孔）、压力穿孔和推杆穿孔（PPM）三种，如图3-33所示。

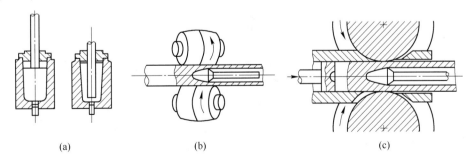

Figure 3-33 Schematic diagram of perforation method
图3-33 穿孔方法示意图
(a) Pressure perforation; (b) Oblique rolling perforation; (c) Push rod perforation
(a) 压力穿孔；(b) 斜轧穿孔；(c) 推杆穿孔

3.3.4.4 Rolled Pipe
3.3.4.4 轧管

Tube rolling is the process of rolling hollow capillary tubes into tubes that are close to the finished size.

轧管是将空心毛管轧成接近成品尺寸的荒管。

3.3.4.5 Sizing and Reducing of Steel Pipe
3.3.4.5 钢管定径与减径

The sizing, reducing and tension reducing of steel tubes are all the processes of continuous rolling hollow tubes without mandrel, which is the last hot deformation process in the production of hot rolled seamless steel tubes, also called hot finishing. The main task of sizing is to control the outside diameter accuracy and true roundness of the finished tube. The number of racks is usually 3~12. The number of frames is generally 9~24. Steel pipes less than 60mm in diameter are generally reduced, in addition to reducing the size of the role of sizing, and the diameter of the tube. In addition to the reducing effect of tension reducing, the wall is reduced by the tension between the racks. The number of the racks is usually 12~24, up to 28. At present there are commonly used two-high, three-high and four-high sizing machine, as shown in Figure 3-34. The roll axes of the two-roll type front and rear adjacent stands are perpendicular to each other 90°. There are two basic types of the three-roll mill: the micro-tension constant reducing mill and the tension constant reducing mill.

钢管定径、减径和张力减径均为无芯棒连轧空心管体的过程，是热轧无缝钢管生产中最后的热变形工序，也称为热精整。定径的主要任务是控制成品管的外径精度和真圆度，机架数一般为3~12架。减径除了起定径作用外，还使管径减小，机架数一般为9~24架。直径小于60mm的钢管一般需经过减径加工。张力减径除了具有减径作用外，还可通过机架间建立张力实现减壁，机架数一般为12~24架，最多达28架。目前常用的有二辊式、三辊式和四辊式定（减）径机，如图3-34所示。二辊式前后相邻机架的轧辊轴线互垂90°，三辊式轧辊轴线互错60°，定（减）径机包括微张力定（减）径机和张力定（减）径机两种基本形式。

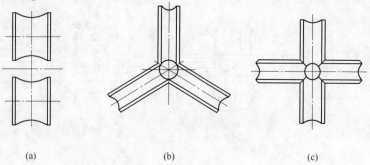

Figure 3-34 Classification of sizing (reducing) mill according to the number of rolls
图3-34 定（减）径机按辊数分类
(a) Two-roll; (b) Three-roll; (c) Four-roll
(a) 二辊式；(b) 三辊式；(c) 四辊式

3.3.4.6　Steel Pipe Finishing
3.3.4.6　钢管精整

Steel pipe finishing including steel pipe cooling, cutting head and tail, subsection, straightening, inspection, manual inspection, spray printing, packing and other basic processes.

钢管精整包括钢管冷却、切头尾、分段、矫直、探伤、人工检查、喷标打印、打捆包装等基本工序。

Exercises
思考题

(1) What is metal pressure working, and what are its methods?
(1) 什么是金属压力加工，有哪些加工方法？
(2) What is the nature of plastic deformation of metals?
(2) 金属塑性变形的本质是什么？
(3) How will the structure and properties of metals change during plastic deformation?
(3) 金属塑性变形时材料组织与性能会如何变化？
(4) What are the laws of metal plastic flow?
(4) 金属塑性流动规律有哪些？
(5) Describe the production process of hot rolled strip steel.
(5) 阐述热轧板带钢生产工艺。
(6) Describe the production process of hot rolled section steel.
(6) 阐述热轧型钢生产工艺。
(7) Describe the production process of hot rolled seamless steel tube.
(7) 阐述热轧无缝钢管生产工艺。

Project 4 Environmental Protection and Utilization of Resources
项目 4 环境保护与资源综合利用

Metallurgical industry is the basic industry of national economic development. It consumes a large amount of energy and resources in the production process, and also produces a large number of by-products (mainly waste gas, waste water and solid waste). If these by-products are discharged directly without treatment, they will have an impact on the environment and cause the waste of resources.

冶金工业是国民经济发展的基础产业,其生产过程中消耗大量的能源和资源,同时也产生大量的副产品(主要为废气、废水、固体废物),这些副产品如果不进行处理就直接排放,将对环境产生影响,同时造成资源的浪费。

Task 4.1 Pollution and Treatment of Waste Gas from Metallurgical Industry
任务 4.1 冶金工业废气的污染与治理

4.1.1 Classification of Waste Gases from Metallurgical Industry
4.1.1 冶金工业废气的分类

Metallurgical industrial waste gas (referred to as waste gas) refers to toxic and harmful gases containing pollutants produced by rock drilling, blasting, ore crushing, screening and transportation, metal smelting and processing, and fuel combustion during non-ferrous metal mining, ore dressing, smelting, processing and related processes.

冶金工业废气(简称废气)是指在有色金属采矿、选矿、冶炼和加工生产及其相关过程中,因凿岩、爆破、矿石破碎、筛分和运输,金属冶炼和加工,以及燃料燃烧等产生的含污染物质的有毒有害气体。

According to the nature of the main pollutants contained in metallurgical industry, waste gas can be broadly divided into three categories: the first category is mining and mineral processing waste gas containing mainly industrial dust; the second category is mainly containing toxic gases containing fluorine, sulfur, chlorine and smoke dust of the metal smelting waste gas; and the third category is mainly containing acid, alkali and oil mist of the metal processing industry waste gas.

冶金工业废气按其所含主要污染物的性质，大体上可分为三大类：第一类为主要含工业粉尘的采矿和选矿工业废气；第二类为主要含有毒有害气体（含氟、硫、氯）与烟尘的金属冶炼废气；第三类为主要含酸、碱和油雾的金属加工工业废气。

In the production process of steel works, there are various kilns emitting a lot of smoke and dust such as sintering, pelletizing, coking, iron making, steelmaking, rolling, forging, metal products and ferroalloy, refractory materials, carbon products and power. In the process of smelting, a large amount of ore, fuel and other auxiliary raw materials are consumed. Every 1t steel needs 6 ~ 7t raw materials, including iron ore, fuel, limestone and manganese ore. These raw materials of 80% are turned into waste.

钢铁厂的烧结、球团、炼焦、炼铁、炼钢、轧钢、锻压以及金属制品与铁合金、耐火材料、碳素制品和动力等生产环节，拥有排放大量烟尘的各种窑炉。冶炼加工过程中，消耗大量的矿石、燃料和其他辅助原料，每生产1t钢需要消耗6~7t原料，其中包括铁矿石、燃料、石灰石、锰矿等，这些原料中的80%变为废物。

The waste gas emitted by iron and steel enterprises can be divided into three types: the first type is the waste gas emitted by chemical reactions in the production process, such as smelting, coke burning, chemical products and the smoke and harmful gases produced in the process of steel pickling; the second type is the flue gas and harmful gas produced by the combustion of fuel in the furnace; and the third type is the dust produced during the transportation, handling and processing of raw materials and fuel.

钢铁企业排放的废气大体可分为三类：第一类是生产工艺过程中化学反应排放的废气，如冶炼、烧焦、化工产品和钢材酸洗过程中产生的烟尘和有害气体；第二类是燃料在炉窑中燃烧产生的烟气和有害气体；第三类是原料和燃料运输、装卸及加工等过程中产生的粉尘。

4.1.2 Characteristics of Waste Gas from Metallurgical Industry
4.1.2 冶金工业废气的特点

4.1.2.1 Characteristics of Waste Gas from Non-ferrous Metallurgical Industry
4.1.2.1 有色冶金工业废气的特点

The characteristics of waste gas from non-ferrous metallurgical industry are as follows:
有色冶金工业废气具有以下特点：

(1) The discharge quantity is big, and the pollution area is wide.

(1) 排放量较大，污染面较广。

(2) The composition of waste gas is complex, and its treatment is difficult.

(2) 废气成分复杂，治理难度较大。

(3) The pollutants in exhaust gas are mainly inorganic, and the environmental pollution has potential influence.

(3) 废气中污染物以无机物为主，环境污染具有潜在的影响。

4.1.2.2　Characteristics of Waste Gas from Iron and Steel Metallurgical Industry
4.1.2.2　钢铁冶金工业废气的特点

The characteristics of waste gas from iron and steel metallurgical industry are as follows:
钢铁冶金工业废气具有以下特点：

(1) The exhaust gas discharge quantity is big, the pollution area is wide. The waste gas pollution source of iron and steel enterprise is concentrated in the furnace of iron-making, steel-making, sintering and coking.

(1) 废气排放量大，污染面广。钢铁企业的废气污染源集中在炼铁、炼钢、烧结、焦化等冶炼工业窑炉，设备集中，规格庞大，废气排放量大。

(2) The soot particles are fine and have strong adsorbability. Most of the iron oxide dust emitted in the smelting process of iron and steel enterprises is easy to be used as the carrier for adsorbing harmful gas because of its fine particle, large specific surface area and strong adsorption.

(2) 烟尘颗粒细，吸附力强。钢铁企业冶炼过程中排放的多为氧化铁烟尘。由于该烟尘尘粒细、比表面积大、吸附力强，其易成为吸附有害气体的载体。

(3) The temperature of flue gas is high, so it is difficult to control.

(3) 烟气温度高，治理难度大。

(4) The volatile gas is strong.

(4) 烟气挥发性强。

(5) Waste gas has recycling value. The waste heat of high-temperature flue gas from iron and steel production can be converted into steam or electric energy through heat energy recovery devices. Coal gas produced in coking, iron-making and steel-making has become the main fuel in iron and steel enterprises and can be used outside. Most of the dust and mud collected in the process of waste gas purification contain iron oxide, which can be recycled in various ways.

(5) 废气具有回收价值。钢铁生产排出的废气中，高温烟气的余热可以通过热能回收装置转换为蒸气或电能。炼焦及炼铁、炼钢过程中产生的煤气，已成为钢铁企业的主要燃料，并可外供使用。各废气净化过程中所收集的尘泥，绝大部分含有氧化铁成分，可采用各种方式回收利用。

4.1.3　Treatment of Waste Gas from Metallurgical Industry
4.1.3　冶金工业废气的治理

4.1.3.1　Treatment of Waste Gas from Metallurgical Industry
4.1.3.1　冶金工业废气的治理方法

There are some pollutants in the exhaust gas from various production processes and people's daily life. The purification technologies used can be divided into two categories: separation method and transformation method. Separation method is the use of external force and other physical methods, the pollutants from the exhaust gas separation. The conversion method is to make the pollutants in the exhaust gas occur some chemical reactions, and make it separate or convert into other substances, and then use other methods for purification.

各种生产工艺过程和人们日常生活中排出的废气含有某些污染物，所采用的净化技术基本上可以分为两大类，即分离法和转化法。分离法是利用外力等物理方法，将污染物从废气中分离出来。转化法是使废气中的污染物发生某些化学反应，然后使其分离或转化为其他物质，再用其他方法进行净化。

For the particulate pollutants such as smoke and fog drops, we can use all kinds of deduster and demister to separate them from the waste gas. For gaseous pollutants, we can use their different physical and chemical properties, using condensation, absorption, adsorption, combustion, catalytic transformation and other methods for purification.

对于烟尘、雾滴之类的颗粒状污染物，可利用其质量较大的特点，用各种除尘器、除雾器使之从废气中分离出去。对于气态污染物，可利用其不同的理化性质，采用冷凝、吸收、吸附、燃烧、催化转化等方法进行净化处理。

Condensation

冷凝法

The condensation method uses the fact that different substances have different saturated vapor pressures at the same temperature, and the same substance has different saturated vapor pressures at different temperatures to cool or pressurize the gas mixture, condense one or more of these contaminants into a liquid or solid and separate them from the gas mixture.

冷凝法是利用不同物质在同一温度下具有不同的饱和蒸气压，以及同一物质在不同温度下具有不同的饱和蒸气压这一性质，将混合气体冷却或加压，使其中某种或几种污染物冷凝成液体或固体，从而由混合气体中分离出来。

Absorption Method

吸收法

The absorption method is the most common method for purifying gaseous pollutants, which can be used for purifying SO_2, NO_x, HF, SiF_4, HCl, NH_3 gaseous pollutants as well as mercury vapor, acid fog, asphalt smoke and multi-component organic vapor. The commonly used absorbent is water, alkaline solution, acid solution, oxidation solution and organic solvent.

吸收法是净化气态污染物最常用的方法，可用于净化含有 SO_2、NO_x、HF、SiF_4、HCl、NH_3 的气态污染物以及汞蒸气、酸雾、沥青烟和多种组分有机物蒸气。常用的吸收剂有水、碱性溶液、酸性溶液、氧化溶液和有机溶剂。

Adsorption Method

吸附法

The adsorption method is mainly used to purify the low concentration pollutant in the waste gas, and to recover the organic vapor and other pollutants in the waste gas. Adsorption method is to make waste gas contact with porous solid adsorbent, so that the pollutants adsorbed on the solid surface and separated from the gas flow. When the concentration of the adsorbate in the gas phase is lower than the equilibrium concentration of the adsorbate on the adsorbent, or when more easily adsorbed substances reach the surface of the adsorbent, the original adsorbate will be separated from the surface of the adsorbent and enter the gas phase. This phenomenon is called desorption. The failed adsorbents can be regenerated to regain adsorption capacity, and the regenerated adsorbents can be reused.

吸附法主要用于净化废气中的低浓度污染物质，并用于回收废气中的有机蒸气及其他污染物。吸附法是使废气与多孔性固体（吸附剂）接触，使其中的污染物（吸附质）吸附在固体表面上而从气流中分离出来。当吸附质在气相中的浓度低于吸附剂上的吸附质平衡浓度时，或者有更容易被吸附的物质到达吸附剂表面时，原来的吸附质会从吸附剂表面上脱离而进入气相，这种现象称为脱附。失效的吸附剂经过再生可重新获得吸附能力，再生后的吸附剂可重新使用。

Burning Method

燃烧法

Combustion is through the combustion of pollutants in the exhaust gas (combustible gas, organic steam and fine dust particles) into harmless substances or easily removed substances. Because this method is often put at the end of all the process, also known as after the burning method, the equipment used is called after the burner. Compared with other treatment methods, combustion method characterized by low concentration of pollutants can be treated waste gas, and a high degree of purification. The combustible waste gas is purified by different methods according to its concentration and oxygen content.

燃烧法是通过燃烧将废气中的污染物（可燃气体、有机蒸气、微细的尘粒等）转变成无害物质或容易除去的物质。由于这种方法常常放在所有工艺流程的最后，又称为后烧法，所用设备称为后烧器。与其他处理方法相比，燃烧法的特点是可以处理污染物浓度很低的废气，净化程度很高。可燃废气根据其浓度和氧含量的不同，采用不同方法进行净化。

Catalytic Conversion

催化转化法

Catalytic combustion (essentially a catalytic combustion process) refers to using the catalytic action of a catalyst to convert pollutants in exhaust gas into harmless compounds (or substances that can be removed more easily than they were before). The catalytic conversion method can be divided into catalytic oxidation method and catalytic reduction method.

催化转化法（催化燃烧法实质也属于催化转化法）就是利用催化剂的催化作用，将废气中的污染物转化成无害的化合物（或比原来存在状况更易除去的物质）。因工作原理不同，催化转化法可分为催化氧化法和催化还原法。

Catalysts used in catalytic conversion should have good activity, selectivity, sufficient mechanical strength and good thermal stability. Gases passing through the catalyst layer shall be free from dust and other substances that may poison the catalyst.

催化转化法所用的催化剂应具备很好的活性和选择性、足够的机械强度和良好的热稳定性。通过催化剂层的气体，应无粉尘及其他可使催化剂中毒的物质。

The catalyst is generally composed of an active substance supported on a carrier. In order to improve the performance of some catalysts, cocatalysts are also added.

催化剂一般由活性物质载于载体上组成。有些催化剂中还加入助催化剂，以改善催化剂性能。

4.1.3.2 Recovery of SO_2 Flue Gas
4.1.3.2 二氧化硫烟气的净化回收

The technology of removing SO_2 from flue gas is simply called flue gas desulfurization. For the flue gas with high concentration of SO_2, the contact self-heating process can be used to produce sulfuric acid. But the treatment of the flue gas with low concentration of SO_2 is more complicated.

从排烟中去除 SO_2 的技术简称为排烟脱硫。对高浓度 SO_2 烟气，可用接触法自热生产硫酸，而低浓度 SO_2 烟气的处理则比较复杂。

Purification and Recovery of a High Concentration Sulfur Dioxide Flue Gas
高浓度二氧化硫烟气的净化回收

The SO_2 flue gas which can meet the requirement of SO_2 (SO_2 more than 2%) is called high concentration SO_2 flue gas. This kind of flue gas often uses the contact method to produce sulfuric acid.

凡是能满足接触法自热生产硫酸的 SO_2 烟气（SO_2 的浓度达 2% 以上），均称为高浓度 SO_2 烟气。此类烟气常采用接触法生产硫酸。

The high concentration SO_2 basic process of sulfuric acid production by flue gas contact method is shown in Figure 4-1.

高浓度 SO_2 烟气接触法生产硫酸的基本流程如图 4-1 所示。

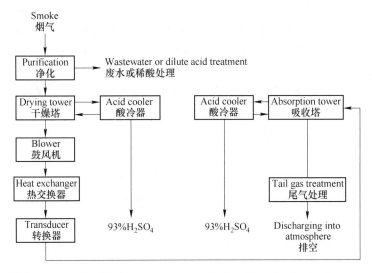

Figure 4-1　High concentration SO_2 basic process of sulfuric acid production by flue gas contact method
图 4-1　高浓度 SO_2 烟气接触法生产硫酸的基本流程

Purification and Recovery of Low Concentration Sulfur Dioxide Flue Gas
低浓度二氧化硫烟气的净化回收

The SO_2 flue gas which can not meet the requirement of contact self-heating sulphuric acid production is called low concentration SO_2 flue gas if its concentration is below 2%. Its control methods usually include the use of low-sulfur fuels, flue gas desulfurization and high stack gas dilution methods.

凡是不能满足接触法自热生产硫酸的 SO_2 烟气（SO_2 的浓度在2%以下），均称为低浓度 SO_2 烟气。其控制方法通常包括采用低硫燃料、烟气脱硫和烟气的高烟囱排放稀释等方法。

Flue gas desulfurization methods can be divided into wet and dry methods. The method using water or aqueous solution as absorbent to absorb SO_2 in flue gas, called wet desulfurization. The method using solid absorbent or adsorbent to absorb or absorb SO_2 in flue gas is called dry desulfurization.

烟气脱硫的方法可分为湿法和干法两种。用水或水溶液作吸收剂吸收烟气中 SO_2 的方法，称为湿法脱硫；用固体吸收剂或吸附剂吸收或吸附烟气中 SO_2 的方法，称为干法脱硫。

4.1.3.3　Dust Removal Techniques
4.1.3.3　除尘技术

Emission Coefficient of a Dust Particle and Its Properties
尘粒的排放系数及其性质

The nature and quantity of dust particles discharged from various pollution sources in metallurgical industry are different due to different production processes, equipments and scales. The parameter of 'Emission Factor' is generally used for rough estimation. The emission coefficient is the dust emission corresponding to the unit quantity of product or the unit quantity of raw material when the pollution control equipment is used or not used under certain conditions of raw material and production process.

冶金工业中的各种污染源因生产工艺、设备和规模不同，所排放尘粒的性质和数量也各异。一般为了粗略估计，可利用"排放系数"这一参数。排放系数是指在一定的原料和生产工艺条件下，采用或不采用污染控制设备时，与单位数量产品或单位数量原料所对应的尘粒排放量。

For dust particles, several physical properties affecting the separation mechanism can be described. The most important of these properties are particle size and density, in addition to resistivity, adhesion, particle shape, hydrophilicity, corrosiveness, toxicity, explosiveness, etc. Since dust particles are often composed of particles of varying sizes, it is important to know their size distribution in order to characterize them. The particle size distribution refers to the particle size distribution of all the particles in the grading. The density of dust particles has a great influence on the performance of the dust removal device. The higher the density, the better the dust removal. The dust particles can be described by several physical properties that affect the separation mechanism. Dust is often a collection of particles between many gaps, so the volume density of dust is much smaller than the true density. The ratio of the interstitial volume to the total dust volume is called the porosity. The larger the porosity is, the smaller the volume density is, and the more difficult the dust collection is. For the dust with the ratio of true density to bulk density more than 10, special attention should be paid to prevent re-flying phenomenon.

对于尘粒，可以用影响分离机理的几种物理特性进行描述。这些性质中最重要的是粒子尺寸和密度，此外，还有电阻率、附着性、粒子形状、亲水性、腐蚀性、毒性、爆炸性等。由于尘粒常常是由大小不等的粒子所组成，为了表示它的特性，还要知道它的粒径分布。粒径分布是指各种粒径的粒子在全部粒子中的分级分布率。尘粒的密度对于除尘装置的性能影响很大，密度越大，对除尘越有利。特别是对于利用重力、惯性力或离心力的机械除尘装置，密度的影响更大。粉尘常常是粒子的集合体，粒子之间有很多间隙，因而粉尘的体积密度要比真密度小得多。间隙体积占粉尘总体积之比称为孔隙率。孔隙率越大，则体积密度越小，粉尘的捕集就越困难。对于真密度与体积密度之比超过 10 的粉尘，要特别注意防止发生再飞扬现象。

Classification of Dust Removal Devices
除尘装置的分类

The resistivity of dust particles has a great influence on the performance of electrostatic precipitator and filter precipitator. Generally speaking, the resistivity of industrial dust is between $10^{-14}\Omega \cdot cm$ (carbon black) and $10^{-3}\Omega \cdot cm$. Dust particles with too low or too high resistivity are not suitable for electrostatic precipitators. The most suitable resistivity range is $10^4 \sim 2\times10^{10}\Omega \cdot cm$. Beyond this range, dust particles can be treated appropriately to change their resistivity.

尘粒的电阻率对于电除尘和过滤除尘装置的性能有很大影响。一般来说，工业尘粒的电阻率介于 10^{-14}（炭黑）$\sim 10^{-3}\Omega \cdot cm$ 之间。电阻率太低或太高的尘粒均不适于采用电除尘器，最适宜的电阻率范围是 $10^4 \sim 2\times10^{10}\Omega \cdot cm$。在此范围之外，可对尘粒进行适当处理以改变其电阻率。

4.1.3.4 Nitrogen Oxide
4.1.3.4 氮氧化物的净化技术

Nitrogen oxide includes N_2O, NO, NO_2, N_2O_3, N_2O_4 and N_2O_5, collectively known as nitrogen oxide and denoted by NO_x.

氮氧化物包括 N_2O、NO、NO_2、N_2O_3、N_2O_4 和 N_2O_5，总称氮氧化物，用 NO_x 表示。

At present, the methods of flue gas denitrogenation include non-selective catalytic reduction, selective catalytic reduction, absorption and adsorption.

目前，排烟脱氮的方法有非选择性催化还原法、选择性催化还原法、吸收法和吸附法等。

Non-selective Catalytic Reduction
非选择性催化还原法

The non-selective catalytic reduction (NCR) method is used to reduce NO_x from flue gas to N_2 by using hydrogen or methane as reductant and platinum as catalyst. The so-called non-selectivity means that the temperature condition during the reaction not only controls the reduction of NO_x in flue gas to N_2, but also has a certain amount of reducing agent and excess oxygen in flue gas. The temperature range selected by this method is $400 \sim 500°C$.

非选择性催化还原法是应用铂作催化剂，以氢或甲烷等还原性气体作还原剂，将烟气

中的 NO_x 还原成 N_2。非选择性是指反应时的温度条件不仅仅控制烟气中的 NO_x 还原成 N_2,而且还控制反应过程中能有一定量的还原剂与烟气中的过剩氧作用。此法选取的温度范围为 400~500℃。

In addition to platinum and other precious metals, cobalt, nickel, copper, chromium and manganese oxides can also be used as catalysts. The waste heat recovery device is necessary in the practical device of non-selective catalytic reduction denitrification.

该法所用的催化剂除铂等贵金属外,还可使用钴、镍、铜、铬、锰等金属的氧化物。在非选择性催化还原法脱氮的实际装置中,要有余热回收装置。

Selective Catalytic Reduction
选择性催化还原法

The selective catalytic reduction method is based on the oxides (with bauxite as the carrier) of precious metals such as platinum or copper, chromium, iron, vanadium, molybdenum, cobalt and nickel as catalysts, ammonia, hydrogen sulfide, chloro-ammonia and carbon monoxide as the reducing agent, and chooses the most suitable temperature range for denitrification reaction. This optimum temperature range varies with the catalyst, reducing agent and flue gas velocity, and is generally 250~450℃.

选择性催化还原法是以贵金属铂或铜、铬、铁、钒、钼、钴、镍等的氧化物(以铝矾土为载体)为催化剂,以氨、硫化氢、氯—氨及一氧化碳等为还原剂,选择最适宜的温度范围进行脱氮反应。这个最适宜的温度范围随着所选用的催化剂、还原剂以及烟气流速的不同而不同,一般为 250~450℃。

According to the different reducing agents, the selective catalytic reduction can be divided into ammonia catalytic reduction, hydrogen sulfide catalytic reduction, chlorine-ammonia catalytic reduction and carbon monoxide catalytic reduction.

根据所选用还原剂的不同,选择性催化还原法可分为氨催化还原法、硫化氢催化还原法、氯—氨催化还原法及一氧化碳催化还原法等。

Absorption Method
吸收法

According to the different absorbents used, absorption methods can be divided into alkali absorption method, molten salt absorption method, sulfuric acid absorption method, magnesium hydroxide absorption method and oxidation absorption method, introduced as follows:

按所使用吸收剂的不同,吸收法可分为碱液吸收法、熔融盐吸收法、硫酸吸收法、氢氧化镁吸收法及氧化吸收法等。其介绍如下:

(1) Alkali absorption method. The method can also remove SO_2 from flue gas. When $n(NO)/n(NO_2) = 1(N_2O_3)$ in the flue gas (i.e. NO and NO_2 are equal), the absorption rate of Alkali liquor is about 10 times faster than that of only 1% NO. The absorption solution is usually 30% sodium hydroxide or 10%~15% sodium carbonate.

(1) 碱液吸收法。该法也可同时除去烟气中的 SO_2。当烟气中 $n(NO)/n(NO_2) = 1(N_2O_3)$(即 NO 和 NO_2 是等物质的量存在)时,碱液的吸收速度比只有1%NO 时的吸收速度快大约10倍。通常采用30%的氢氧化钠溶液或10%~15%的碳酸钠溶液作为吸收液。

(2) Molten salt absorption method. The method is a method of absorbing NO_x from flue gas by molten salts of alkali or alkaline-earth metals. It can also remove SO_2 from flue gas.

(2) 熔融盐吸收法。该法是以熔融状态的碱金属或碱土金属的盐类吸收烟气中的 NO_x 的方法,此法也可同时除去烟气中的 SO_2。

(3) Sulfuric acid absorption method. The method can also remove SO_2 from flue gas.

(3) 硫酸吸收法。该法也可同时除去烟气中的 SO_2。

(4) Magnesium hydroxide. The method is to remove SO_2 from flue gas with $Mg(OH)_2$.

(4) 氢氧化镁吸收法。该法就是用 $Mg(OH)_2$ 脱除烟气中 SO_2 的方法。

(5) Oxidation absorption method. Since NO is difficult to be absorbed, the method of NO oxidation to NO_2 firstly and then to be absorbed by oxidant is proposed. The oxidants used are sodium chlorite, sodium hypochlorite, potassium permanganate, ozone, etc..

(5) 氧化吸收法。鉴于 NO 很难被吸收,因而提出用氧化剂将 NO 先氧化为 NO_2,然后再吸收下来的方法。所用的氧化剂有亚氯酸钠、次氯酸钠、高锰酸钾、臭氧等。

Adsorption Method
吸附法

The adsorbents used to remove nitrogen oxide are activated carbon, silica gel and molecular sieve. These adsorbents oxidize NO to NO_2 and then adsorb it in the form of NO_2. The adsorption capacity of activated carbon for NO_x is only a fraction of that of adsorbed SO_2, so the amount of activated carbon needed is too much and it is difficult to be applied. The adsorption capacity of NO_x on activated carbon is only a fraction of the adsorption capacity of SO_2, so the amount of activated carbon required is too much. There are practical difficulties. The effect is better, heating regeneration to get NO_2 can be used. However, if there is moisture in the exhaust gas, the moisture must be removed first, so the adsorption method is not suitable for the flue gas containing a lot of moisture.

用于脱除氮氧化物的吸附剂有活性炭、硅胶和分子筛等。这些吸附剂都是将 NO 氧化成 NO_2 后,以 NO_2 的形式加以吸附。活性炭对 NO_x 的吸附容量仅为吸附 SO_2 的几分之一,因而所需活性炭的数量太多,实用上有困难。以硅胶和分子筛吸附 NO_x 的效果较好,加热再生时得到的 NO_2 可以利用。但如废气中含有水分,则必须先除去水分,所以吸附法对于含大量水分的烟道气并不适用。

Task 4.2　Pollution and Treatment of Metallurgical Industrial Waste Water
任务 4.2　冶金工业废水的污染与治理

The natural water body receives from the waste water, the atmosphere and the solid waste material pollutant pollution, called the water pollution. The pollution of waste water to water body, atmosphere, soil and organism is called waste water pollution.

自然水体受到来自废水、大气、固体废料中污染物的污染,称为水污染;废水对水体、大气、土壤、生物的污染,称为废水污染。

4.2.1 Classification of Metallurgical Industrial Waste Water
4.2.1 冶金工业废水的分类

The classifications of metallurgical industrial waste water are as follows:
冶金工业废水的分类包括：

(1) According to the degree of pollution, the waste water can be divided into pure waste water and waste water. There is a considerable amount of production drainage that is not actually dirty, such as cooling water, so the term of 'Wastewater' is appropriate for all drainage. In the case of dirty water, the terms of 'Waste Water' and 'Sewage' can be used interchangeably.

(1) 根据污染程度，废水可分为净废水和污水两种。有相当数量的生产排水其实并不脏（如冷却水），因而用"废水"一词统称所有的排水比较合适。在水质污浊的情况下，"废水"与"污水"两种术语可以通用。

(2) According to the source, the wastewater can be divided into domestic sewage and industrial wastewater. Domestic sewage refers to the waste water discharged from people's life, mainly including fecal water, bath water, washing water and flushing water, etc. Industrial waste water refers to the waste water discharged from industrial production. The wastewater discharged from cities and towns is called urban wastewater, including domestic sewage and industrial wastewater.

(2) 根据来源，废水可分为生活污水和工业废水两大类。生活污水是指人们生活过程中排出的废水，主要包括粪便水、洗浴水、洗涤水和冲洗水等。工业废水是指工业生产中排出的废水。由城镇排出的废水称为城市废水，包括生活污水和工业废水。

(3) According to the chemical types of pollutants, wastewater can be divided into organic wastewater and inorganic wastewater. The former mainly contains organic pollutants with biodegradability, while the latter mainly contains inorganic pollutants without biodegradability.

(3) 根据污染物的化学类型，废水可分为有机废水和无机废水两种。前者主要含有机污染物，具有生物降解性；后者主要含无机污染物，无生物降解性。

(4) According to the different kinds of poisons, the waste water can also be divided into phenol-containing waste water, mercury-containing waste water and cyanide-containing waste water. It should be noted that mercury-containing wastewater only indicates that the main toxicant is mercury, but does not mean that mercury is the most abundant or that it is the only pollutant.

(4) 根据毒物的种类不同，也可把废水分为含酚废水、含汞废水、含氰废水等。应当注意，含汞废水仅表明其中主要毒物是汞，但并不意味汞含量最多或者汞是唯一的污染物。

In addition, it can also produce waste water sector or production process to name, such as cooling waste water, electroplating waste water and so on.

此外，还可以根据产生废水的部门或生产工艺来命名，例如冷却废水、电镀废水等。

4.2.2 Pollution of Metallurgical Industrial Waste Water
4.2.2 冶金工业废水的污染

With the development of modern industry, industrial wastewater has become the main source of pollution. Slag and leaching slag from smelting process and tailings from mines are leached by rain water, and all kinds of heavy metals are absorbed into surface water and groundwater. Some heavy metals and their compounds can cause harm by accumulating and accumulating in fish and other aquatic organisms, as well as in crop tissues.

随着现代工业的发展，工业废水已成为主要污染源。冶炼过程产生的熔渣和浸出渣以及矿山产出的尾矿，经雨水淋溶，将各种重金属带入地表水和地下水中。某些重金属及其化合物能在鱼类及其他水生物体内以及农作物组织内积累富集，从而造成危害。

4.2.3 Treatments of Metallurgical Industrial Waste Water
4.2.3 冶金工业废水的治理

4.2.3.1 Physical Methods
4.2.3.1 物理方法

Gravity Sedimentation
重力沉降法

Under the action of gravity, the suspended matter in wastewater with density greater than 1 sinks to remove it from the wastewater. This method is called gravity settling method. The gravity sedimentation method can not only separate the original suspended solids such as silt, iron filings and coke powder in the wastewater. It can also separate the secondary suspended solids, such as chemical sediment, chemical floc and microbial floc, which are generated in the process of wastewater treatment. Because this method is simple, easy to operate and has good separation effect, and the separation of suspended solids is often an indispensable pre-treatment or follow-up process of water treatment system, it is widely used.

重力沉降法是指在重力作用下，废水中密度大于1的悬浮物下沉，使其从废水中去除的方法。重力沉降法既可分离废水中原有的悬浮固体（如泥沙、铁屑、焦粉等），又可分离在废水处理过程中生成的次生悬浮固体（如化学沉淀物、化学絮凝体以及微生物絮凝体等）。由于这种方法简单、易行、分离效果较好，而且分离悬浮物又往往是水处理系统不可缺少的预处理或后续工序，因此应用十分广泛。

There are four basic types of sedimentation, according to the concentration of sediment and the degree of flocculation:

根据废水中可沉物质的浓度高低和絮凝性能的强弱，沉降有下述四种基本类型：

(1) Free Settlement. Free settling, also known as discrete settling, refers to the settling of solid particles in dilute solution with no or weak tendency to flocculation. Because of the low concentration of suspended solids and the non-fusion of particles, the shape, diameter and density of particles remain unchanged during the settling process. The initial precipitation of particles in sediment basin and primary sedimentation basin belongs to free sedimentation.

(1) 自由沉降。自由沉降也称离散沉降，是指一种无絮凝倾向或有弱絮凝倾向的固体颗粒在稀溶液中的沉降。由于悬浮固体浓度低，且颗粒间不发生融合，因此在沉降过程中颗粒的形状、粒径和密度都保持不变，各自独立地完成沉降过程。颗粒在泥沙池及初次沉淀池内的初期沉淀属于自由沉降。

（2）Flocculation sedimentation. Flocculation sedimentation is the sedimentation of a kind of flocculent particles in dilute suspension. Although the concentration of suspended solids in wastewater is not high, the physical properties and settling velocity of particles change continuously due to the adhesion of particles to each other to form larger flocs. The later sedimentation in primary sedimentation tank and the initial sedimentation in secondary sedimentation tank belong to flocculation sedimentation.

（2）絮凝沉降。絮凝沉降是指一种絮凝性颗粒在稀悬浮液中的沉降。虽然废水中的悬浮固体浓度不高，但在沉降过程中各颗粒之间互相黏合成较大的絮体，因而颗粒的物理性质和沉降速度不断发生变化。初次沉淀池内的后期沉淀及二次沉淀池内的初期沉降属于絮凝沉降。

（3）Layer settlement. Stratified settlement is also called group settlement. When the concentration of suspended matter in wastewater is high and the particles are close to each other, the settlement of each particle is disturbed by the force of the surrounding particles. But the relative position of the particles is constant, and the whole covering layer sinks together. At this point, a clear interface is formed between the water and the particles, and the subsidence process is actually the subsidence process of the interface. Because the sinking overburden must replace the water of the same volume below, there is relative movement between the two, and the water forms a non-negligible resistance to the particle group. The settling of floc and the later settling of active water sludge in secondary sedimentation tank belong to layered settling.

（3）成层沉降。成层沉降也称集团沉降。当废水中的悬浮物浓度较高、颗粒彼此靠得很近时，每个颗粒的沉降都受到周围颗粒作用力的干扰，但颗粒之间相对位置不变，成为一个整体的覆盖层共同下沉。此时，水与颗粒群之间形成一个清晰的界面，沉降过程实际上就是这个界面的下沉过程。由于下沉的覆盖层必须把下面同体积的水置换出来，两者之间存在相对运动，水对颗粒群形成不可忽视的阻力，因此成层沉降又称为受阻沉降。化学混凝中絮体的沉降及活性水淤泥在二次沉淀池中的后期沉降属于成层沉降。

（4）Compression. When the concentration of suspended solids in wastewater is very high, the particles touch and support each other. Under the action of the gravity of the upper particles, the water in the gap between the lower particles is squeezed out of the interface, the relative position of the particles is changed, and the particle group is compressed. The concentration of activated sludge in the sludge hopper of secondary sedimentation tank and in the thickening tank is compression.

（4）压缩。当废水中的悬浮固体浓度很高时，颗粒之间便互相接触、彼此支承。在上层颗粒的重力作用下，下层颗粒间隙中的水被挤出界面，颗粒相对位置发生变化，颗粒群被压缩。活性污泥在二次沉淀池泥斗中及浓缩池内的浓缩属于压缩。

Sedimentation Tank

沉淀池

A device for separating suspended solids from water by gravitational sedimentation is called a sedimentation tank. There are ordinary sedimentation tank and inclined plate, inclined tube sedimentation tank, introduced as follows:

用重力沉降法分离水中悬浮固体的设备称为沉淀池。沉淀池有普通沉淀池和斜板、斜管沉淀池之分。其介绍如下:

(1) Common sedimentation tank. According to the flow direction of water in the sedimentation tank, the common sedimentation tank is divided into three types: horizontal flow, radial flow and vertical flow, which are introduced as follows:

(1) 普通沉淀池。根据水在沉淀池内的流向,普通沉淀池又分为平流式、辐流式和竖流式三种。其介绍如下:

1) Horizontal sedimentation tank. The sedimentation tank is rectangular in shape, with waste water flowing in from the head of the tank, horizontally flowing through the tank, and flowing out fromits wake. The bottom of the pool head is provided with a mud storage bucket to concentrate the removal of mud scraped by the mud scraping equipment. Mud scraper equipment chain belt scraper, bridge road scraper and so on. In addition, multi-bucket gravity sludge discharge can also be used.

1) 平流式沉淀池。其平面呈矩形,废水从池首流入,水平流过池身,从池尾流出。池首底部设有储泥斗,集中排除刮泥设备刮下的污泥。刮泥设备有链带刮泥机、桥式行车刮泥机等。此外,也可以采用多斗重力排泥。

2) Vertical flow sedimentation tank. The sedimentation tank is generally round or square, and the waste water is allocated from the bottom of the contral tube, rising uniformly, and discharged from the top periphery. The cone at the bottom of the pool is a mud bucket, and the sludge is discharged by the static pressure of water.

2) 竖流式沉淀池。其平面一般呈圆形或正方形,废水由中心筒底部配入,均匀上升,由顶部周边排出。池底锥体为储泥斗,污泥靠水的静压力排除。

3) Radial flow sedimentation tank. The pond body plane takes the circle as many, also having the square. The waste water enters the pool from the central inlet pipe and flows slowly around the pool along the radius direction. The suspended solids settle in the flow and enter the sludge hopper along the slope of the pool bottom to clarify water flowing out of the channel around the pool. However, there are also radial flow sedimentation tanks with peripheral water inflow and central discharge.

3) 辐流式沉淀池。池体平面以圆形为多,也有方形的。废水自池中心进水管进入池内,沿半径方向向池周缓缓流动。悬浮物在流动中沉降,并沿池底坡度进入污泥斗,澄清水从池周溢流出水渠。但也有周边进水、中心排出的辐流式沉淀池。

(2) Inclined plate (inclined tube) sedimentation tank. A parallel (inclined plate) pipe is arranged in the clarification area of the sedimentation tank to improve the treatment capacity of the sedimentation tank.

(2) 斜板(斜管)沉淀池。在沉淀池澄清区设置平行的斜板(斜管),以提高沉淀池的处理能力。

Examples of Treatment of Wastewater Containing Iron Sheet
含铁皮废水处理实例

The examples of treatment of wastewater containing iron sheet are as follows:

以下介绍含铁皮废水处理实例：

(1) Wastewater characteristics. The waste water containing iron sheet mainly comes from rolling and continuous casting. This kind of waste water contains a large number of iron sheet, and contains a variety of lubricants and other impurities. The quantity of wastewater and the composition of wastewater vary widely, which depends on the performance, operation, maintenance level and management experience of process equipment. The results showed that the iron sheet contained 1000~8000mg/L and oil 30~1200mg/L, the iron sheet contained 50~2000mg/L and oil 5~40mg/L respectively. The size of iron sheet particles in steel rolling wastewater varies with factors such as steel rolling, ranging from a few centimeters to tens of centimeters, and small only a few microns. The particle size of iron sheet in continuous casting wastewater varies with the type of continuous casting machine and the size of billet. the big one is over 5mm, and the small one is only a few microns. The density of iron sheet is generally $5~5.3t/m^3$.

(1) 废水特性。含铁皮废水主要来自轧钢、连续铸锭等车间。这类废水中除含有大量铁皮外，还含有各种润滑油及其他杂质。废水数量和废水中所含成分的变化范围较大，这取决于工艺设备性能、操作情况、维护水平和管理经验等。轧钢废水含铁皮 1000~8000mg/L，含油 30~1200mg/L；连铸废水含铁皮 50~2000mg/L，含油 5~40mg/L。轧钢废水中铁皮颗粒的大小随轧钢种类等因素的不同而不同，大的从几厘米到几十厘米，小的仅几微米；连铸废水中铁皮的粒度也随连铸机种类、铸坯尺寸的不同而不同，大的在 5mm 以上，小的仅几微米。铁皮的密度一般为 $5~5.3t/m^3$。

(2) Treatment. At present, there are four common methods to treat iron sheet wastewater at home and abroad：

(2) 处理方法。目前国内外处理含铁皮废水常用的方法有以下四种：

1) The first method, as shown in Figure 4-2, has been widely used in domestic large and medium rolling mills. The settling efficiency is 60%~75% when the wastewater stays in the primary sedimentation tank for 1~2min, and the effluent quality is only 100~200mg/L when it stays in the secondary advection sedimentation tank for 40~60min. The advantages of this method are less equipment and simple operation. Its disadvantages are large area, slag removal difficulties, and poor water quality. In order to improve the quality of effluent water ensure a higher recycling rate, and two improvement methods can be adopted. One is to add a filter after the secondary sedimentation tank, at this time the treatment of water quality can generally reach about 10mg/L. The filtering device can adopt gravity filter or pressure fast filter, and the latter type is more commonly used. The other is that the secondary sedimentation tank is changed to coagulation sedimentation, which has the advantages of high adaptability, good effluent quality, high operation cost and trouble management.

1) 第一种方法如图 4-2 所示，大中型轧机上采用得较多。废水在一次沉淀池内停留 1~2min，沉淀效率达 60%~75%；在二次平流沉淀池内停留 40~60min，出水水质一般仅能达到 100~200mg/L。这种方式的优点是设备少、操作简单，其缺点是占地面积大、清渣困难、处理水质较差。为了提高出水水质，保证更高的循环率，可采用两种改进方式。其一是在二次沉淀池后增设过滤器，此时处理水质一般可达 10mg/L 左右。过滤装置可采用重力滤池或压力快速过滤器，后一种形式采用较多。其二是二次沉淀池改用混凝沉淀方式，其优点是适应性强、出水水质好，缺点是运行费较高、管理麻烦。

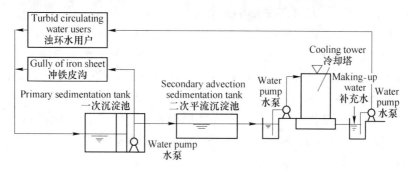

Figure 4-2　Sheet metal wastewater treatment process
图 4-2　含铁皮废水处理流程

2) In the second method, the primary sedimentation tank and the secondary sedimentation tank are replaced by the cyclone sedimentation tank. The efficiency of the cyclone sedimentation tank is higher than that of the advection sedimentation tank, reaching about 95%. Under the condition of large load per unit area, compared with two advection sedimentation tanks, the cyclone sedimentation tank has the advantages of good water quality, 40% saving in investment, 35% reducing in outlay, and only 15% of the advection sedimentation tank in floor area. Because of the advantages mentioned above, the cyclone sedimentation tank has been widely used in the treatment of wastewater containing iron sheet from steel rolling workshop. In recent years, there is a tendency to adopt the down-swirl sedimentation tank, which avoids the shortcoming that the up-swirl sedimentation tank is easily blocked due to the deep inlet pipe. The quality of the effluent from the cyclone sedimentation tank can be reduced to about 10mg/L by the pressure filter treatment.

2) 第二种方法以旋流沉淀池代替一次沉淀池和二次沉淀池。旋流沉淀池比平流沉淀池效率高，可达 95% 左右。在单位面积负荷较大的情况下，旋流沉淀池与两次平流沉淀池相比，出水水质好，投资节省 40%，经费减少 35%，占地面积仅为平流沉淀池的 15% 左右，清渣方便。由于旋流沉淀池具有上述优点，目前其已广泛用于轧钢车间含铁皮废水的处理。近年来，趋向于采用下旋式旋流沉淀池，从而避免了上旋式旋流沉淀池因进水管过深而容易堵塞的问题。旋流沉淀池出水经压力过滤器处理，出水水质可降到 10mg/L 左右。

3) In the third method, high gradient magnetic filter is used to replace the pressure fast filter in the above method, which has the advantages of faster filtering speed and higher efficiency. High Gradient magnetic filtration (hgmf) has been used in some factories abroad, but it is still in trial production in China. According to the test data of a steel rolling plant in China, the effect of High gradient magnetic filtration on iron sheet wastewater treatment is better. The concentration of suspended matter in influent was 150~350mg/L, the magnetic field intensity was 2000 Gs, the filtration rate was 500m/h, the concentration of suspended matter in effluent was

less than 20mg/L, the purification efficiency was 97%, the filtration period was 45~60min, the rinsing time was 1.5min, and the ratio of rinsing water to treated water was 1.8%~2.7%. The advantages of this method are high treatment efficiency and small area, but the concentration of suspended solids in influent should not be too high.

3) 第三种方法是以高梯度磁过滤器代替上述方法中的压力快速过滤器,从而使得滤速更快,效率更高。高梯度磁过滤在国外部分工厂已用于有关废水的处理,但国内尚处于生产试用阶段。根据国内某轧钢厂的试验资料,用高梯度磁过滤处理轧钢含铁皮废水的效果较好。进水悬浮物含量为150~350mg/L,磁场强度为2000Gs,滤速为500m/h,出水悬浮物含量低于20mg/L,净化效率达97%,过滤周期为45~60min,冲洗时间约为1.5min,冲洗水占处理水的比率为1.8%~2.7%。这种方法的优点是处理效率高、占地面积小,但进水悬浮物含量不宜过高。

4) The fourth method is called DSD waste water treatment equipment in foreign countries, which is adopted in a domestic factory when the continuous casting workshop waste water treatment system is introduced. The advantages are: the load of the cyclone sedimentation tank can be increased, and the diameter of the cyclone sedimentation tank can be reduced; and the suspended matter content of the outlet water can be reduced to 300~400mg/L, but the suspended matter content of the outlet water can be up to 100mg/L by the hydrocyclone treatment, which can still meet the requirement of the pressure filter for the quality of the inlet water. Its main shortcoming is the equipment increase, the operation management is more complex.

4) 第四种方法在国外称为 DSD 废水处理设备,国内某厂引进连铸车间废水处理系统即采用这种方法。其优点是:可以提高旋流沉淀池负荷,缩小旋流沉淀池的直径;虽然相应地降低了出水水质(出水悬浮物含量为300~400mg/L),但通过压力式水力旋流器处理,出水悬浮物含量可达到100mg/L左右,仍可满足压力过滤器对进水水质的要求。其主要缺点是设备增多、操作管理较复杂。

Examples of Blast Furnace Gas Cleaning Wastewater Treatment
高炉煤气清洗废水处理实例

The examples of blast furnace gas cleaning wastewater treatment are as follows:
高炉煤气清洗废水处理实例介绍如下:

(1) Wastewater characteristics. The composition of blast furnace gas cleaning wastewater is very unstable, which mainly depends on the composition of raw material and fuel, and the smelting operation conditions. In the process of gas cleaning, the hardness is increased temporarily because the gas and CaO dust particles are easily soluble in water. The hardness of calcium and magnesium in waste water increases with each cleaning of gas. The phenomenon of increasing hardness is an important characteristic of this kind of wastewater.

(1) 废水特性。高炉煤气清洗废水的成分很不稳定,主要取决于原料和燃料的成分以及冶炼操作条件。在煤气清洗过程中,由于气体和 CaO 尘粒易溶于水中,暂时使得硬度升高。每清洗煤气一次,废水中钙、镁硬度均增加。具有增加硬度的现象是这种废水的重要特点。

(2) Treatment. Because this kind of wastewater is easy to scale after recycling, so in the

treatment method, there should be not only removal of suspended solids facilities, but also measures to control scaling. There are three common approaches as follows:

（2）处理方法。由于这类废水循环使用后容易结垢，因此在处理方法中，不仅应有去除悬浮物的设施，而且应有控制结垢的措施。常用的处理方法有以下三种：

1) Natural precipitation method. The amount of blast furnace gas is about $6.8 \times 10^5 \text{m}^3/\text{h}$, the gas cleaning equipment includes a washing tower and a venturi tube (treatment water volume is $4000\text{m}^3/\text{h}$), two radial flow sedimentation tanks with a diameter of 30m, two 400m^2 hyperbolic natural ventilation cooling towers, two concentration tanks with a diameter of 12m, four 18m^2 disk-type vacuum filters (treatment capacity is $700\text{m}^3/\text{h}$, accounting for 17.5%), a lime softening station, a set of equipment for adding carbon dioxide gas and other auxiliary supporting equipment. After being put into production, the operation performance is good, and the main technical indicators are:

1）自然沉淀法。某厂高炉煤气发生量约为 $6.8 \times 10^5 \text{m}^3/\text{h}$，煤气清洗设备有洗涤塔和文氏管（处理水量为 $4000\text{m}^3/\text{h}$），设有 2 座直径为 30m 的辐流式沉淀池、2 座 400m^2 双曲线自然通风冷却塔、2 座直径为 12m 的带斜板的浓缩池、4 台 18m^2 盘式真空过滤机（处理能力为 $700\text{m}^3/\text{h}$，占总水量 17.5%）、1 座石灰软化站、1 组加二氧化碳气的装置及其他辅助配套设备。投产后运行情况较好，其主要技术指标为：

①The circulation rate was more than 94%, the emission rate was less than 1.76%, and the concentration multiple n was no less than 1.88.

①循环率大于 94%，排污率小于 1.76%，浓缩倍数 $n \geq 1.88$。

②The surface load of the sedimentation tank was $1.93 \text{m}^3/(\text{m}^3 \cdot \text{h})$.

②沉淀池表面负荷为 $1.93\text{m}^3/(\text{m}^3 \cdot \text{h})$。

③The content of suspended matter at the inlet was 400~4000mg/L, and that at the outlet was less than 100mg/L.

③进口悬浮物含量为 400~4000mg/L，出口悬浮物含量小于 100mg/L。

④The circulating water temperature was about 55℃ at the entrance and 40℃ at the exit of the cooling tower in summer.

④循环水水温，夏季冷却塔进口为 55℃ 左右，出口为 40℃ 左右。

⑤Lime softening facility could reduce the temporary hardness (equivalent) by 2~3mg/L.

⑤石灰软化设施可降低暂时硬度（当量）2~3mg/L。

⑥When adding the smoke well, the effluent CO_3^{2-} concentration was zero, and the free CO_2 concentration was 1~3mg/L.

⑥加烟井，出水 CO_3^{2-} 浓度为零，游离 CO_2 浓度为 1~3mg/L。

⑦The processing load of the disk vacuum filter was $0.2~0.35\text{t}/(\text{m}^2 \cdot \text{h})$, the mud concentration before dehydration was 35%~55%, and the moisture content of the filter cake was about 20% after dehydration.

⑦盘式真空过滤机处理负荷为 $0.2~0.35\text{t}/(\text{m}^2 \cdot \text{h})$，脱水前泥浆浓度为 35%~55%，脱水后滤饼含水率为 20% 左右。

The system uses lime carbonization method to stabilize water quality and basically realizes recycling. It can save 15 million tons of water and recover about 40,000 tons of sludge every year.

该系统采用石灰—碳化法进行水质稳定，基本上实现了循环利用，每年可节约用水1500万吨，回收泥渣约4万吨。

2）Coagulation sedimentation. The effect of coagulation and sedimentation of blast furnace gas cleaning waste water is better when polyacrylamide is used alone or combined with iron trichloride, and the concentration of suspended matter in effluent can be reduced to less than 50mg/L.

2）混凝沉淀法。高炉煤气清洗废水混凝沉淀法，以聚丙烯酰胺单独使用或与三氯化铁复合使用的效果较好，出水悬浮物浓度可降到50mg/L以下。

3）Slag filtration method. The slag filtration method uses the blast furnace water slag as the filter material to filter the blast furnace gas to clean the waste water. The other part of the biological filter tower in the cyanide, water for washing slag or discharge. Using water residue filter to treat blast furnace gas cleaning wastewater can save a set of gas cleaning wastewater treatment structures and sludge treatment facilities, so it occupies a small area, infrastructure investment and low operating costs.

3）渣滤法。渣滤法用高炉水渣作滤料来过滤高炉煤气清洗废水，滤过水的一部分经凉水池降温后，可循环使用；另一部分在生物滤塔中脱氰，出水供冲渣用或者排放。利用水渣滤池来处理高炉煤气清洗废水，可以节省一套煤气清洗废水处理构筑物和泥渣处理设施，因此占地面积小，基建投资和经营费用低。

Filtration

过滤法

The filtration includes as follows：

过滤法包括：

（1）Sieving. Sieving refers to the method of separating liquid from solid by means of mesh and lattice devices such as grids or sieves. The Grille is a frame made of a series of parallel steel bars, which are tilted in front of a wastewater treatment structure or in a channel at the inlet of a water sump at a pumping station to intercept large floating objects in the wastewater, to prevent blocking the structure of the holes, gates and pipes or damage to water pumps and other mechanical equipment. Therefore, the Grille is actually a protective role of security facilities. The bars of a grating are usually made of flat or round steel. The flat steel mostly adopts the section of 50mm×10mm or 40mm×10mm, which is characterized by high strength, not easy to bend and deform, but large loss of head. Round steel diameter commonly used 1omm, and its characteristics is just the opposite of flat steel. The spacing of the bars varies with the size of the floating object to be intercepted, mostly between 15mm and 50mm. There are two ways to remove the grid slag on the grid bar: Manual and mechanical. When the daily cutting slag quantity is more than $0.2m^3$, mechanical cleaning is adopted. For grates with a daily cut-off capacity of more than 1t, a crusher is often attached to crush the grates, which are conveyed to the sludge treatment system by hydraulic means.

（1）筛滤。筛滤是指通过网目状和格子状设备（如格栅或筛子等）进行液固分离的方法。格栅是由一组平行的钢质栅条制成的框架，倾斜架设在废水处理构筑物前或水泵站集水池进口处的渠道中,用以拦截废水中的大块漂浮物,以防阻塞构筑物的孔洞、闸门和管道或损坏水泵等机械设备。因此,格栅实际上是一种起保护作用的安全设施。格栅的栅条多用扁钢或圆钢制成,扁钢大多采用50mm×10mm或40mm×10mm的断面,其特点是强度大、不易弯曲变形,但水头损失较大;圆钢直径多用10mm,其特点恰好与扁钢相反。栅条间距随拦截的漂浮物尺寸而定,大多在15~50mm之间。被拦截在栅条上的栅渣有人工和机械两种清除方法。一般日截渣量大于0.2m³时,采用机械清渣;对日截渣量大于1t的格栅,常设有破碎机,以便将栅渣粉碎,再用水力输送到污泥处理系统进行处理。

(2) Granular media filtration. When the waste water passes through granular media such as quartz sand bed, the suspended matter and colloid are trapped in the surface and inner space of the filter media. It can be used as pretreatment before activated carbon adsorption and ion exchange, and as final treatment after chemical coagulation and biochemical treatment. The filtration process includes two basic stages: filtration and backwashing. Filtration is to intercept pollutants, and backwashing is to wash the pollutants from the filter layer, so that the recovery of filtering capacity. The time from the beginning to the end of the filter is called the filter cycle or work cycle, and from the beginning of the filter to the end of the backwashing is called a filter cycle. There are many kinds of granular media filter, which can be divided into three types according to the filter speed: slow filter speed is $0.04 \sim 0.4 m^3/(m^2 \cdot h)$, fast filter speed is $4 \sim 8 m^3/(m^2 \cdot h)$, and high speed filter speed is $10 \sim 16 m^3/(m^2 \cdot h)$. According to the flow direction of the water, it can be divided into three types: downward flow filter, upward flow filter and bidirectional flow filter.

(2) 粒状介质过滤。废水通过粒状滤料(如石英砂)床层时,其中的悬浮物和胶体就被截留在滤料的表面和内部空隙中,这种通过粒状介质层分离不溶性污染物的方法称为粒状介质过滤。该方法既可用于活性炭吸附和离子交换等深度处理过程之前作为预处理,也可用于化学混凝和生化处理之后作为最终处理过程。过滤工艺包括过滤和反洗两个基本阶段。过滤阶段主要是截留污染物;反洗阶段是把污染物从滤料层中洗去,使之恢复过滤能力。从过滤开始到结束所延续的时间称为过滤周期(或工作周期),从过滤开始到反洗结束称为一个过滤循环。粒状介质滤池的种类很多,按过滤速度,可分为慢滤池[滤速为$0.04 \sim 0.4 m^3/(m^2 \cdot h)$]、快滤池[滤速为$4 \sim 8 m^3/(m^2 \cdot h)$]和高速滤池[滤速为$10 \sim 16 m^3/(m^2 \cdot h)$]三种;按作用水头(即过滤推动力),分为重力式滤池(作用水头4~5m)和压力式滤池作用水头(15~20m)两类;按水的流动方向,又分为下向流滤池、上向流滤池和双向流滤池三种。

Buoyancy Floatation
浮力上浮法

The separation of solid or liquid primary suspended pollutants with a density of less than (or close to 1) from the wastewater by the buoyancy of the water. It is also possible to separate secondary aerosols with a density greater than 1 and a density less than 1 after a certain physical chemistry treatment, known as buoyancy floatation. Generally, there are three kinds of buoyancy floatation: natural floatation, bubble floatation and reagent floatation.

借助于水的浮力,使废水中密度小于1(或接近于1)的固态或液态原生悬浮污染物浮出水面而加以分离，也可以分离密度大于1而在经过一定的物理化学处理后转为密度小于1的次生悬浮物，这种处理方法称为浮力上浮法。一般浮力上浮法分为三种，即自然上浮法、气泡上浮法和药剂浮选法。

Centrifugation
离心分离法

Centrifugal force is produced when a body rotates at a high speed. In a centrifugal force field, all particles are subjected to centrifugal forces many times greater than their own weight. This centrifugal force separation of suspended matter in waste water method is called centrifugal separation.

物体做高速旋转时将产生离心力。在离心力场内，所有质点都将受到比其本身重量大许多倍的离心力的作用。用这一离心力分离废水中悬浮物的方法，称为离心分离法。

Under the condition of a certain rotating speed, the centrifugal force of the particle in the centrifugal force field depends on the mass of the particle. Therefore, when suspended wastewater moves in a high-speed circular motion, the suspended solids with a mass greater than water are thrown to the periphery and the suspended solids with a mass smaller than water (such as emulsified oil) are pushed to the inner layer due to their different centrifugal forces to the different masses of suspended solids and water. Thus, if the respective outlets of suspended matter and water are properly arranged, the suspended matter can be separated from the water. It can be seen that centrifugal sedimentation and centrifugal buoyancy can be performed in the centrifugal force field.

在转速一定的条件下，离心力场内质点所受到的离心力的大小取决于质点的质量。所以，当含悬浮物的废水做高速圆周运动时，由于悬浮物的质量与水不同，它们受到的离心力也不相同，质量比水大的悬浮物固体被甩到外围，而质量比水小的悬浮物（如乳化油）则被推向内层。这样，如果适当地安排悬浮物和水的各自出口，就可以使悬浮物与水分离。由此可见，在离心力场中能够进行离心沉降和离心浮升两种操作。

According to the different ways of producing centrifugal force, the centrifugal separation equipment can be divided into two categories. One is the water cyclone separation equipment, which is characterized by the container is fixed, and the waste water itself rotates along the tangential high speed into the device to generate centrifugal force. The other is the device, which is characterized by high-speed rotating container driven by waste water rotation to generate centrifugal force, such equipment is actually a variety of centrifuges.

按产生离心力方式的不同，离心分离设备可分为两大类。一类是水旋分离设备，其特点是容器固定不动，而由沿切向高速进入器内的废水本身旋转来产生离心力；另一类是器旋分离设备，其特点是由高速旋转的容器带动器内废水旋转来产生离心力，这类设备实际上就是各种离心机。

Magnetic Separation
磁力分离法

Magnetic separation is a method to extract magnetic suspended solids from wastewater by means of magnetic field. The method has the advantages of high treatment capacity, high efficiency, low energy consumption and compact equipment. It can be used for the purification of blast furnace gas washing water, steel-making dust purification wastewater, steel-rolling wastewater and sintering wastewater.

磁力分离法是借助磁场的作用，将废水中具有磁性的悬浮固体吸出的方法。此法具有处理能力强、效率高、能耗少、设备紧凑等优点，可用于高炉煤气洗涤水、炼钢烟尘净化废水、轧钢废水和烧结废水的净化。

4.2.3.2 Physical Chemistry
4.2.3.2 物理化学方法

Adsorption Method
吸附法

Adsorption is a method of using the surface of porous solid adsorbents to adsorb one or more pollutants in wastewater. A solid substance that has the ability to adsorb solute is called an adsorbent, and the solute adsorbed is called an adsorbate. This method is often used in the treatment of low concentration industrial wastewater.

吸附法是利用多孔性固体吸附剂的表面，吸附废水中一种或多种污染物溶质的方法。对溶质有吸附能力的固体物质称为吸附剂，而被吸附的溶质称为吸附质。这种方法常用于低浓度工业废水的处理。

The commonly used adsorbents are activated carbon, zeolite, diatomite, coke, charcoal, wood chips, slag, bauxite, macroporous adsorption resin and humic acid adsorbent, among which activated carbon is the most widely used. After activated carbon adsorption treatment of wastewater, it can not contain color, smell, foam and other organic matter. It can also meet the water quality standards and recycling requirements.

常用的吸附剂有活性炭、沸石、硅藻土、焦炭、木炭、木屑、矿渣、炉渣、矾土、大孔径吸附树脂以及腐殖酸类吸附剂等，其中以活性炭使用最为广泛。经过活性炭吸附处理后的废水，可以不含色度、气味、泡沫和其他有机物，能达到水质排放标准和回收利用的要求。

- Mechanism of Adsorption Process
- 吸附过程的机理

In wastewater treatment, adsorption occurs at the liquid-solid interface, and the adsorption of absorbents occurs due to the surface force of the solid adsorbent. At present, the nature of the surface force is not very clear, so the nature of adsorption is still under further study. Surface energy has been used to explain that the reduction in surface energy of an adsorbent can only be achieved by a reduction in surface tension. In other words, the adsorbent can adsorb certain solute, because this kind of solute can reduce the surface tension of the adsorbent. So the surface of the adsorbent can adsorb those substances that can reduce its surface tension.

在废水处理中，吸附发生在液—固两相界面上，由于固体吸附剂表面力的作用，才产生对吸附质的吸附。目前对表面力的性质还不是十分清楚，因此吸附的本质尚在进一步研究中。有人用表面能来解释，吸附剂要使其表面能减少，只有通过表面张力的减少来达到。也就是说，吸附剂之所以能吸附某种溶质，是因为这种溶质能降低吸附剂的表面张力。所以，吸附剂的表面可以吸附那些能降低其表面张力的物质。

There are three kinds of forces between adsorbents and adsorbate: intermolecular force, chemical bond force and electrostatic force. The adsorption produced by the intermolecular attraction (the Johannes Diderik van der Waals Force), is called physical adsorption. Since molecular gravity is the force between various adsorbents and adsorbents, physical adsorption has no selectivity. Physical adsorption and desorption speed are faster, and easy to reach the equilibrium state. In general, the adsorption at low temperature is mainly physical adsorption. If a chemical reaction takes place between the adsorbent and the adsorbate, a chemical bond is formed and adsorption is initiated. This adsorption is called chemical adsorption. Due to the formation of chemical bonds, chemical adsorption is selective, and not easy to adsorption and desorption, reaching the equilibrium slowly. The heat released from chemical adsorption is also 40~400kJ/mol, similar to the chemical reaction. Chemisorption increases with the increase of temperature, so chemisorption usually takes place at higher temperature. If an ion of adsorbate is adsorbed at the charged point on the surface of the adsorbent due to electrostatic attraction, the resulting adsorption is called ion exchange adsorption. In this adsorption process, there is an exchange of equivalent ions. If the concentration of the adsorbate is the same, the more the charge of the ion band, the stronger the adsorption. For the ions with the same charge, the smaller the hydration radius, the closer to the adsorption point, and the more conducive to adsorption. These three kinds of adsorption can occur together with the change of the external conditions, because of the influence of comprehensive factors, in a system may showing some adsorption play a leading role. In wastewater treatment, most of the adsorption is the comprehensive performance of several adsorption, which is mainly physical adsorption.

吸附剂和吸附质之间的作用力可分为三种，其分别为分子间力、化学键力和静电引力。通过分子间的引力（即范德华力）而产生的吸附，称为物理吸附。由于分子引力是普遍存在于各种吸附剂与吸附质之间的力，物理吸附无选择性。物理吸附的吸附速度和解吸速度都较快，易达到平衡状态。一般在低温下进行的吸附主要是物理吸附。如果吸附剂与吸附质之间产生了化学反应，生成化学键而引起吸附，这种吸附称为化学吸附。由于生成了化学键，化学吸附是有选择性的，而且不易吸附和解吸，达到平衡慢，化学吸附放出的热量也大（40~400kJ/mol），与化学反应相近。化学吸附随温度的升高而增加，所以，化学吸附常在较高温度下进行。如果一种吸附质的离子由于静电引力，被吸附在吸附剂表面的带电点上，由此产生的吸附称为离子交换吸附。在这种吸附过程中，伴随着等当量离子的交换。如果吸附质的浓度相同，离子带的电荷越多，吸附就越强。对于电荷相同的离子，水化半径越小，越能紧密地接近吸附点，越有利于吸附。这三种吸附随着外界条件的改变可以相伴发生，由于综合因素的影响，在一个系统中可能表现出某种吸附起主导作用。在废水处理中，大部分的吸附是几种吸附的综合表现，其中主要是物理吸附。

- Adsorption Process
- 吸附工艺过程

Adsorption operations are divided into static intermittent and dynamic continuous, also known as static adsorption and dynamic adsorption. Wastewater treatment is adsorption under continuous flow condition, so it is mainly dynamic adsorption. Static adsorption is generally only used for experimental studies or small-scale wastewater treatment. There are three kinds of dynamic adsorption methods: fixed bed adsorption, moving bed adsorption and fluidized bed adsorption. Among them, fixed bed adsorption is the most commonly used method in wastewater treatment.

吸附操作分为静态间歇式和动态连续式两种,也称为静态吸附和动态吸附。废水处理是在连续流动条件下的吸附,因此主要是动态吸附。静态吸附一般仅用于实验研究或小型废水处理。动态吸附有固定床吸附、移动床吸附和流化床吸附三种方式,其中,固定床吸附是废水处理工艺中最常用的一种方式。

- Regeneration of Activated Carbon
- 活性炭再生

In the case that the structure of the activated carbon does not change or rarely change, a special method is used to remove the adsorbed material from the pores of the activated carbon, so that the activated carbon has a performance close to the new activated carbon, known as activated carbon regeneration. The regeneration methods of activated carbon mainly include steam stripping, solvent regeneration, acid and alkali washing and roasting.

活性炭再生是指在活性炭本身结构不发生或极少发生变化的情况下,用特殊的方法将其上被吸附的物质从活性炭的孔隙中去除,以便活性炭重新具有接近新活性炭的性能。活性炭再生的方法主要有:水蒸气吹脱法,溶剂再生法,酸、碱洗涤法,以及焙烧法。

- Main Factors Affecting Adsorption
- 影响吸附的主要因素

The main factors affecting adsorption are as follows:

影响吸附的主要因素如下:

(1) Nature of the adsorbent. The adsorbent should meet the requirements of large adsorption capacity, high adsorption rate, high mechanical wear resistance and long service life.

(1) 吸附剂本身的性质。吸附剂应满足吸附容量大、吸附速率高、机械耐磨强度高和使用寿命长的要求。

(2) The restriction of pollutant property in waste water. For example, the smaller the solubility of a contaminant in water, the easier it is to be adsorbed and the harder it is to be desorbed. The solubility of organic compounds decreases with the increase of molecular chain length.

(2) 废水中污染物性质的制约。例如,污染物在水中溶解度越小,越容易被吸附,越不易解吸。有机物的溶解度随分子链长的增加而减少。

- Applications of Adsorption Method in Metallurgy
- 吸附法在冶金中的应用

The applications of adsorption method in metallurgy:

吸附法在冶金中的应用有:

(1) Treatment of scrubbing water from zinc fluidized roaster flue gas with humic acid adsorbent. The production process of zinc smelter is as follows: Zinc sulfide concentrate Fluidized roasting—Vertical retort—Rectification. The waste water mainly comes from the scrubbing water of the flue gas of the fluidized roaster, which contains not only sulfuric acid, but also many kinds of

heavy metals. The pH value can be controlled from 7.3 to 11, and the heavy metal ions such as Pb, Zn, Cd and Hg in the wastewater can basically meet the discharge standard.

（1）用腐殖酸吸附剂处理锌沸腾焙烧炉烟气洗涤水。某炼锌厂的生产工艺流程为：硫化锌精矿—沸腾焙烧—竖罐蒸馏—精馏生产精锌。其中废水主要来自沸腾焙烧炉烟气洗涤水，除含硫酸外，还含有多种重金属。应用石灰中和腐殖酸煤吸附处理制酸废水，pH 值可控制在 7.3~11 之间，废水中 Pb、Zn、Cd、Hg 等重金属离子基本达到排放标准。

（2）Treatment of radioactive element wastewater with activated carbon adsorbent. Firstly, after removing the oil in the sump, it is pumped into the waste water buffer pool to mix and quantitatively add barium chloride to remove radium. After settling for 4h, the supernatant, which accounts for 1/2~2/3 of the waste water, is discharged into the clear water tank. The sludge is pumped into a suspension clarifier and flocculants are added. The supernatant flows into the clear water tank at the top of the suspension clarifier. The sludge is sent from the bottom of the suspension clarifier to the plate and frame filter press, the filtrate is returned to the buffer pool, and the filtrate is sent to the waste sludge tank.

（2）用活性炭吸附剂处理高浓度铀、钍等放射性元素废水。首先在集水池中去除油类后，由泵送入废水缓冲池混合，定量加入氯化钡除镭。废磷碱液用氢氧化钠调节至 pH=8~9，沉降 4h 后，占废水量 1/2~2/3 的上清液流入清水池排放。污泥由泵送入悬浮澄清器，投加絮凝剂，上清液在悬浮澄清器顶部流入清水池。污泥从悬浮澄清器底部送至板框压滤机压滤，滤液返回缓冲池，滤渣送入废渣库。

Ion Exchange Method
离子交换法

The method of using ion exchange agent to exchange ion pollutants in waste water equally is called ion exchange method. The ion exchanger that can replace cation is called cation exchanger, and the ion exchanger that can replace anion is called anion exchanger. Industrial applications include organic ion exchangers (such as sulphonated coal and ion-exchange resin) and inorganic ion exchangers (such as zeolite and zirconium phosphate). The inorganic ion exchanger with compact particle structure, can only exchange on the surface, small exchange capacity and little application. Among the organic ion exchangers, sulfonated coal is the first one, which is prepared by sulfonating coal with sulfuric acid and introducing active genes. Sulphonated coal was replaced by ion-exchange Resin coal because of its low cost and low price, but easy crushing and poor chemical stability. The most widely used in industry is the ion-exchange resin.

离子交换法是指利用离子交换剂，等当量地交换废水中离子态污染物的方法。能置换阳离子的离子交换剂称为阳离子交换剂，能置换阴离子的离子交换剂称为阴离子交换剂。工业应用的离子交换剂包括有机离子交换剂（如磺化煤和离子交换树脂）和无机离子交换剂（如沸石、磷酸锆等）。无机离子交换剂的颗粒结构致密，仅能进行表面交换，交换容量小，应用不多。有机离子交换剂中，磺化煤是最初使用的交换剂，它是利用煤质本身空间结构为骨架，用硫酸进行磺化，引入活性基团制得。磺化煤成本低、价格便宜，但易粉碎、化学稳定性差，所以被离子交换树脂所取代。在工业中应用最广泛的还是离子交换树脂。

There are many active groups on the resin matrix, which are composed of fixed ions and active ions (also known as counter-ions or exchange ions). The fixed ion is fixed on the resin skeleton, and the active ion is combined with the fixed ion by the electrostatic gravitation.

树脂母体上有很多活性基团，活性基团由固定离子和活动离子（也称反离子或交换离子）组成。固定离子固定在树脂骨架上，活动离子则依靠静电引力与固定离子结合在一起，两者电性相反、电荷相等，处于电中和状态。

Liquid Membrane Separation
液膜分离法

Liquid membrane separation is essentially a method of desalting. A concentrated solution of salt can be obtained by reducing the total salt content in the solution. There are four main methods used in wastewater treatment: reverse osmosis, ultrafiltration, electrodialysis and liquid membrane separation.

液膜分离法实质上是一种除盐的方法。在降低溶液中盐的总含量的同时，可以得到一种盐的浓缩液。在废水处理中主要应用的方法有四种，其分别为反渗透、超滤、电渗析和新近发展起来的液膜分离法。

Liquid membrane separation method is characterized by simple structure, easy operation, and it can work at ambient temperature, continuous process and easy automation. But the main disadvantage is to solve the problem of concentrated liquid brine treatment.

液膜分离法的特点是：设备结构比较简单，操作方便，可以在周围环境温度下工作，过程能连续，便于实现自动化。但其缺点是要解决生成的浓缩液（盐水）的处理问题。

Task 4.3　Pollution and Treatment of Solid Waste in Metallurgical Industry
任务4.3　冶金工业固体废物和污染与治理

4.3.1　Classification of Solid Waste in Metallurgical Industry
4.3.1　冶金工业固体废物的分类

The classifications of solid waste in metallurgical industry are as follows:
以下对冶金工业固体废物进行分类：

(1) Solid waste from mining. Solid waste of mining industry mainly refers to various wall rocks stripped off from the main ore when mining metal ore. The amount of this kind of waste rock is huge. Tailings are the process of extraction of concentrate after the remaining tailings. The amount is also quite large, and general concentrator is set up specially tailings storage. The tailings from the non-ferrous metal concentrate also contain valuable metals such as Cu, Ni, Zn and Pb, as well as sulfur and various useful oxides, which can be recycled.

(1) 矿业固体废物。矿业固体废物主要指开采金属矿时从主矿上剥离下来的各种围岩，这类废石数量巨大，从工业应用角度来看，利用价值不大，多在采矿现场就地堆放。其次是尾矿，尾矿是选矿过程中经过提取精矿后剩余的尾渣，数量也相当大，一般选厂都专门设置尾矿库堆放。有色金属矿精选后的尾矿中还含有 Cu、Ni、Zn、Pb 等有价金属以及硫、各种有用的氧化物等，可以回收利用。

(2) Solid waste of iron and steel metallurgical industry. The solid waste of iron and steel metallurgy industry, except the waste rock and tailings from mining and mineral processing, mainly refers to the waste residue from iron-making and steel-making, which can be collectively called metallurgical residue. It mainly includes blast furnace slag and steel slag. In addition, there is smoke and dust produced in the production process, introduced as follows:

(2) 钢铁冶金工业固体废物。钢铁冶金工业固体废物除了包括采矿和选矿生产过程中产出的废石和尾矿外，还包括炼铁、炼钢冶炼过程中排出的废渣，这些废渣可以统称为冶金渣。其中，冶金渣主要包括高炉渣和钢渣，还有在生产过程中产生的烟尘。现分别介绍如下：

1) Blast furnace slag. According to the smelting method, it can be divided into cast pig iron slag, steel pig iron slag and special pig iron slag. According to the chemical composition, it can be divided into basic slag ($r>1$), neutral slag ($r=1$) and acid slag ($r<1$). According to the physical properties and forms, it can be divided into granular slag, pumice or ball slag, lump slag, and powder slag.

1) 高炉渣。按冶炼生产方法其可分为铸造生铁矿渣、炼钢生铁矿渣和特种生铁矿渣；按化学成分，其可分为碱性矿渣（$r>1$）、中性矿渣（$r=1$）和酸性矿渣（$r<1$）；按物理性质及形态，其可分为粒状矿渣、浮石状或球状矿渣、块状矿渣和粉状矿渣。

2) Steel Slag. According to the smelting method, it can be divided into open-hearth steel slag, converter steel slag and electric furnace steel slag. According to the chemical composition, it can be divided into low alkalinity steel slag ($r<1.8$), medium alkalinity steel slag ($r=1.8\sim2.5$), and high alkalinity steel slag ($r>2.5$). According to physical form, it can be divided into water-quenched granular steel slag, massive steel slag, and powder steel slag.

2) 钢渣。按冶炼方法，其可分为平炉钢渣、转炉钢渣和电炉钢渣；按化学成分，其可分为低碱度钢渣（$r<1.8$）、中碱度钢渣（$r=1.8\sim2.5$）和高碱度钢渣（$r>2.5$）；按物理形态，其可分为水淬粒状钢渣、块状钢渣和粉状钢渣。

3) Soot. In the process of iron and steel smelting, the high temperature flue gas discharged from metallurgical furnace contains a lot of smoke and dust. For example, $50\sim100$kg of iron-containing dust is produced for 1t of iron smelting in blast furnace, and $15\sim20$kg of iron-containing dust is produced for 1t of steel smelting in converter. The dust collected by wet-process dedusting of blast furnace, converter or dry-process dedusting of open hearth furnace is called dust mud. According to the production method, it can be divided into converter dust mud, open hearth dust mud and blast furnace dust mud. The content of iron and alkaline oxide is high, the content of harmful impurities is low, and the composition is close to that of iron ore powder, so these dust and mud have great utilization value.

3) 烟尘。在钢铁冶炼过程中，冶金炉排出的高温烟气中含有大量烟尘。例如，高炉每炼 1t 铁要产出 $50\sim100$kg 含铁烟尘，转炉每炼 1t 钢要产生 $15\sim20$kg 含铁烟尘。用湿法除尘（高炉、转炉）或干法除尘（平炉）收集的烟尘统称为尘泥。按生产方法，其可分为转炉尘泥、平炉尘泥和高炉尘泥。这些尘泥中,铁和碱性氧化物含量较多,有害杂质含量少,成分接近铁矿粉,因此这些尘泥具有很大的利用价值。

4.3.2 Principles of Solid Waste Treatment in Metallurgical Industry and Significance of Comprehensive Utilization
4.3.2 冶金工业固体废物处理的原则和综合利用的意义

Solid waste of metallurgical industry is the final state of various pollutants. Under the influence of natural conditions, these solid wastes can also spread into the atmosphere or penetrate into the ground, causing long-term harm to soil and water bodies.

冶金工业固体废物是各种污染物的终态。而这些固体废物在自然条件的影响下,同样会扩散到大气中或渗透到地下,对土壤和水体造成长期危害,因此必须引起高度重视。

The principles of solid waste disposal in metallurgical industry are as follows. First of all, it is necessary to realize the optimal control of solid waste discharge and reduce the discharge to the minimum. Secondly, the solid waste discharged must be used comprehensively, so that it can be used as a secondary resource; those that can not be used under the existing conditions must be treated innocuously, and finally be rationally restored to the natural environment. Solid waste discharged should be properly treated to make it safe, stable, harmless to minimize its quantity. Therefore, the solid waste should be treated by physical, chemical and biological methods. In the process of treatment and disposal, attention should be paid to the prevention of secondary pollution.

对冶金工业固体废料处理的原则是:首先要实现固体废物排放量的最佳控制,把排放量降到最低程度。其次,必须排放的固体废物要进行综合利用,使它们成为二次资源加以利用;在现有条件下不可能利用的要进行无害化处理,最后合理地还原于自然环境中。对必须排放的固体废物应妥善处理,使其安全化、稳定化、无害化,并尽可能减少其数量。为此,对固体废物要采取物理的、化学的和生物的方法进行处理,在处理和处置过程中,要注意防止二次污染的产生。

4.3.3 Management of Solid Waste in Metallurgical Industry
4.3.3 冶金工业固体废物的治理

4.3.3.1 Treatment and Utilization of Metallurgical Slag
4.3.3.1 冶金渣的处理和利用

Comprehensive Utilization of Blast Furnace Slag
高炉矿渣的综合利用

Most of blast furnace slag is close to neutral blast furnace slag, and the quantity of high basicity and acid blast furnace slag is less. The chemical composition of blast furnace slag fluctuates greatly due to the difference of ore grade and pig iron smelting. The comprehensive utilization technology of blast furnace slag has a history of several decades in China, and the utilization rate of blast furnace slag has reached over 90%. Of these, 90% are washed into water-quenched slag, most of which are used as raw materials for cement mixing and cement without clinker, and a few are used to produce slag bricks and tiles, etc. The rest is used as road bed slag, railway ballast, concrete aggregate, production of slag wool, and expanded slag beads.

高炉矿渣大部分接近中性矿渣，高碱性及酸性高炉渣数量较少。由于矿石的品位和冶炼生铁的种类不同，高炉矿渣的化学成分波动范围很大。高炉矿渣的综合利用技术在我国已有几十年的历史，高炉矿渣的利用率已达到90%以上。其中90%冲成水淬矿渣，大部分用作水泥的混合原料和无熟料水泥的原料，少部分用来生产矿渣砖瓦等，其余用作道路路基渣、铁路道砟、混凝土骨料以及生产矿渣棉、膨胀矿渣珠等。

Comprehensive Utilization of Steel Slag
钢渣的综合利用

Generally, 200~300kg of steel slag is produced for every 1t of steel smelted. Steel slag returned for sintering, iron-making and steel-making, can also be used for highway subgrade, cement, railway ballast, asphalt mixture and agricultural fertilizer, etc., introduced as follows:

一般每冶炼1t钢产生200~300kg的钢渣。钢渣可以返回，供烧结、炼铁、炼钢使用，也可用于公路路基、水泥、铁路道砟、沥青拌和料和农肥等方面。现分别介绍如下：

(1) Application of steel slag in sintering production. By adding 5%~10% steel slag less than 8mm into sinter instead of flux, the macroscopical and microscopical structure of sinter can be improved, the drum index and the caking rate can be increased, and the weathering rate decreases and the rate of finished products increases. The use of sinter with steel slag in blast furnace can make the operation of blast furnace run smoothly to increase the output and reduce the coke ratio.

(1) 钢渣在烧结生产中的应用。在烧结矿中配入5%~10%的小于8mm的钢渣代替熔剂使用，可利用渣中钢粒及其有益成分，显著改善烧结矿的宏观及微观结构，提高转鼓指数及结块率，使风化率降低、成品率增加。高炉使用配入钢渣的烧结矿，可使高炉操作顺行，产量提高，焦比降低。

(2) Steel slag can be used as flux. The steel slag with low phosphorus content can be used as the flux of blast furnace and cupola, and it can also be used in converter. The return of steel slag to blast furnace can not only save consumption of flux limestone, dolomite and fluorite, but also improve the fluidity of blast furnace slag. The converter slag can be used to replace the flux of lime stone and part of fluorite.

(2) 钢渣可以作为熔剂使用。含磷低的钢渣可作为高炉、化铁炉的熔剂，也可返回转炉利用。钢渣返回高炉，既可节约熔剂（石灰石、白云石、萤石）消耗，又可利用其中的钢粒和氧化铁成分，改善高炉渣的流动性。用转炉钢渣代替化铁炉石灰石和部分萤石熔剂，效果也比较好。

(3) Extracting rare elements from steel slag to make use of secondary resources. Niobium and vanadium in steel slag can be extracted by chemical leaching.

(3) 从钢渣中提取稀有元素，发挥二次资源的利用价值。用化学浸取的办法可以提取钢渣中的铌、钒等稀有金属。

(4) Steel slag brick. Steel slag brick is a kind of building brick which is made of powder steel slag (or water quenched steel slag) as main raw material, mixed with part of blast furnace

water slag (or fly ash, activator lime) and gypsum powder and mixed with water. Steel slag brick can be used in civil building masonry wall, column structure, etc.

（4）钢渣制砖。钢渣砖是以粉状钢渣或水淬钢渣为主要原料，掺入部分高炉水渣（或粉煤灰）和激发剂（石灰、石膏粉）加水搅拌，经轮碾、压制成型、蒸养而制成的建筑用砖。钢渣砖可用于民用建筑中砌筑墙体、柱子构造等。

(5) Making steel slag slag cement. Steel slag cement has formed a new series of cement in China, including steel slag cement, steel slag pumice cement, steel slag fly ash cement and so on.

（5）制作钢渣矿渣水泥。钢渣矿渣水泥在我国已形成一种新的水泥系列，包括钢渣矿渣水泥、钢渣浮石水泥、钢渣粉煤灰水泥等，其生产工艺和主要性能大致相近。

(6) Application of steel slag in agriculture. In the process of steelmaking with medium and high phosphorus hot metal, the phosphorus slag in the initial stage is recovered without adding fluorite to make slag, and it can be directly crushed and ground to make steel slag phosphate fertilizer. In addition, steel slag such as calcium and silicon more, can be used as calcium silicon fertilizer.

（6）钢渣在农业中的应用。采用中、高磷铁水炼钢时，在不加萤石造渣的情况下回收初期含磷渣，将其直接破碎、磨细可制成钢渣磷肥。此外，钢渣中如含钙和硅较多，可作钙硅肥料。

4.3.3.2 Treatment and Utilization of Metallurgical Dust
4.3.3.2 冶金粉尘的处理和利用

The mechanical dust produced in the course of metallurgical production, such as the dust produced in the crushing process, the dust produced in the feeding process or the dust produced in the transportation process of materials, does not change during the production process, and the physical chemistry of the mechanical dust produced in the course of metallurgical production does not change. The composition is basically the same as that of the material before dust formation. In order to improve the recycling efficiency of resources in metallurgical enterprises, this kind of dust is usually used as return powder and returned to the main smelting process to recover the valuable metals.

冶金生产过程中产生的机械性粉尘（如破碎过程产生的粉尘、加料或物料运输过程产生的粉尘等），在生产过程中不发生物理化学变化，其成分基本上与成尘前的物料成分相同。为了提高冶金企业内部的资源循环利用率，通常将这类粉尘作为返粉，返回主体冶炼流程以回收其中的有价金属。

The volatile dust produced in metallurgical production is caused by the processes of oxidation, reduction, sublimation, evaporation and solidification. The physical chemistry changes during the dust forming process and its composition is not necessarily the same as that of the material before the dust forming process. The valuable metals that enter the soot during the dust formation process often accumulate in significant quantities and must be recovered. In particular, the dispersed metals, are no single mineral in nature to extract the metal, only from the dust or other materials enriched with the metal. So the comprehensive utilization of the dust is of special significance.

在冶金生产过程中产生的挥发性烟尘,是由于发生氧化、还原、升华、蒸发和凝固等过程而形成的。其在成尘过程中发生了物理化学变化,其成分与成尘前的物料成分不一定相同。在成尘过程中进入烟尘的有价金属常常富集到相当多的数量,因此必须予以回收。特别是稀散金属,在自然界中没有可供提取该金属的单独矿物,因此只能从富集有该金属的烟尘或其他物料中提取。所以,烟尘的综合利用具有特别重要的意义。

In the process of iron and steel smelting, the high temperature flue gas discharged from metallurgical furnace contains a lot of smoke and dust. For example, 15~20kg of iron-containing dust is produced for every 1t of steel smelted in converter, and 50~100kg of iron-containing dust is produced for every 1t of iron smelted in blast furnace. In order to prevent air pollution and make use of combustible components in flue gas, wet dust removal (or dry dust removal) is usually used. The sludge produced by wet dedusting and the dust collected by dry dedusting are collectively referred to as iron-containing dust in the wastewater treatment process as sludge.

在钢铁冶炼过程中,冶金炉排出的高温烟气中含有大量烟尘。例如,转炉每冶炼 1t 钢要产生 15~20kg 的含铁烟尘,高炉每冶炼 1t 铁要产生 50~100kg 的含铁烟尘。为防止大气污染并利用烟气中的可燃成分,一般采用湿法(或干法)除尘。这种由湿法除尘排出的污水经处理后产生的污泥和干法除尘收集的烟尘,统称为含铁尘泥(在废水处理工艺中称为泥渣)。

Chemical Composition, Mineral Composition and Physical Properties of Iron-bearing Dust Mud

含铁尘泥的化学成分、矿物组成及物理性质

The main components of the iron-containing mud are iron and iron oxides, as well as calcium oxide and silicon dioxide, etc. The chemical composition of the mud is shown in Table 4-1.

含铁尘泥的主要成分是铁和铁的氧化物以及氧化钙、二氧化硅等,其化学成分见表 4-1 所示。

Table 4-1 Chemical composition of iron dust sludge
表 4-1 含铁尘泥的化学成分

名称	Components (mass fraction) /% 化学成分(质量分数)									
	TFe	FeO	Fe_2O_3	CaO	SnO_2	Al_2O_3	MgO	P	S	MnO
转炉尘泥	50~62	36~65	13~16	8~14	2~5	0.6~1.3	1~6	0.55	0.2~0.5	0.8~3
高沪尘泥	30~40	5~10	40	8~12	10~15	5~7	2~3		0.4~0.5	

The mineral composition of open hearth dust is mostly hematite, magnetite and metallic iron. It is round or ring-shaped under a microscope and covered with metal. The mineral composition of converter dust is mainly magnetite, which is mostly round, with a small amount of magnetite enclosed by iron and hematite with irregular shape.

平炉烟尘的矿物组成大部分为赤铁矿,其次为磁铁矿和金属铁。其在显微镜下呈圆形或环状,并包裹着金属。转炉尘泥的矿物组成主要为磁铁矿(大多呈圆形,有少量磁铁矿被铁包裹着)和无规则形状的赤铁矿。

Utilization of Iron-bearing Dust

含铁尘泥的利用

The content of iron and basic oxide in the mud containing iron dust are more, and the content of harmful impurity is less. The composition is close to iron ore powder, and it has great utilization value.

含铁尘泥中铁和碱性氧化物的含量较多,有害杂质的含量少,成分接近铁矿粉,有很大的利用价值。

The slag of mixed system of FeO·CaO was made by adding basic materials to the bottom-blown oxygen converter sludge, which was consolidated and caked at low temperature and returned directly to the converter. The technology of treating and balling iron-bearing dust mud is as follows: After sending the mud from the sedimentation tank to the mud distribution tank, dewatering by the filter, the water content of the dust mud is reduced to 25%~30%; Then it is mixed with the waste lime powder in the mixer to make the moisture of the sludge be controlled at 18%~20%, and the moisture is reduced to below 15%; The finished ball is consolidated by low temperature 150~250℃, and the moisture content is less than 1%. The pellet containing iron dust was used as slag-forming agent and coolant in the converter. And the smelting effect of fast slag melting, good dephosphorization and stable operation was obtained.

将氧气顶吹转炉除尘污泥配加碱性物料,制成 FeO·CaO 混合系渣料,在低温下固结造块,直接返回转炉。含铁尘泥处理及成球工艺为:将沉淀池的泥浆送至泥浆分配槽后,经过滤机脱水,使尘泥含水率降到 25%~30%;然后将其与废石灰粉在搅拌机内强制混合消化,使泥料水分控制在 18%~20%,并静放使水分降至 15% 以下;再由压球机压制成球,成品球经低温(150~250℃)固结,含水率小于 1%。含铁尘泥球团返回转炉作造渣剂和冷却剂,取得了化渣快、脱磷好和操作稳定的冶炼效果。

The mixture of 25%~30% (by mass) of iron dust with water content and the return ore from sintering plant is made into a ball with water content less than 10%, and then the mixture is added into the sinter to improve both the permeability of the mixture and the sintering process.

将含水(质量分数)25%~30%的含铁尘泥与烧结厂的返矿混合成球(含水率小于10%),然后加入烧结料中配料。这样既能提高混合料的透气性,又能改善烧结过程。

Exercises

思考题

(1) What are the common types of pollutants in the metallurgical production process?

(1) 冶金生产过程常见的污染物有哪几大类?

(2) What are the treatment methods for metallurgical industrial waste gas?

(2) 冶金工业废气常见的处理方法有哪些?

(3) What are the common treatment methods for metallurgical industrial wastewater?

(3) 冶金工业废水常见的处理方法有哪些？

(4) What are the pollution hazards of solid waste from metallurgical industry?

(4) 冶金工业固体废物的污染危害性有哪些？

(5) What are the principles for the treatment of metallurgical solid waste?

(5) 冶金固体废物处理的原则是什么？

References
参 考 文 献

[1] 杜长坤. 冶金工程概论 [M]. 北京：冶金工业出版社, 2018.
[2] 王庆义, 等. 钢铁技术概论 [M]. 北京：冶金工业出版社, 2006.
[3] 鲍燕平, 等. 钢铁冶金学教程 [M]. 北京：冶金工业出版社, 2008.
[4] 黄聪玲. 现代钢铁生产概论 [M]. 北京：冶金工业出版社, 2011.
[5] 薛正良. 钢铁冶金概论 [M]. 北京：冶金工业出版社, 2008.
[6] 李慧. 高炉炼铁基础知识 [M]. 北京：冶金工业出版社, 2005.
[7] 张训鹏. 冶金工程概论 [M]. 长沙：中南大学出版社, 2005.
[8] 傅杰. 现代电炉炼钢理论与应用 [M]. 北京：冶金工业出版社, 2009.
[9] 曲英. 炼钢学原理 [M]. 北京：冶金工业出版社, 1980.
[10] 周孝信. 冶金与材料制备工程科学 [M]. 北京：科学出版社, 2006.
[11] 李生智, 等. 金属压力加工概论 [M]. 北京：冶金工业出版社, 2014.
[12] 李荣, 等. 转炉炼钢操作与控制 [M]. 北京：冶金工业出版社, 2012.
[13] 宫娜. 型钢孔型设计与螺纹钢生产 [M]. 北京：冶金工业出版社, 2016.
[14] 张秀芳, 等. 热轧无缝钢管生产实训指导 [M]. 北京：冶金工业出版社, 2017.
[15] 贾寿峰. 中厚板生产与实训 [M]. 北京：冶金工业出版社, 2017.
[16] 宫娜. 冶金生产过程检测技术 [M]. 北京：冶金工业出版社, 2015.

冶金工业出版社部分图书推荐

书 名	作 者	定价(元)
冶金专业英语（第3版）	侯向东	49.00
电弧炉炼钢生产（第2版）	董中奇 王 杨 张保玉	49.00
转炉炼钢操作与控制（第2版）	李 荣 史学红	58.00
金属塑性变形技术应用	孙 颖 张慧云 郑留伟 赵晓青	49.00
自动检测和过程控制（第5版）	刘玉长 黄学章 宋彦坡	59.00
新编金工实习（数字资源版）	韦健毫	36.00
化学分析技术（第2版）	乔仙蓉	46.00
冶金工程专业英语	孙立根	36.00
连铸设计原理	孙立根	39.00
金属塑性成形理论（第2版）	徐春阳 辉 张 弛	49.00
金属压力加工原理（第2版）	魏立群	48.00
现代冶金工艺学——有色金属冶金卷	王兆文 谢 锋	68.00
有色金属冶金实验	王 伟 谢 锋	28.00
轧钢生产典型案例——热轧与冷轧带钢生产	杨卫东	39.00
Introduction of Metallurgy 冶金概论	宫 娜	59.00
The Technology of Secondary Refining 炉外精炼技术	张志超	56.00
Steelmaking Technology 炼钢生产技术	李秀娟	49.00
Continuous Casting Technology 连铸生产技术	于万松	58.00
CNC Machining Technology 数控加工技术	王晓霞	59.00
烧结生产与操作	刘燕霞 冯二莲	48.00
钢铁厂实用安全技术	吕国成 包丽明	43.00
炉外精炼技术（第2版）	张士宪 赵晓萍 关 昕	56.00
湿法冶金设备	黄 卉 张凤霞	31.00
炼钢设备维护（第2版）	时彦林	39.00
炼钢生产技术	韩立浩 黄伟青 李跃华	42.00
轧钢加热技术	戚翠芬 张树海 张志旺	48.00
金属矿地下开采（第3版）	陈国山 刘洪学	59.00
矿山地质技术（第2版）	刘洪学 陈国山	59.00
智能生产线技术及应用	尹凌鹏 刘俊杰 李雨健	49.00
机械制图	孙如军 李 泽 孙 莉 张维友	49.00
SolidWorks实用教程30例	陈智琴	29.00
机械工程安装与管理——BIM技术应用	邓祥伟 张德操	39.00
化工设计课程设计	郭文瑶 朱 晟	39.00
化工原理实验	辛志玲 朱 晟 张 萍	33.00
能源化工专业生产实习教程	张 萍 辛志玲 朱 晟	46.00
物理性污染控制实验	张 庆	29.00
现代企业管理（第3版）	李 鹰 李宗妮	49.00